D0276735

MANAGEMENT OF RESEARCH AND DEVELOPMENT ORGANIZATIONS

WILEY SERIES IN ENGINEERING & TECHNOLOGY MANAGEMENT

Series Editor: Dundar F. Kocaoglu, Portland State University

Badiru / PROJECT MANAGEMENT IN MANUFACTURING AND HIGH TECHNOLOGY OPERATIONS, SECOND EDITION

Badiru / INDUSTRY'S GUIDE TO ISO 9000

Baird / MANAGERIAL DECISIONS UNDER UNCERTAINTY: AN INTRODUCTION TO THE ANALYSIS OF DECISION MAKING

Edosomwan / INTEGRATING INNOVATION AND TECHNOLOGY MANAGEMENT

Eisner / ESSENTIALS OF PROJECT AND SYSTEMS ENGINEERING MANAGEMENT

Eschenbach / CASES IN ENGINEERING ECONOMY

Gerwin and Kolodny / MANAGEMENT OF ADVANCED MANUFACTURING TECHNOLOGY: STRATEGY, ORGANIZATION, AND INNOVATION

Gōnen / ENGINEERING ECONOMY FOR ENGINEERING MANAGERS

Jain and Triandis / MANAGEMENT OF RESEARCH AND DEVELOPMENT ORGANIZATIONS: MANAGING THE UNMANAGEABLE, Second Edition

Lang and Merino / THE SELECTION PROCESS FOR CAPITAL PROJECTS

Martin / MANAGING INNOVATION AND ENTREPRENEURSHIP IN TECHNOLOGY-BASED FIRMS

Martino / R & D PROJECT SELECTION

Messina / STATISTICAL QUALITY CONTROL FOR MANUFACTURING MANAGERS

Morton and Pentico / HEURISTIC SCHEDULING SYSTEMS: WITH APPLICATIONS TO PRODUCTION SYSTEMS AND PROJECT MANAGEMENT

Niwa / KNOWLEDGE BASED RISK MANAGEMENT IN ENGINEERING: A CASE STUDY IN HUMAN-COMPUTER COOPERATIVE SYSTEMS

Porter et al. / FORECASTING AND MANAGEMENT OF TECHNOLOGY

Riggs / FINANCIAL AND COST ANALYSIS FOR ENGINEERING AND TECHNOLOGY MANAGEMENT

Rubenstein / MANAGING TECHNOLOGY IN THE DECENTRALIZED FIRM

Sankar / MANAGEMENT OF INNOVATION AND CHANGE

Streeter / PROFESSIONAL LIABILITY OF ARCHITECTS AND ENGINEERS

Thamhain / ENGINEERING MANAGEMENT: MANAGING EFFECTIVELY IN TECHNOLOGY-BASED ORGANIZATIONS

MANAGEMENT OF RESEARCH AND DEVELOPMENT ORGANIZATIONS

Managing the Unmanageable

Second Edition

R. K. JAIN
Associate Dean for Research
and International Engineering
University of Cincinnati
Cincinnati, Ohio

H. C. TRIANDIS
Professor of Psychology and Labor & Industrial Relations
University of Illinois
Urbana, Illinois

A WILEY-INTERSCIENCE PUBLICATION
JOHN WILEY & SONS, INC.
NEW YORK CHICHESTER BRISBANE TORONTO SINGAPORE WEINHEIM

This text is printed on acid-free paper.

This publication is designed to provide accurate and authoritative information in regard to the subject matter covered. It is sold with the understanding that the publisher is not engaged in rendering legal, accounting, or other professional services. If legal advice or other expert assistance is required, the services of a competent professional person should be sought.

Library of Congress Cataloging in Publication Data:

Jain, R. K. (Ravinder Kumar), Date

Management of research and development organizations : managing the unmanageable / R.K. Jain, H.C. Triandis. — 2nd ed.

p. cm. — (Wiley series in engineering and technology management)

Includes bibliographical references and index.

ISBN 0-471-14613-7 (cloth : alk. paper)

1. Research, Industrial—Management. I. Triandis, Harry Charalambos, 1926– . II. Title. III. Series.

T175.5.J35 1997

658.5′7—dc20 96-26912

Printed in the United States of America

10 9 8 7 6 5 4 3 2 1

To Terumi, Avra, Anna, Andrew, Pola and Louisa

CONTENTS

PREFACE

"Teams of professionals, armed with laptop computers, fax-modems, E-mail, voice mail, videoconferencing, interactive databases, and frequent-flyer memberships, are being sent to conduct business in this global arena" [Armstrong and Cole, 1995, p. 187].

This picture may well become typical for many R&D teams in the twenty-first century, because the new technology allows individuals to do some of their work at home, or to collaborate with specialists located on different continents, or to work on complex problems that require the input of dispersed professionals. Even in the past, R&D professionals rarely were "managed" the way most people think of management. In fact, when asked, "How do you manage research conducted by the many faculty members and staff in your school," a Dean at Harvard replied, "We don't." He asked, "Have you ever tried to manage prima donnas?"

Managing R&D organizations and focusing on their *productivity* and *excellence* presents unique problems and unusual challenges which get augmented when the team is dispersed. This uniqueness stems from two basic reasons: (1) the character of the enterprise and (2) the type of people involved in R&D.

John Naisbitt and Daniel Bell have suggested that one of the major trends of this century is the transformation of an industrial society (i.e., manufacturing) into an informational society. Nobel Laureate Ken Arrow has stated, "The central economic fact about the processes of invention and research is that they are devoted to the production of information" [Arrow, 1974, p. 152]. The generation of information requires research; therefore, research is going to be one of the most important jobs in the society of the future. Just as farmers in preindustrial society were central players, particularly in periods of famine, so the researchers will be the central players in the future, especially in the advanced industrialized economies.

In addition to the R&D organizations' focus on information, the work itself involves considerable uncertainty since the output can never be predicted perfectly from the various inputs used. Ken Arrow stated "even human consciousness itself would disappear in the absence of uncertainty" [Arrow, 1974, p. 1]. Uncertainty, therefore, gives R&D enterprises a unique quality, and people involved in research and development have some unique characteristics. The obvious ones are postgraduate training and high aptitude. Perhaps, more importantly, people working in R&D have been socialized differently from others. This process occurred during their graduate training, since in order to do well they had to work autonomously and show some initiative and curiosity. To some degree, it was a self-selection process.

Commenting on managing R&D organizations, an eminent R&D manager (Keith Williams, Industrial Fellow, Churchill College) stated, "It is more difficult to manage R&D organizations because of the nature of their activities and people involved—mostly the people. People are more independent and articulate . . . so they need to be handled differently."

How differently? Well as an example, a progressive manager would be well advised to use few sticks and every available carrot. Also, accepting "odd" behavior and granting considerable autonomy to the researcher are highly desirable. In Chapter 3 we discuss "the Amadeus complex," which reminds us that genius is associated with behaviors that some people would call immature. The R&D manager has to learn to tolerate a broad range of behavior from subordinates and colleagues.

In managing an R&D organization, one also has to understand the ethos of a scientific community with its focus on universalism and sharing of scientific knowledge.

Managing an R&D organization, then, is essentially the art of integrating the efforts of diverse, creative, intelligent and autonomous individuals. Paraphrasing John D. Rockefeller, Jr., good management consists of showing superior people how to do the work of near-geniuses.

WHY SUCH A BOOK?

This book, as the reader will see from the background of the authors, is a collaboration between an engineer/scientist and a social and organizational psychologist.

Besides the interest and experience of the authors in R&D, which encompasses managing and directing significant R&D programs, teaching, and writing about many technical, social, and behavioral issues related to organizations, we feel R&D is a very important activity of a modern technological society. For R&D, the United States alone will spend about $182 billion in 1995 [Battelle Memorial Institute, C&EN, February 1995]. The important role R&D plays in the effectiveness of a technology-based organization, the profitability of a business enterprise, and in the economic well-being of a nation, can hardly be overstated. These and other related science policy issues are explored more fully in the book.

Much has been written recently about this country's loss of competitiveness. Most of this loss can be traced to the spending pattern of R&D funds. While our competitors have researched the U.S. market and developed products that can sell here, we have spent the overwhelming portion of our R&D budget on military hardware that becomes obsolete every few years. No one questions the importance of national defense. The question is, Is the United States carrying a disproportionate share of this burden? Thus, a word about our science policy in a book discussing the management of R&D is in order. The managers of R&D must influence science policy and redirect it in the twenty-first century.

With the end of the cold war, the need for major defense expenditures is somewhat lower than it has been. Yet, increased global competitiveness requires even more efficient operations, and the development of such operations requires research. R&D is a major component required for survival in the world of global economic competition.

There is evidence [Nadiri, 1980] that return on R&D investment in industry is higher than on other activities. Some studies showed that the *average* return on this investment is 30 to 1. Furthermore, there is evidence that technology stimulates science [Bondi, 1967], science stimulates technology [Gibbons and Johnston, 1974], and both stimulate the economy [Freeman, 1982].

A few conclusions about R&D funding and researchers' productivity are worth stating at the outset:

1. Basic research is more likely to be supported by government or foundations than by private enterprises.
2. Some of the cost of basic research is covered by the researcher, because researchers are people of great talent and dedication who put in considerable extra time for which they are rarely reimbursed.
3. The most productive researchers do some basic as well as applied research (Chapter 1).
4. R&D investment in the United States has slipped in recent years, relative to Japan and Germany, particularly when we examine the non-defense-related R&D funding.

Looking ahead to the next century, U.S. industry will have to continue developing products for the global industrial society [Brown and Kay, 1987]. The unified European market, the North American market (soon to include, in addition to Canada and Mexico, Chile), and an opening Japanese market together will form a world market of one billion people interested in products that current R&D might develop. In addition, a middle class is emerging in China and India. It is estimated that at the end of the 1990s about a quarter of a billion people will have reached European level incomes in those countries. For instance, one can already see bus loads of tourists from India in many parts of Europe. The countries of central and Eastern Europe and the former Soviet Union will eventually supply another quarter of a billion people with middle class aspirations. Thus, a global market of about 1.75 billion people will emerge. Corporations that sell only in the United States will be at a disadvantage, because they will be dividing their R&D costs with only 270 million instead of 1.75 billion potential customers.

To design a product for a global industrial society, however, will require familiarity with the needs of the customers and, in turn, will require a much broader cross-cultural education. People working in R&D will also need a global perspective on research and innovation and the ability to organize and manage commercialization activities for international markets.

This book has resulted from the experiences of the authors in actually managing R&D organizations, teaching courses in R&D management, and conducting research in theories of management, cross-cultural organizational psychology, and organization psychology. Many of the topics are evolving, as new research brings in new theoretical perspectives. Knowing the needs of physical scientists and engineers for practical suggestions, rather than for a review of the pros and cons of various theories, we used our best judgment of what constitutes the most appropriate answer to a theoretical controversy, rather than burden the reader with such controversies. This means that some managers and scholars in the field would disagree with some of our positions, but that is inevitable in a fast-changing field where precise answers are rarely available.

FOR WHOM IS THIS BOOK?

This book focuses on ways one can improve R&D organization productivity and foster excellence in such organizations. Thus, it is written for (a) principal investigators (P.I.s) and their colleagues and supervisors in research and development organizations and (b) faculty members, department heads, and research administrators at academic institutions. While the profile of such individuals frequently includes a Ph.D. in a physical, biological, or social science or engineering, whatever they learned in behavioral science courses they took has long since been forgotten. Basically, they have been given the job of managing people without much training in how to do this. We assume that individuals in such a role will want to have a short, easily accessible guide to the literature on the best way to manage a research enterprise or a small research group.

Since the first edition of the book was also used for courses related to R&D management, this book has been augmented with new chapters and new topics to make it more useful as a text. Much of the inspiration concerning what topics to cover in the book came from a needs assessment that David Day and Harry Triandis, of the University of Illinois, Urbana–Champaign, carried out in R. K. Jain's laboratory. Over a period of several months, we spoke to groups of P.I.s about their management problems in order to assess what they needed to know. Although these P.I.s did not want to become social psychologists or organizational theorists, they expressed a need to learn something about the behavioral sciences. The needs assessment determined that topics such as how to resolve conflicts, how to change attitudes, how to motivate subordinates, how to design the best work environment, how to make decisions about priorities, and leadership theory were of the greatest concern and probable utility to the P.I.s.

In addition, R. K. Jain, as supervisor of many P.I.s, noted that they would do their job better and get more funds to support their research if they had a broader perspective about science policy. So, in addition to the "micro" topics identified in the needs assessment, we have included some "macro" topics that will help the P.I.s in their search for research funding. University experience as a Research Dean (R. K. Jain) and research related to diversity and cross-cultural issues (H. C. Triandis) provided a stimulus for including chapters focusing on these topics.

In writing this book we have kept principal investigators in mind. However, there are others—for example, university department heads and research administrators, consulting engineers, managers responsible for sponsoring research, and policy-makers concerned with science and technology—who should find the information presented here of interest. Some reviewers suggested that the book provides information that is quite relevant and useful to all managers of creative and rather autonomous personnel.

We will cover some topics that will already be familiar to some of the readers. To help the reader go through the book most efficiently, and skip sections in the book of insufficient interest, we will provide, at the beginning and end of each chapter, an introduction and a summary.

In writing this book we thought of it as opening doors for further study and discussion. Consequently, at the end of each chapter, we have provided a list of Questions for Class Discussion and Suggested Further Readings. The list of questions can be used for paper topics, group projects, or homework assignments, as well as for developing case studies related to R&D organizations.

WHAT THE BOOK IS ALL ABOUT

Managing a research and development (R&D) organization is to a great degree the art of coordinating and integrating the efforts of highly trained and rather autonomous participants. The manager has to provide order, purpose, and foresight and do this while dealing intelligently with the uncertainty inherent in an R&D enterprise. It is hoped that discussions and ideas presented in this book focus on ways one can improve the productivity of R&D organizations and foster excellence in such organizations. Based on needs assessment and the experience of the authors, topics that we thought to be most helpful to R&D managers and their colleagues have been covered. As the book outline shows, the topics range from the motivation of individuals to science policy.

The first edition of this book was well received, but the literature that was covered in that edition included material only through 1987. In this second edition we have incorporated literature to 1996. In addition, the organizations that describe R&D activities in various countries issue new reports almost every year. This second edition utilizes the most recent reports on the status of R&D activities in many parts of the world.

Since R&D organizations are becoming more diverse—that is, women, East Asians, and others frequently comprise the R&D teams—we included a chapter that explicitly covers the topic, "Dealing with Diversity in R&D Organizations."

In a global society, with extreme specialization, more and more R&D teams are dispersed, sometimes on different continents. While such heterogeneity increases the creativity of these teams, it does create problems of communication. Misunderstandings are more common than in face-to-face homogeneous teams. We have included, in the second edition, discussions that may help managers in dealing with such teams.

In the earlier edition, there was little discussion about the university research enterprise and strategic planning for R&D organizations. These topics are covered in two new chapters in this edition.

Chapter 1 develops a typology of R&D activities and the people who engage in them. What is research and development and what is unique about managing R&D organizations are discussed. A section examines the question, What to research? To some extent this is a key question for an R&D manager and for the organization.

Chapter 2 covers basic elements needed for an R&D organization: people, ideas, and funds. It examines communication networks and the innovation process. The discussion of the R&D organizational culture includes avoiding the not-invented-here syndrome, fit of the person and the job, and managing antithesis and ambiguity. The discussion has implications for the selection of people in R&D organizations and the shaping of the culture of such organizations.

The key role for a manager is to create a productive and effective R&D organization, which is the topic of Chapter 3. We ask questions such as, What is organizational effectiveness? Who are the inventors and innovators? How are new ideas generated? Formation of the teams and the ethos of a scientific community that are likely to result in an effective organization are discussed.

Chapter 4 focuses on the design of jobs, careers, and organizational hierarchies and on keeping researchers as innovative as possible throughout their careers. Chapter 5 covers influencing people, peoples' attitude, and how attitudes can be changed. A behavioral science case and its analysis are also presented. Chapter 6 examines what is relevant about human motivation, with special emphasis on rewards, communication, and social and organizational structures that are likely to motivate R&D personnel. It also examines how to develop a sense of control and community for a research organization.

We have added Chapter 7 in response to the need for a better understanding of how to deal with diversity in a global society. Joint research projects with scientists who are different in culture, gender, discipline, organization level, and function are becoming more common than ever before. As research organizations become culturally diverse, there is a greater need for dealing with this diversity.

Leadership is the topic of Chapter 8. We examine a number of theories of leadership and the leadership styles that are likely to be effective in R&D organizations.

Chapter 9 provides a discussion of conflict in organizations. Three kinds of conflict (within a person, between individuals, and between groups) are discussed. Conflict is not always undesirable. There is a productive as well as destructive conflict. We explore how to take advantage of productive conflict, how to reduce destructive conflict, and the ethics that are likely to achieve such ends in R&D organizations.

Chapter 10 is on performance appraisal. We make suggestions concerning how to successfully structure a performance appraisal system in R&D organizations. To do that, the discussion takes into account the different goals and activities of scientists and engineers. Monetary rewards, status, and other rewards can be associated with the results of the appraisal, but some of the dangers of too close a connection between appraisal and monetary rewards are also explored. A performance appraisal implementation strategy and example performance appraisal systems at research organizations are presented.

To be effective, an R&D organization must also be successful in technology transfer, which is the subject of Chapter 11. We ask, What are the stages of such transfer? What factors affect technology transfer? What is the optimal strategy for such transfer?

Chapter 12 provides the manager with an overview of organizational change, what goes on in organizational change, and how to evaluate it.

Chapter 13 focuses on the university research enterprise. Issues ranging from the basis for the university research activities to the importance of the university–industry linkage and the role of the academic institutions in the innovation process are discussed.

Strategic planning has become an important consideration for industrial and academic organizations with major research and development components. Various strategic planning elements unique to an R&D organization along with a case study are presented in Chapter 14.

Finally, a discussion of research and development and society as well as issues important for developing science policy is provided in the Appendix. It examines R&D expenditures and their effect on economic development. A discussion of the need and the level of resource allocation for basic research is included.

CONFESSIONS AND ACKNOWLEDGMENTS

When all is said and done, one reflects on one's completed work and finds many shortcomings. In other words, reality sets in. One could argue that there is not much here that has not already occurred to, or been postulated by, others. We hope our attempt to integrate and formalize some of the concepts will be of value to our readers.

Many of the concepts discussed here were developed during Jain's stay as a Fellow at Churchill College, Cambridge University, a distinct honor and a memorable experience.

The cases that appear in this book were developed by Harry Triandis and David Day, both of the University of Illinois, Urbana–Champaign, in the course of a training needs assessment at Jain's laboratory, organized and directed by David Day. The cases have been distorted, exaggerated, and changed sufficiently so that no one can recognize the players, least of all Jain!

We benefited immensely from the review of the manuscript and interaction with the following colleagues and eminent managers of R&D organizations: Sir Hermann Bondi, Master, Churchill College, Cambridge University; Mr. Keith Williams, Industrial Fellow Commoner, Churchill College (formerly of Shell International Petroleum); The Rt. Hon. Aubrey Jones, Industrial Fellow Commoner, Churchill College (former Minister in the British government and author of many books); Sir William Hawthorne, former Master, Churchill College, Cambridge University and former Hunsaker Professor of Aeronautical Engineering at MIT; Professor David Day, Industrial and Labor Relations, University of Illinois, Urbana–Champaign; Professor David Marks, Professor and Department Head, Civil Engineering, Massachusetts Institute of Technology (MIT); Professor Andrew Schofield, Engineering Department, Cambridge University.

We are grateful to Pola Triandis, who improved the writing style of the book by meticulously editing the entire manuscript, and to Susan Bill, Elle Mengon and Suellen Fortine for their conducting background research and for preparing figures and tables for the manuscript. Many individuals at John Wiley & Sons were most generous with their assistance in finalizing the manuscript and producing the text. Working with Bob Argentieri, Editor, Wiley–Interscience Division, was a pleasure; he tactfully provided many critical comments to improve the manuscript. Personal attention provided to this project by Bob Argentieri made the crucial difference in effectively completing this long, demanding, and exciting journey and finally producing a published volume.

Although the assistance and support provided by our organizations and colleagues are gratefully acknowledged, the responsibility for what is presented in the book is solely that of the authors.

R. K. JAIN
H. C. TRIANDIS

Cincinnati, Ohio
Urbana, Illinois

1

R&D ORGANIZATIONS AND RESEARCH CATEGORIES

Clockmakers were the first consciously to apply the theories of mechanics and physics to the making of machines. Progress came from the collaboration of scientists—Galileo, Huygens, Hooke, and others—with craftsmen and mechanics.

DANIEL J. BOORSTIN
The Discoverers

The historic collaboration between scientists and craftsmen to create the clock, which Boorstin calls "the mother of machines," represents a rudimentary R&D organization.

Today the complexity of the technology has created correspondingly complex organizations, with sometimes hundreds of employees. Many disciplines have to be coordinated and it is the manager who brings the many components together so they can function smoothly, each making an optimal contribution to the R&D organization. Thus, today, as in the past, progress requires collaboration.

Managing a research and development (R&D) organization is, to a great degree, the art of integrating the efforts of its many participants. Beyond this, the manager has to provide order, purpose, and foresight and do this while dealing intelligently with the uncertainty inherent in an R&D enterprise. Considering the important role R&D plays in the economic well-being of a nation, the profitability of a business enterprise, the effectiveness of a technology-based governmental agency (e.g., the Department of Defense), and the enormous investment nations make in R&D activities ($182 billion in 1995 in the United States), effective R&D management can have profound and far-reaching consequences. Effective management, coupled with a vigorous research and science policy, is necessary for a nation to sustain economic growth, provide a strong national defense at an affordable cost, and maintain a position of leadership in the international community. It is, therefore, important to understand R&D organizations and their relationship to society. For this reason, the first chapter provides some basic definitions of research categories and research organizations and the Appendix covers macro issues related to R&D and science policy. This information should be useful to those who conduct and manage research, and especially to those who seek funding support for research and who want to develop allies in influencing science policy.

This chapter first provides a perspective on R&D management and then discusses research and development definitions and categories. Sections that follow examine the question: What to research? This is in some respects a key question for a R&D manager. To what extent, for instance, should the manager allow basic research to be done in addition to the applied research needed by the organization? What is the best way to establish priorities among competing research projects? There are numerous suggestions in the literature on how to do that, and we provide a guide to that literature in the form of an annotated bibliography at the end of each chapter. Since a question is often raised as to what is so unique about an R&D organization management, a discussion of this issue is included in this introductory chapter.

1.1 HOW INFORMATION CAN BE USED

Some readers may want to take a cursory look at the information presented in this chapter and keep in mind how some of it may help them. In addition to having important implications for R&D management, this information has other possible uses as well. Some examples follow.

As a principal investigator (PI), if you are interested in being involved primarily in basic research, in what kind of an organization should you be seeking employment? If you are working in industry you should not be too surprised if you are required to focus your efforts on "products and profits." As shown in Figure 1.1, on the average, 72% of industrial R&D is focused on product development and only

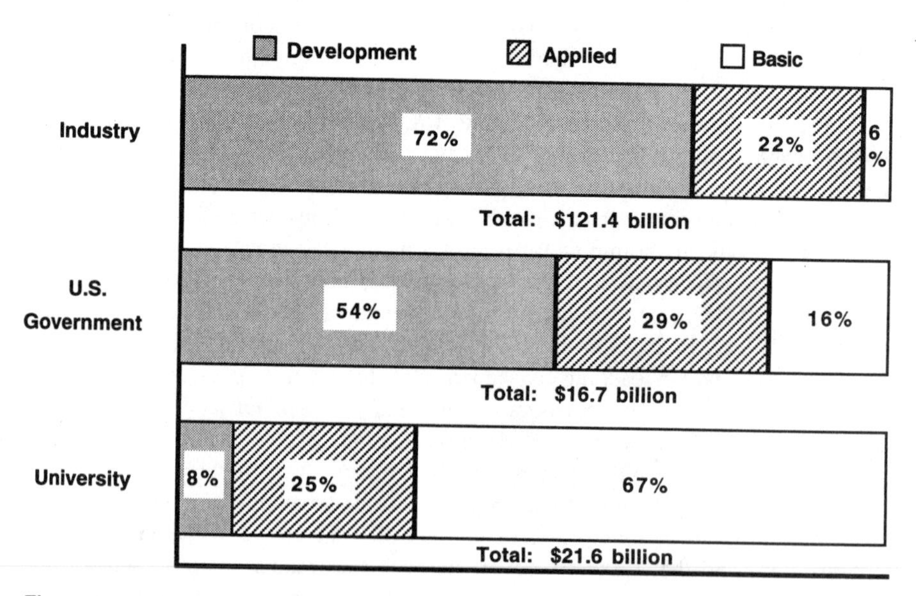

Figure 1.1. U.S. R&D spending by category for 1995. (Source: *Science and Engineering Indicators,* 1996.)

6% of the total industrial R&D is devoted to basic research. Expenditures on R&D by source, performer, character of work and sector are shown in Figure 1.2. During the last five years, trends in terms of character of R&D work have not changed much except there is somewhat more of an emphasis on applied research than on other areas.

In this chapter, indeed in this book, we argue that, in a productive and effective research organization, a researcher should have a mix of activities including basic, applied, and product development research. Examples of successful organizations and results of studies conducted are provided to support this assertion. For a manager of an R&D organization interested in productivity and effectiveness, understanding this issue is crucial and has important managerial implications. If we are successful in persuading you to include basic research in your mix of activities, even if your organization focuses on product development, would you not use the information in this chapter to persuade corporate decision-makers to allow this flexibility?

Is there any R&D manager who has not been accused of being unresponsive to customer needs and of focusing on esoteric, nonproductive research activities? Throughout this book a strong case has been made for customer participation in needs assessment and in the innovation process. The issue is much broader.

Let us consider an R&D organization that works only on those research needs identified by the customer. Would such an organization not be working on yesterday's, or, at best, today's problems in a very narrow framework? Using this approach, during World War II, would research have been working on bigger and better binoculars to detect incoming airplanes rather than on developing radar?

We propose a two-tier model, which includes an economic index model and a portfolio model, that should overcome some of these difficulties. Further, a systematic and a conceptual approach for prioritizing potential projects is presented. De-

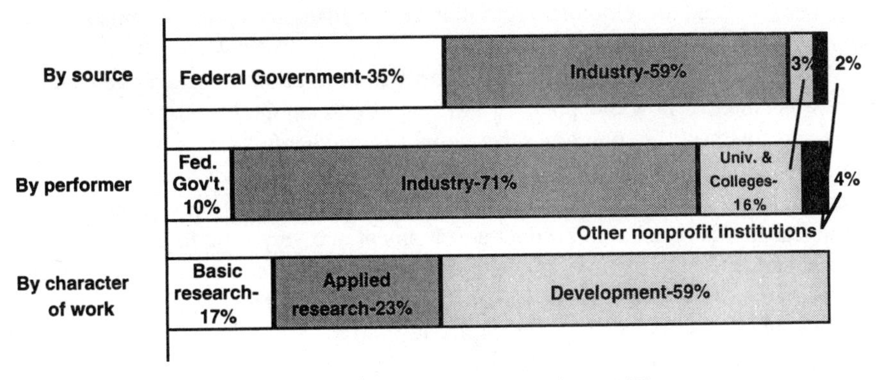

Figure 1.2. The national R&D effort. (Source: *Science and Engineering Indicators*, 1996.)

pending on the organizational setting and the decision-makers involved, this approach provides a crucial mechanism for research project selection and effective decision-making. By being systematic, it also gives psychological comfort to the decision-makers.

Oh yes, how about these mundane definitions! Anyone involved in research knows them, or should know them. Maybe so. Careful reading would show that there are some key points brought out that are not commonly appreciated. For example, what really differentiates basic research from applied research? Basic research is not inevitably unapplied. Differences lie elsewhere. If nothing else, these definitions may facilitate communication among the various actors involved in conducting and sponsoring research.

1.2 A PERSPECTIVE ON R&D MANAGEMENT

The ideas presented in this book focus on ways to improve the productivity of R&D organizations and foster excellence in such organizations. The book is primarily aimed at principal investigators, their colleagues, and supervisors. As indicated, others may also find the information presented here interesting.

In mathematics or physics, most concepts can be readily judged as useful or worthless. Management concepts, on the other hand, are more difficult to evaluate. The following example might illustrate the case.

One well-known scientist was recruited to be vice-president of a biotechnology company. In trying to prepare for this important new position he took a course at the California Institute of Technology on "Managing Research and Development." After completing the course the scientist felt that the course had failed to teach him how to prioritize and manage research projects. On his evaluation he stated that the course had been "expensive and worthless." In response to this criticism, the course program director pointed out that the scientist had "completely misunderstood the goals of the course." According to the director, the course was geared toward planning research and development activities rather than managing scientists [*Wall Street Journal,* November 10, 1986].

Managing researchers is one of the most daunting tasks a manager can undertake. It is not clear how one plans or anticipates a "scientific breakthrough." If this is the case, is there any point in undertaking extensive efforts in strategic planning or doing any planning at all? Scientists are thought to be dedicated to ideas and research. However, as shown in Figure 1.1, except at universities, a great majority of the research is devoted to product development and very little to basic research. The challenge then is to provide a mix of activities to achieve organizational goals and sustain the researcher's motivation and curiosity, which are essential to scientific breakthroughs and product development.

The effect public policy and management decisions have on the resources available for R&D is well understood; one needs to consider and understand, also, the important role engineers and scientists can and should play in developing science policy. Of the approximately 450,000 doctoral scientists and engineers employed

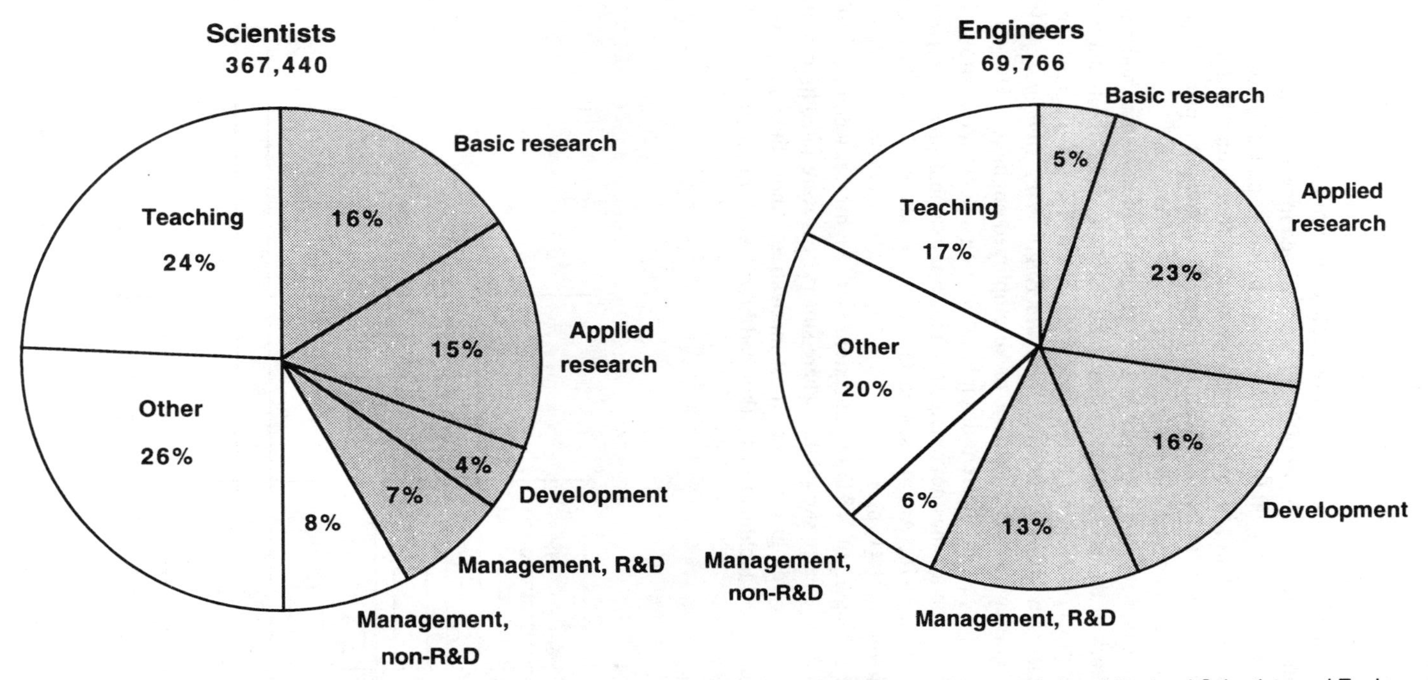

Figure 1.3. Employed doctoral scientists and engineers, by primary work activity: 1991. (Source: Characteristics of Doctoral Scientists and Engineers in the United States, NSF 94-307, 1994.)

in the United States as of 1991, approximately 160,000 work in R&D and another 33,000 in management of R&D [*Characteristics of Doctoral Scientists and Engineers in the United States,* 1991, p. 37]. The remaining doctoral scientists and engineers are involved in many forms of professional practice, in addition to the substantial number who teach (99,000). Those involved in professional services and consulting number nearly 60,000. Consulting engineers and scientists undertake creative activities that are, in many ways, responsible for closing the loop between research and development and application. Figure 1.3 shows the primary work activities of doctoral scientists and engineers.

A doctorate is a research degree, and the majority of scientists and engineers with PhDs work in research, development, and teaching. It is significant that relatively few engineers, as compared to scientists, hold doctoral degrees. In 1991, 69,800 engineers held doctorates [*Science and Engineering Indicators,* 1993, p. 76], which represented only about 3% of all employed engineers. The percentage of employed engineers in different disciplines holding a doctorate is shown in Figure 1.4. This percentage has not changed much. For example, the same proportion held doctorates in 1976. Among scientists, however, about 20% hold doctorates [*Science Indicators,* 1985, p. 61].

We favor managers of R&D organizations with high-level technical skills, because studies have clearly shown that where supervisors were rated highest in technical skills the research groups were most innovative. And where supervisors did not possess excellent technical skills (but had high-level administrative skills), the

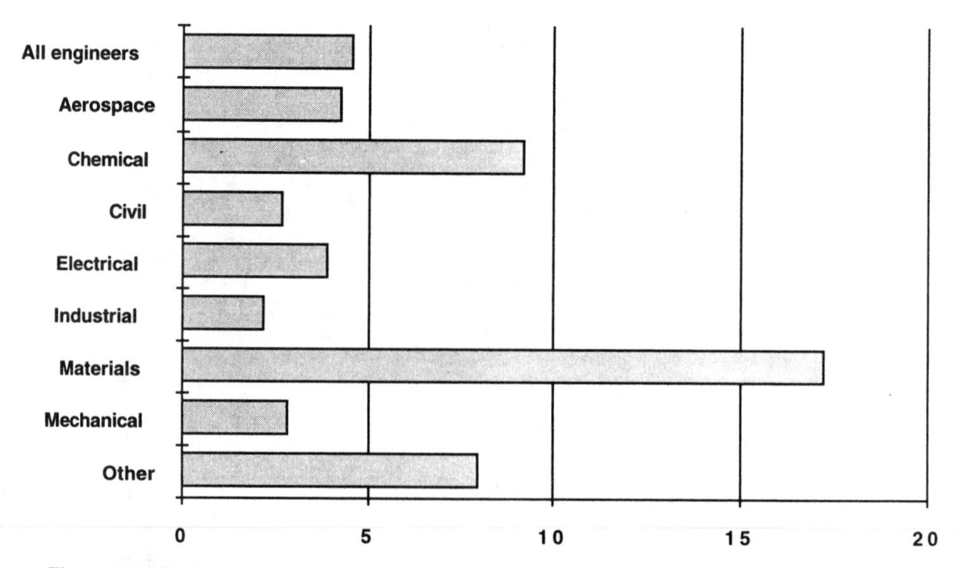

Figure 1.4. Percentage of employed engineers holding a doctorate: 1993. (Source: *NSF Division of Science Resources Studies*, 1996.)

research groups were least innovative [Farris, 1982, p. 340]. These findings in no way minimize the importance of administrative skills, but rather point to a fundamental need for a supervisor in an R&D organization who possesses excellent technical skills. Ideally, both kinds of skills should be available to a manager. Consequently, the role of a scientist* in managing R&D organizations has and will continue to be an important one.

To make sure we communicate effectively, we must first define some basic terms. We will do this in the next section.

1.3 WHAT IS RESEARCH AND DEVELOPMENT?

The National Science Foundation (NSF) classifies and defines research as follows [*Science and Engineering Indicators,* 1993, p. 94]:

Basic Research. Basic research has as its objective "a more complete knowledge or understanding of the subject under study, without specific applications in mind." To take into account industrial goals, NSF modifies this definition for the industry sector to indicate that basic research advances scientific knowledge "but does not have specific immediate commercial objectives, although it may be in fields of present or potential commercial interest."

Applied Research. Applied research is directed toward gaining "knowledge or understanding to determine the means by which a specific, recognized need may be met." In industry, applied research includes investigations directed "to discovering new scientific knowledge that has specific commercial objectives with respect to products, processes, or services."

Development. Development is the "systematic use of the knowledge or understanding gained from research, directed toward the production of useful materials, devices, systems or methods, including design and development of prototypes and processes."

The Organization of Economic Co-operation and Development (OECD) in its publication, *The Measurement of Scientific and Technical Activities* [1993], defines some research activities as follows:

Basic research is experimental or theoretical work undertaken primarily to acquire new knowledge of the underlying foundations of phenomena and observable facts, without any particular application or use in view. Basic research analyzes properties, structures, and relationships with a view to formulating and testing hypotheses, theories or laws. The results of basic research are not generally sold but are usually published in scientific journals or circulated to interested colleagues. *Pure basic research*

*Whenever we are considering engineering, technology, or pure science for the purpose of this book, the word *scientist* is used to apply to a person (engineer or scientist) who possesses the technical knowledge and skills that are essential to the work of an R&D organization.

> is carried out for the advancement of knowledge, without working for long-term economic or social benefits and with no positive efforts being made to apply the results to practical problems or to transfer the results to sectors responsible for its applications. *Oriented basic research* is carried out with the expectation that it will produce a broad base of knowledge likely to form the background to the solution of recognized or expected current or future problems or possibilities. *Applied research* is also original investigation undertaken in order to acquire new knowledge. It is, however, directed primarily towards a specific practical aim or objective. Applied research develops ideas into operational form. *Experimental development* is systematic work, drawing on existing knowledge gained from research and practical experience, that is directed to producing new materials, products and devices; to installing new processes, systems and services; or to improving substantially those already produced or installed.

Research and development covers many of these activities. The OECD defines R&D as "creative work undertaken on a systematic basis in order to increase the stock of knowledge of man, culture and society, and the use of this stock of knowledge to devise new applications."

In order to provide functional and understandable definitions for various research activities, *Science Indicators* categorizes R&D activities as efforts in science and engineering as follows:

- Producing significant advances across the broad front of understanding of natural and social phenomena—*basic research.*
- Fostering inventive activity to produce technological advances—*applied research and development.*
- Combining understanding and invention in the form of socially useful and affordable products and processes—*innovation.*

Many United States governmental agencies have categorized research and development activities to provide a better focus on these activities and, ostensibly, to facilitate technology transfer. One such categorization for the U.S. Department of Defense (DOD) is depicted in Table 1.1. Since DOD accounts for approximately 60% of the Federal Government's R&D expenditures, some understanding of its research program categorization would be helpful to those seeking research support from the DOD.

1.4 RESEARCH CATEGORIES

Harvey Brooks [1968, p. 46] has suggested a general set of dimensions and categories of research:

- The degree to which the research is fundamental or applied—for example, basic research versus applied research and development. The term "fundamental" refers to an intellectual structure, a hierarchy of generality, while the term "applied" refers to a practical objective. It is true that fundamental research is generally less closely related to practical application, but not inevitably so.

TABLE 1.1 U.S. Department of Defense Research Program Categorization

6.1 *Research: Directed to the Development of Fundamental Knowledge.* Includes scientific study and experimentation directed toward increasing knowledge and understanding in those fields of the physical, engineering, environmental, biological–medical, and behavioral–social sciences related to long-term national security needs. It provides fundamental knowledge for the solution of identified military problems. It also provides part of the base for subsequent exploratory and advanced developments in defense-related technologies and of new or improved military functional capabilities in areas such as communications, detection, tracking, surveillance, propulsion, mobility, guidance and control, navigation, energy conversion, materials and structures, and personnel support.

6.2 *Exploratory Development: Directed to the Development of New Techniques, Methodologies, and Criteria.* Includes all effort directed toward the solution of specific military problems, short of major development projects. This type of effort may vary from fairly fundamental applied research to quite sophisticated breadboard hardware, study, programming, and planning efforts. It would thus include studies, investigations, and minor development effort. The dominant characteristic of this category of effort is that it be pointed toward specific military problem areas with a view to developing and evaluating the feasibility and practicability of proposed solutions and determining their parameters.

6.3 *Advanced Development: Concerned with Design and Development and Hardware (Material) Items for Experimentation.* Includes all projects that have moved into the development of hardware for experimental or operational test. It is characterized by line item projects and program control is exercised on a project basis. A further descriptive characteristic lies in the design of such items being directed toward hardware for test or experimentation as opposed to items designed and engineered for eventual service use.

6.4 *Engineering Development: Directed to Testing and Demonstration of New Techniques or Methodologies, and to Technical Systems Equipment.* Includes those development programs being engineered for service use but that have not yet been approved for procurement or operation. This area is characterized by major line item projects and program control will be exercised by review of individual projects.

6.5 *Management and Support: Directed to the Support of Installations for Their Operations and Maintenance and for the Procurement of Special Purpose Equipment.* Includes research and development effort directed toward support of installations or operations required for general research and development use. Included would be test ranges, military construction, maintenance support of laboratories, operation and maintenance of test aircraft and ships, and studies and analyses in support of the R&D program. Costs of laboratory personnel, either in-house or contract operated, would be assigned to appropriate projects or as a line item in the research, exploratory development, or advanced development program areas, as appropriate. Military construction costs directly related to a major development program will be included in the appropriate element.

Source: AR70–9 Army Research Information Systems and Reports, May, 1981, NTIS, Springfield, VA.

- The scientific discipline—for example, physics, chemistry, or biology.
- The function of the research, or its primary focus—for example, defense, health, or environment.
- The institutional character of research—for example, academic (university), governmental laboratory, or industrial.
- The scale of research or style of research—for example, big science versus little science.
- The extent to which the research is multidisciplinary focusing on a single class of objects—for example, environment, space science, oceanography, or requiring multiple disciplines.

For planning purposes, Brooks [1968, p. 57] has suggested three broad categories of research organizations: mission-oriented research, scientific institutional research, and academic research.

Mission-Oriented Research Organizations

The term "mission" refers to an objective defined in terms of the long-range goals of the organization rather than a specific technical objective. Examples of such organizations include Department of Defense research laboratories and industrial research laboratories. Such research laboratories are vertically integrated organizations that conduct both basic and applied research and may provide technical support for operation or manufacturing. While their research may be of the most sophisticated and fundamental type, it is directed to fulfilling the objectives and the mission of the organization rather than to the development of science per se.

Scientific Institutional Research Organizations

This covers organizations whose mission is defined primarily in scientific terms—for example, advancement of high-energy physics or molecular biology. Such research organizations follow some sort of a coherent program adapted to changing frontiers in their area of interest.

Academic Research Organizations

Academic research is usually small-scale basic research carried out in academic departments of universities by students or research associates under the direction of university professors who also teach.

1.5 WHAT TO RESEARCH

There are few discussions of research funding, research program planning, and execution that do not include comments about what really ought to be researched.

Governmental agency and industry management hierarchies constantly talk about the need for a better focus on research programs so that research will meet agency and organization needs. Users in production departments, operational personnel in agencies, and consumers often complain about the lack of relevance of the research program and about the lack of timeliness of research results.

Let us take the case of a research laboratory where sponsors, though quite satisfied with the research output of the laboratory, nonetheless provided these kinds of comments about the research program:

- Research takes too long.
- Our need to solve the groundwater contamination problem is now, not three years from now. We just can't wait for years for researchers to study the problem.
- We need answers quicker than researchers provide them.
- The research program is too esoteric. We need solutions that are practical.
- Researchers study the problem to death to find a 100% solution. What is wrong with a quicker solution which is not quite 100%?
- This problem seems to go on forever. Five years ago I worked at the Department of the Interior. We thoroughly studied the problem of land disposal of hazardous toxic waste. I thought we solved the problem or at least put the issues to bed. When asked whose bed and what were the results, the sponsor did not know.
- We always hear about your previous accomplishments. How about the future? What can we expect from you next year and the year after? Be specific.

First and foremost, R&D managers need to understand the sponsor's perspective and then develop a strategy for effective communication. Recall (Figure 1.1) that 96% of industry R&D expenditure is for applied research and development and that 83% of U.S. government R&D expenditure also is for applied research and development. Consequently, the focus of such research is rather "specific," "commercial" and "product-oriented." For the sponsors to raise questions, as exemplified in the preceding quotes, is to some degree understandable. Consequently, the response of the R&D manager or the PI need not be defensive. For basic research, however, issues are likely to be of a different nature.

How, then, should one respond? One could take each question and provide extensive documentation to refute the sponsor's assertion. For example, one could prove that studying and solving the groundwater contamination problem, which was created through decades of neglect, would take some time. Solutions, especially cost-effective and environmentally safe solutions, may well take three years, or even longer, to find. One could also ignore sponsor assertions and go on with the research activity since the sponsor is not likely to find any other researcher who could do the work any faster anyway.

Another approach that an R&D manager could utilize would be a two-part strategy:

- First, empathize with the sponsor's needs and be responsive in a genuine manner. This would translate to providing interim solutions, to the degree possible, for critical problems. Explain to the sponsor the limitations and uncertainties involved.
- Second, educate the sponsor regarding the nature of the research enterprise. Focus on why it is in his/her best interest to follow a systematic, though time-consuming, process of research and development so that solutions developed are scientifically valid, are appropriate to the problem at hand, and truly provide a more advantageous solution to the problem than the existing technology does. This could involve undertaking a mix of research activities ranging from basic research that might take three to five years to applied research that might provide some solutions within one to two years.

What to research is also affected by what our adversaries or competitors are doing. Some governmental agencies (for example, the Department of Defense) and some industries (for example, high technology) often are concerned about being surprised by a technological development by an adversary or competitor. This is simply because the payoff or effectiveness of the defense establishment of a nation, or profitability of an industry, depends on its own capabilities and also on the capabilities of its adversaries or competitors. New technological developments of an adversary or a competitor can have a profound effect on the security of a nation and on the competitive success of an enterprise.

Other questions and issues related to the issue of what to research often include the following:

- How should user needs be considered?
- Who are the real users?
- How should a comprehensive and responsive research program be formulated?
- How should the tradeoffs between long-range research needs and short-range or immediate requirements be made?

Many approaches to formulating research programs have been proposed. For example, Merten and Ryu [1983, pp. 24–25] have proposed dividing an industrial laboratory's research activities into five categories:

- Background research
- Exploratory research
- Development of new commercial activities
- Development of existing commercial activities
- Technical services

Schmitt [1985] has discussed generic versus targeted research and market-driven versus technology-driven research. Shanklin and Ryans [1984] contend that high-technology companies can make a successful transition from being innovation-driven to being market-driven by linking R&D and marketing efforts.

A considerable literature is available related to R&D project selection. The proper approach applicable to an organization would clearly vary depending on the needs of an organization. Publications relevant to this are included under Further Readings at the end of the chapter.

Two criteria seem most important in deciding what to research: (1) What will advance the science? and (2) What do the customers of our research need? Once we have answered those questions, we need to ask: What are the prospects for a solution?

There are other considerations that may override them. Other criteria may apply in the solution of very specific problems. For example, in oil exploration, safety considerations may be a top research priority. Such problems may have to be solved regardless of cost because the organization would be wrong to ignore them. Research needed to protect human health and the environment from improper disposal of hazardous waste falls in the same category.

One of the most difficult problems is deciding when to abandon a problem that does not seem to be solvable. There is always the hope that with a few more months of work the problem will be solved. Yet, one usually has some sense of what is likely to happen. If one researcher is sure that the problem can be solved and no one else is so convinced, it is necessary to determine whether the one researcher is a "genius" or a "neurotic." People do get attached to hopeless causes, and when that happens they exhibit a variety of such symptoms as extreme tension and the inability to be self-critical. Managers must be sensitive to clues that indicate that the optimism about a project is unjustified. Since stopping such a project without destroying the motivation of the scientist is important, some suggested approaches to achieve this follow.

A manager may agree to give the scientist short deadlines and establish mutually agreed-upon milestones to ascertain whether tangible progress toward the goal is being made. If the project indeed is hopeless, lack of project progress during the milestone review would reveal the problem. In most cases, the scientist would, on his own initiative, agree to drop the project.

Should the scientist still request to continue the project, the manager should consider allowing the scientist to spend some time (say 20%) on the project and again establish agreed-upon milestones to review progress. If results again are not very promising and the scientist still perseveres and wants to continue, two options are possible. One, the manager may direct that the project be stopped. The other possibility is to still allow the scientist to spend some time on the project but strip away all support, such as for laboratory equipment, computer expenses, and technicians. In time the project will fade away.

The manager, however, should not be too surprised when some researchers supposedly pursuing unpromising theories or projects thought to be nonproductive in their early stages end up producing promising results. It is good for all concerned, especially for the manager, to keep in mind that predictions about the success or failure of research projects are most unreliable. Two examples come to mind, one dealing with fundamental research and the other with applied research.

Astrophysicist S. Chandrasekhar was working on the theory of black holes and white dwarfs. He sought to calculate what would happen in the collapse of larger stars when they burn out. He theorized that if the mass of a star was more than 1.4 times that of the sun, the dense matter resulting from the collapse could not withstand the pressure and thus would keep on shrinking. He wrote that such a star "cannot pass into the white dwarf stage." His paper on this theory was rejected by the *Astrophysical Journal,* of which he was later to become a well-respected editor.

As reported in the *New York Times* (October 20, 1983), Sir Arthur Edington, rejecting Dr. Chandrasekhar's theory, stated that "there should be a law of nature to prevent the star from behaving in this absurd way." Chandrasekhar was urged by other scientists to drop his research project because it did not seem very promising. Dr. Chandrasekhar persisted and in 1983 won the Nobel Prize for his discovery. His research led to the recognition of a state even more dense than that of a white dwarf: the neutron star. The so-called Chandrasekhar limit has now become one of the foundations of modern astrophysics.

As another example, a group of researchers developing a complex environmental impact analysis system and associated relational databases chose to pursue this research project by using a higher-order computer language instead of the traditional FORTRAN. They also wanted to experiment using an operating system developed by the Bell Laboratories. Management attitudes ranged from enthusiastic support to tepid support, opposition, and downright hostility. The less technically knowledgeable managers were opposed; and the further removed they were from the research group, the more opposed they were to the continuation of this research project. Because of the creativity of the researchers and with some degree of support and acquiescence of the management, the project was allowed to continue in parallel with other activities. On completion the project was one of the most successful and one of the most widely used systems in the agency. It received the agency's highest R&D achievement award and became an archetype for future systems development research activities.

No one approach for categorizing or organizing research and for identifying the research needs of an agency or an industrial enterprise may satisfy the complex and, at times, unique needs of an organization. We propose a two-tier model for identifying "what to research," in an effort to develop an approach that provides a flexible, systematic framework for integrating various requirements that at times seem in conflict with each other. The model includes an economic index model and a portfolio model. This two-tier model may apply more readily to mission-oriented research than to scientific institutional or academic research. Further discussions of this model follow.

Economic Index Model

Under this model, research needs are defined as those needs designed to improve the operation or manufacturing efficiency of the organization or the enterprise. The emphasis is on building a "better mousetrap" to reduce the cost of doing things. Inputs for such needs come from the users, operation units, and scientists, as well as from looking at competitive products and operations.

Portfolio Model

Under this model, normative, comparative, and forecasted research needs are considered. *Normative needs* are those of the user (a user being the primary or follow-on beneficiary of the research product). *Comparative needs* relate to research needs derived from reviewing comparable organizations, competitive product lines, and related enterprises. *Forecasted research needs* focus on trend analysis in terms of consumer or organization needs derived from new requirements, changed consumer behavior, new technological developments, new regulations (e.g., environmental, health, and safety regulations), and new operational requirements. Often the effectiveness of a commercial enterprise or of a national defense effort depends not only on how well the organization itself does but also on how well the organization does in comparison with its competitor or adversary. Consequently, it is necessary to have effective intelligence concerning the portfolio of a competitor in order to focus properly on comparative and forecasted research needs.

After defining research needs using these two models, some research projects would be essentially modifying, adapting, or adopting existing scientific knowledge and would correspond to applied research and development; other research projects would fill technology gaps and would correspond to basic or fundamental research.

Inevitably, there are more projects to be researched than there are funds available. This is a normal and a healthy situation. A model derived from the work of Keeney and Raiffa [1976], which takes into account multiple objectives, preferences, and value tradeoffs, is suggested for deciding which projects to select among competing requirements. The main problem in using such an approach is the tendency on the part of many technical users to quantify items that do not lend themselves to quantification.

In developing a policy (at higher levels) or in making specific project choices among competing demands (at lower levels), the decision-maker can assign utility values to consequences associated with each path instead of using explicit quantification. The payoffs are captured conceptually by associating to each path of the tree a consequence that completely describes the implications of the path. It must be emphasized that not all payoffs are in common units and many are incommensurate. This can be mathematically described as follows [Keeney and Raiffa, 1976, p. 6]:

$$a' \text{ is preferred to } a'' \Leftrightarrow \sum_{i=1} P'_i U'_j > \sum_{j=1} P''_j U''_j$$

where a' and a'' represent choices, P probabilities, and U utilities; the symbol $\Leftrightarrow$ reads "such that."

Utility numbers are assigned to consequences, even though some aspects of a choice are not in common units or are subjective in nature. This, then, becomes a multiattribute value problem. This can be done informally or explicitly by mathematically formalizing the preference structure. This can be stated mathematically [Keeney and Raiffa, 1976, p. 68] as

$$\nu(x_1, x_2, \ldots, x_n) \geq \nu(x_1', x_2', \ldots, x_n')$$
$$\Leftrightarrow (x_1, x_2, \ldots, x_n) \gtrsim (x_1', x_2', \ldots, x_n')$$

where ν is the value function that may be the objective of the decision-maker, x_i is a point in the consequence space, and the symbol $\gtrsim$ reads "preferred to" or "indifferent to."

After the decision-maker structures the problem and assigns probabilities and utilities, an optimal strategy that maximizes expected utility can be determined. When a comparison involves unquantifiable elements, or elements in different units, a value tradeoff approach can be used either informally, that is, based on the decision-maker's judgment, or explicitly, using mathematical formulation.

After the decision-maker has completed the individual analysis and has ranked various policy alternatives or projects, then a group analysis can further prioritize the policy alternatives or specific projects. A modified Delphi technique [Jain et al., 1980] is suggested as an approach for accomplishing this.

After research project selection and prioritization, an overall analysis of the research portfolio should be made. The research project portfolio should contain both basic and applied research. The mix would depend on the following:

- Technology of the organization
- Size of the organization
- Research staff capabilities
- Research facilities
- Access to different funding sources

It should be noted that the distinction between basic and applied research can become rather blurred. What is *basic* research to one organization can be *applied* to another, and what is *basic* one year can be *applied* the next. Also, given the same general research project title, different emphases during project execution can affect the nature of research. As will be discussed below, to maximize R&D organizational effectiveness, scientists and work groups should be involved in a mix of basic and applied research.

1.6 EMPHASIS ON BASIC VERSUS APPLIED RESEARCH

We have discussed some research organization categorization and ways of developing an R&D portfolio. For planning purposes, three types of research organization categorization were presented. The emphasis on basic research versus applied research within each organization varies; consequently, there is a certain amount of conflict. The conflict is due to the fact that basic research is often dictated by the questions that science is asking. Such research may require activities that are not compatible with the mission-oriented research that a commercial or government organization is supposed to do. For example, a scientist while reading a scientific journal may have an insight that requires further experimentation. However, his supervisor may have already asked him to develop a particular product that meets particular specifications. Obviously the two activities are incompatible and some of the conflict that occurs within the scientist is due to the conflict between the need to discover and the requirements of the organization.

Some quite successful organizations—for example, 3M in Minnesota—have developed procedures that allow their scientists a certain amount of time to work on topics that are of interest to them. What percent of the scientist's time will be spent on such topics, and when such activities should take place, are matters of negotiation between the scientist and his or her supervisor. A successful scientist, who has had a better track record, may be given more time to discover other things by pursuing his or her own interest than one who does not have a good track record.

Pelz and Andrews [1966a] did a study of 1300 scientists in 11 laboratories. They studied scientists in both industrial and government laboratories and they used five criteria to identify successful scientists: (1) the judgments of their peers, (2) the judgments of their boss, (3) the number of papers they published, (4) the number of patents they were awarded, and (5) the number of reports they issued. They then conducted intensive interviews to identify what discriminated the effective from the less effective scientists. One of the findings was that the more effective scientists did both basic and applied research.

We will return to the study of Pelz and Andrews throughout this book; but for the time being, one basic point that we should keep in mind when thinking about how to structure research and development organizations is that both kinds of research are done by the more effective scientists. It is obvious that if a scientist has an insight while reading a journal that requires an experiment, the inability to do the experiment will be quite frustrating. It is exactly this point that indicates that some sort of freedom to experiment should be allowed by the organization. If reading scientific journals results in frequent frustration, it is very likely that the scientist will become obsolete by giving up such reading. Similarly, the organization should encourage its scientists to publish, since this provides an opportunity for the organization to acquire prestige in the eyes of the scientific community and also tests the capabilities of the individual scientist to become effective in relating to the wider scientific community.

It should be remembered that there are about 7000 journal articles published every day in the sciences. Thus the output of any particular individual is a minute contribution to a very large pool of activity. However, the fact that a person has made a contribution essentially "buys" the ticket that allows him or her to interact with other scientists, to learn from them, and to discover what they are currently doing.

1.7 WHAT IS UNIQUE ABOUT MANAGING R&D ORGANIZATIONS?

R&D organizations are different from other organizations because of the people working in such organizations, the ideas that are generated, the funds or research support that are obtained, and the culture of the organization. These four elements—people, ideas, funds, and culture—are the basic elements of an R&D organization and are discussed in detail in the next chapter. A brief review of each element as related to an R&D organization's uniqueness follows.

People

People in R&D organizations normally would have graduate training and relatively high aptitude. They are socialized during their graduate training to work autonomously and show considerable initiative.

An anecdote will help convey more clearly what is special about R&D personnel. The famous German scientist Hermann Helmholtz put a sign up on his lab: "Do not disturb." This was all that his students and collaborators were able to see for a month. After some 30 days Helmholtz emerged with an important new theory which eventually led to the development of radio and television (related in Boring, 1950).

Ideas

Ideas in an R&D organization are generated through a unique communication network (discussed in the next chapter) and facilitated by the ethos of a scientific community (discussed in Chapter 3).

Funds

In general, funding sources for R&D organizations are different from those for any similar large enterprise. For example, in the United States about 42% (1993) of funds for R&D are provided by the federal government. The federal government spends approximately four times as much on basic research as does industry. Even for academic institutions, the majority of research funding support, 57% (1993), is derived from the federal government. This funding support, coupled with research productivity benefits that accrue to society at large rather than the individual or the sponsoring organization, gives R&D organizations a unique characteristic.

Culture

The culture of an organization relates to both objective and subjective elements. For an R&D organization, objective elements such as research laboratory facilities and equipment and office buildings are different from those of other organizations. Subjective elements such as rules, laws, standard operating procedures and unstated assumptions, values, and norms for an R&D organization are also different. For example, scientific discoveries, whatever their source, are subjected to impersonal judgments, and scientists often participate in organized skepticism and critically evaluate scientific ideas and discoveries. This permeates all aspects of an organization's function. Management decisions affecting individuals are thus critically evaluated and questioned by the researchers. After attending a senior management conference, a newly assigned deputy administrator of a federal research organization stated that he had never worked in an organization where people were so vocal and where management decisions were reviewed and discussed as openly and fully.

The culture and other elements vary from one R&D organization to another; however, as a group, R&D organizations generally possess unique characteristics.

1.8 SUMMARY

We first pointed out that the essence of R&D management is the coordination of the activities of many individuals. An effective R&D organization should have a mix of research activities that are both basic and applied. The chapter provided definitions of terms such as basic and applied research and development, and it reviewed proposals for a system of categories of research. One key issue is "What to research?" A model that deals with this question was presented. Finally, we examined what is unique about managing R&D organizations. One unique aspect is the need for the intricate coordination of people, ideas, funds, and culture. In the next chapter we discuss these elements and their coordination further, and the rest of the book is concerned with how a manager can be most effective and lead an organization that will be most productive.

1.9 QUESTIONS FOR CLASS DISCUSSION

1 How much R&D is too much for a corporation? When is it not enough?

2 How much R&D is too much for a country? When is it not enough?

3 Define and compare basic and applied research.

4 How much basic research is desirable in what kind of an R&D lab?

5 Take the actual case of a government or industry research laboratory. Develop a systematic procedure and a short-term and long-term research plan.

1.10 FURTHER READINGS

Allio, R. J. and D. Sheehan (1984). Allocating R&D resources effectively. *Research Management,* **27**(3), 14. The model proposed was used for R&D programs at the Allied

Corporation. It suggests that R&D expenditures should support the business strategy and be sufficiently intensive and yield marketplace results that can be sustained in the face of vigorous competition. Technological innovations that do not improve competitive position are not cost-effective. R&D results should be converted into a product that is marketable in the face of vigorous competitive research response. Also, the business must have in place an appropriate marketing and distribution system and have adequate financial resources to withstand competitive situations.

Brenner, M. S. (1994). Practical R&D project prioritization. *Research-Technology Management,* **37**(5), 38–42 (September–October). A systematic project solution process used at Air Products & Chemicals Inc. identifies and builds consensus around the key issues for success, communicates these factors to improve project proposals, and helps to extend limited funding to maximize project progress and completion. Decision-makers select and weigh criteria in a structured framework based upon the analytic hierarchy process. Project champions then propose their projects within that framework. Project strengths and weaknesses are clearly identified by using profiles of the project ratings for each criterion.

Ellis, L. W. (1984). Viewing R&D projects financially. *Research Management,* **27**(2), 29 (March–April). For project selection purposes, the paper focuses on R&D projects as investments for prospective return. Internal rate of return is offered as a tool for project selection and evaluation.

Erickson, T. J. (1993). Managing the link to corporate strategy. *Management Review,* **82** (12), 10–18 (December). In the newest generation of R&D management, technology plays an integral role in corporate strategy. Companies articulate why they are investing in technology, identify the importance of technology to the company's long-term success, facilitate an environment in which senior management views technology in a partnership mode, and view technology in a portfolio sense.

Gibson, J. E. (1981). Rational selection of R&D projects. In *Managing Research and Development.* New York: Wiley, p. 289. This chapter discusses issues relating to ranking, scoring, or rating methods, economic rating methods, formal optimization methods, risk analysis, and decision analysis methods. Some examples of scoring methods for R&D projects are also presented. A staged approach to R&D project selection and some interactive decision analysis methods are also discussed.

Jackson, B. (1983). Decision methods for selecting a portfolio of R&D projects. *Research Management,* September–October, p. 210. This approach for selecting a portfolio of research projects is based on using analytical techniques such as linear programming, dynamic programming, and chance constraint programming techniques.

Krawiec, F. (1984). Evaluating and selecting research projects by scoring. *Research Management,* March–April, **27**(2), 21. In this paper, scoring, augmented by subjective probabilistic risk assessment, is seen as the most suitable ranking technique for developing a balanced R&D portfolio. Solar thermal R&D is used as an example.

Schmidt, Robert L. A model for R&D project selection with combined benefit, outcome and resource interactions. *IEEE Transactions on Engineering Management,* 1993, **40**(4), 403–410 (November). Three types of interactions are generally recognized to occur within a set of projects: (1) benefit interactions, (2) resource interactions, and (3) outcome interactions. A model is presented that accounts for the combined effect of benefit, outcome, and resource interactions within a single set of projects. A branch and bound algorithm is presented to solve the resulting nonlinear integer program with multiple quadratic constraints.

Winkofsky, E. P., R. M. Mason, and W. E. Sauder (1980). R&D budgeting and project selection: A review of practices and models. In B. V. Dean and J. L. Goldhar (Eds.), *Management of Research and Innovation. TIMS Studies in the Management Sciences,* Vol. 15, New York: North-Holland, p. 183. This paper describes major aspects of R&D budgeting and project selection practices. This descriptive work is then used to evaluate the state of the art in quantitative models of R&D project selection.

2 ELEMENTS NEEDED FOR AN R&D ORGANIZATION

The basic elements required for an R&D organization are (1) people, (2) ideas, (3) funds, and (4) cultural elements. These four basic ingredients have to be coordinated with skill by the management of R&D organizations in order to achieve high productivity and excellence. In this chapter we will cover some of the introductory topics concerning these basic elements. In later chapters we will focus more specifically on the task of coordinating and managing.

It is obvious that the most important element is creative people. Such people have the bright ideas and skills to do research and then translate research results into useful products. However, these people must be organized into structures that permit effective cooperation. In doing so it is important to keep in mind that certain mixes of people work better than others. To ensure a smoothly functioning organization, one needs unstated assumptions, beliefs, norms, and values—in other words, an organizational culture that will favor creativity and innovation. Last, but not least, one needs funds.

2.1 PEOPLE

The kinds of people who are most likely to succeed in a R&D organization are those who are analytical, curious, independent, intellectual, and introverted and who enjoy scientific and mathematical activities. Such people tend to be complex, flexible, self-sufficient, task-oriented, and tolerant of ambiguity, and they have high needs for autonomy and change and a low need for deference [Winchell, 1984].

However, success in an R&D organization requires joint action; that is, people should not be loners. Thus the extreme introvert may simply not fit.

A person with a graduate degree probably already has many of these attributes. Other important attributes, however, may be lacking. For example, it is necessary to scrutinize very carefully a person's tolerance for ambiguity and need for autonomy and change.

People with internal standards and self-confidence are highly desirable, because in many cases research can be very discouraging. The person who is not easily discouraged and is sure of his goals and how to reach them is more likely to persist. Interaction with peers is also essential, since most new ideas are generated not by reading the literature but by talking with others who are working on similar problems. Finally, and this is admittedly cynical, a successful scientist needs to be able to tolerate what he might consider "bad management." The kind of person who gets upset too easily if the manager is insensitive to his needs may not be able to deal with a research environment. Most managers are technical people, interested in research rather than in managing others, so they are likely to do a less than optimal job. But there is a saving grace: Research has shown that people who enjoy their job can tolerate poor supervisors!

Another desirable attribute is internal locus of control. This is the tendency to think that the causes of events are internal (e.g., ability, hard work) rather than external (e.g., help from others, luck). Research has shown that internals are better than externals at collecting information and at deciding for themselves about the correct course of action [for a review see Spector, 1982].

Creativity is, of course, highly desirable. Unfortunately, there are few reliable and valid tests for this attribute. However, previous creativity is a good predictor of future inventiveness.

Friedman [1992] identified primary activities performed by R&D managers at different supervisory levels. The 48 tasks which managers related as having spent the most time on were factor analyzed, resulting in the identification of three primary activities: project management, personnel, and strategic planning. Strategic planning was rated as requiring significantly higher levels of logical reasoning, originality, fluency of ideas, communication skills, and resistance to premature judgment than was required for the other two primary activities. Strategic planning in R&D organizations is becoming an important activity; consequently, a new chapter on this topic has been added to this text.

In summary, an effective scientist needs to be an individualist [Allen, 1977] who has internal standards, self-confidence [Pelz and Andrews, 1966b], and persistence and who works in the right organizational environment. It is important to stress that even the most creative person will be a failure if the environment is not right. One can think of the analogy of a rectangle. The area of the rectangle depends on the size of both its sides. Similarly, creativity depends on both the attributes of the person and the environment. If either one is missing, creativity can be zero. An R&D manager has to be able to integrate the activities of diverse, autonomous, and talented people and must do well in handling activities associated with R&D project management, personnel, and strategic planning.

2.2 SPECIALIZATION

The question of specialization is also related to both person and environment. Some people enjoy specialization, while others prefer to be generalists. Some environments encourage and some discourage specialization. The literature suggests that successful R&D personnel are not overspecialized. They are interested in several topics and are able to talk with others about *their* problems with ease. Specialization can be tolerated in the early stages of a career, but later one looks for broader interests and the ability to talk constructively with a wide range of colleagues.

In selecting people who have such attributes, the manager can look for specific behaviors. For example, the kind of person who tolerates answers such as "probably," "approximately," and "perhaps" is likely to be tolerant of ambiguity.

Finally, when selecting members of an R&D team, it is desirable to look for managerial talent. Since such talent is generally rare among highly technical people, when it occurs it should receive some special attention. While technical competence is of the utmost importance in managers of R&D organizations, their ability to deal with people makes them especially desirable. They should therefore be selected over their peers who are equally technically competent but lack interpersonal skills.

One more criterion should be kept in mind in putting together the R&D team: It is desirable to choose a diverse workforce. It needs entrepreneurs, project leaders, gatekeepers, coaches, public relations people, and others [Roberts and Fusfeld, 1981]. One should consider the mix of people, as well as the fact that conditions do change and what is popular today may not be popular or fundable in 10 years. With a sufficient mix one may be able to survive during periods of radical change in the environment of R&D organizations.

Diversity in R&D organizations is another crucial issue. The chapter entitled "Dealing with Diversity in R&D Organizations" (Chapter 7) presents key issues related to this topic.

2.3 STAFFING

Often one hears managers say "people are our most important resource." Indeed, in an R&D organization, highly trained, able, and motivated researchers provided with well-equipped laboratories are essential. Excellent and productive R&D organizations are all characterized by such assets.

In staff selection and staff development, some social issues, such as equal employment opportunity and biases against certain ethnic groups and women, are examined briefly in the chapter on diversity. These are important issues and there is clear and ample historical evidence of such biases. These biases first manifest themselves in the way people raise their children or in the initial counseling received in high school. They continue during interviews for the first job and also when decisions are made concerning pay, staff development, and promotion to higher administrative and executive level positions. Other staffing issues such as need identification, interviewing, selection, placement, staff development, promotion, and pay are quite important but occur in R&D organizations just as in manufacturing organizations and are therefore already covered in the standard personnel literature. For this reason they will not be discussed here.

The selection of new employees should be done in collaboration with the people who are going to work with them. They are the ones who are most critical and the most involved. Furthermore, once they have participated, they will have some commitment to making that person a success in the organization. A work team interview is a good way to accomplish this.

The discussion here will focus primarily on the types of skills an R&D organization needs to facilitate the innovation process. These skills are categorized into three major areas:

- Support staff
- Technicians
- Research staff

Support staff includes such functions as financial management, contracting, technical editing, reference library work, typing, and other clerical duties.

Technicians include laboratory technicians, fabricators of experimental models, computer technicians, and laboratory and field experimental support staff.

Making support staff and technicians true members of the team, along with the research staff, is crucial for the success of the innovation process. They make a significant contribution to the innovation process, and their contribution needs to be recognized. Clockmakers were the first to apply scientific theories to the making of machines. Innovations came as a result of the collaboration among scientists, craftsmen, and mechanics. This collaboration, which was necessary for innovation centuries ago, is still required today. It is not uncommon for a clever technician to think of ways to set up an experiment or collect field data more efficiently, or for other support staff to facilitate administrative activities associated with the innovation process, thus saving time and effort. Often, the project sponsor's first contact is with the support staff (e.g., the receptionist or the secretary), and many technical assistance activities are handled by the technicians working closely with the user or the customer. Because of the crucial role support staff and technicians play in the innovation process, recruiting, training, and motivating them are quite important.

The entire staff needs to become integrated, as the following true story suggests. Not long ago, a professor of psychology at the University of Illinois used fruitflies as part of an experiment in behavioral genetics. Several generations of fruitflies had been developed to obtain the particular type needed for the experiment. Then one evening a janitor opened the laboratory windows and the draft killed the fruitflies. Several years of the professor's work had been inadvertently destroyed! Obviously, had the janitor understood the significance of the work he would not have opened the windows.

A similar disaster to an experiment occurred on an oceanographic ship. The crew and the scientists did not get along. One of their disagreements concerned what should be placed in the refrigerator—the scientists' specimens or beer. After 6 months of collecting specimens in the Pacific, the scientists discovered, to their horror, that the crew had thrown the specimens overboard and put the beer into the refrigerator!

For the *research staff,* more than idea-generating personnel are needed. Other critical functions involve entrepreneuring (marketing), communicating, gatekeeping, coaching, and project leading or supervising [Roberts and Fusfeld, 1981, p. 25]. There is some overlap among these functions, and an individual can perform more than one of them.

2.4 IDEAS

For idea-generating, the personnel needs to be technically competent in one or more fields and have the ability to conceptualize. They must be comfortable with abstract thinking and have a real interest in R&D.

In an R&D organization one finds that some people are particularly good at projecting beyond the obvious and thus generating ideas. To foster an idea-generating environment it is important to allow new ideas to be presented without immediately making judgments about their soundness. A group of researchers was once asked to present its ideas regarding some new research initiatives. After listening to the ideas, the managers quickly gave their comments and told the participants why the ideas were not particularly sound and thus could not be considered further. Participation in presenting new research decreased rapidly and after the initial two or three research presentations no one had anything more to offer. The research team finally disbanded because of low morale. Managers should not be too hasty to relegate ideas to the wastebasket.

Successful *entrepreneuring* or marketing requires individuals with the ability to sell or market new ideas to others and obtain resources for R&D projects. These individuals should be technically competent, possess a wide range of interests, and be energetic and willing to take risks. Entrepreneuring has some other important implications for organizational control and for organizational change. An organization that obtains much of its funding through the entrepreneuring activities of its research staff has to allow for more autonomy than others. Initiating new directions in research requires considerable participation by the affected research staff. A case in point is the type of research conducted at universities where much of the research funding is generated by individual faculty members. Consequently there is a strong tradition of faculty autonomy and dominance in academic institutions.

An important function in a laboratory is that of *key communicator* [Chakrabarti and O'Keefe, 1977]. A key communicator reads the literature in the field, particularly the "hard" papers, and talks frequently with outsiders and insiders in the laboratory. Chakrabarti and O'Keefe studied three government laboratories and found that about one-seventh of the professional staff could be described as doing that. Key communicators helped in a number of ways by providing desired information to others, locating written sources, participating in the generation of ideas, putting people in contact with each other, ending the search for nonexisting research in a particular area, evaluating ideas, offering support, selling a new idea, briefing key decisionmakers about recent developments in the field, and making contacts both outside and inside the laboratory to promote an idea. Key communicators were in supervisory positions only half the time. Such people, when identified, deserve an increased personal budget to facilitate travel, release time, formal recognition, and special training and encouragement, since they are invaluable for a laboratory.

Related to the idea of key communicator is the idea of a gatekeeper, or of a person in a "boundary-spanning role" [Keller and Holland, 1975]. Keller and Holland tested the hypothesis that such people might suffer from role conflict and role ambiguity and might be dissatisfied with their positions. Their data suggested that boundary-spanning activity did not produce much conflict, and even was positively correlated with job satisfaction with co-workers, pay, and promotions. However, it was negatively correlated with satisfaction with supervision. Thus, on the whole, this is a useful role that does not adversely affect those in it.

Also related to the role of key communicator is the role of "champion in product innovation" [Chakrabarti, 1974]. Such individuals are technically competent, know both the company and the market, are aggressive, and politically astute.

For gatekeeping,* individuals should possess a high level of technical competence, be personable and approachable, and enjoy contact with people and helping others. These individuals should keep themselves informed of related developments outside the organization via journals, professional conferences, and personal contacts. Gatekeeping is an informal role. Formalizing this role by assigning it to an individual or group would undermine the very purpose it is supposed to serve. In an R&D organization an individual with a high level of technical competence who has contacts with the wider scientific community and the appropriate personality frequently assumes this informal responsibility. Many times one hears the statement "If you have a question about acoustics, check with Dr. X, he can tell you what is the latest." Often the supervisor acts as a gatekeeper, especially for locally oriented projects, which will be discussed later in this section.

It is clear from what was stated above that there is a controversy about the extent to which gatekeepers or key communicators should be identified and rewarded. One view is that formalizing the role will undermine it; the other view is that by encouraging and rewarding the role these key functions could be done even better. We are inclined toward the latter view, but without too much emphasis on the singling out of the individual. Rather, top management might provide some extra travel allowances, some extra encouragement, and some more rewards without formally identifying the position of a key communicator or gatekeeper. It is the offering of support for the activity that is needed rather than a formalization of the role.

One can also make the case that it is not so much *how well* the role is carried out but how the *environment* in which the role is carried out is structured that deter-

*The gatekeeping concept is further discussed in the "Communication Networks" section (Section 2.5) of this chapter and in Chapter 11, "Technology Transfer." A gatekeeper essentially links the organization to external information sources.

mines the effectiveness of gatekeepers [Davis and Wilkof, 1988]. In a bureaucratic organization the best gatekeeper will fail; in an organic organization a moderately good gatekeeper will be effective. As Davis and Wilkof put it: "Most R&D groups long ago discovered that they cannot effectively operate with such a bureaucratic arrangement. They found that the restrictions and loss of autonomy inherent in the mechanistic structure stifled individual creativity and led to a sense of indifference and alienation, especially among those on the lower rungs of the organizational ladder" (p. 51).

In an organic organization, professionals are recognized for their expertise, and not for their position in the organizational hierarchy. That means that an expert at the lowest level of the organization may be heard as much as a nonexpert at the top of the organization. The corporation is viewed as an individual/team initiative system within which ideas are bought and sold, packaged, organized, and implemented. Status is based on technical competence. Top honors go to those who generate, package, and sell ideas. Management screens ideas to ensure their compatibility with overall corporate objectives.

The organic form of organization encourages every team member to be a gatekeeper. Instead of identifying gatekeepers, the organization supports the function of gatekeeping for all team members. Thus, gatekeeping is the rule, not the exception. With more gatekeepers there is a higher probability of tapping broader sources of information. Information is transferred across project lines. There is a relative egalitarian structure and highly participatory project management, and the emphasis is on the group's "collective intelligence." It is the group that might assign the role of gatekeeper to a particular member, for a particular topic. Group meetings are important vehicles for information transfer.

The organic form of organization, desirable as it is, is not without its problems [see Davis and Wilkof, 1988, pp. 56–57], such as the neglect of routine functions and the danger of spending too much time in meetings. Managers will do well to move in the direction of organic organizations with prudence and caution so that routine functions and other essential organization needs are not ignored.

For *coaching,* individuals should be in a more senior position in the organization, be good listeners and helpers, and be technically competent enough to develop new ideas. They should provide encouragement and guidance and should act as sounding boards for others in the research group. Those coaching should have access to higher-level management within and outside the organization and be able to buffer the projects from unnecessary organizational constraints. These individuals should have the ability to coach the members of the research team in a way that will enable them to develop their talents.

Project leading or *supervising* and coaching have some overlap and can often be accomplished by the same person. Project leading calls for individuals who are able to plan and organize the various project activities and can ensure that administrative and coordination requirements are met. They should have the ability to provide leadership and motivation and be sensitive to the needs of others. They must be able to understand the organizational structure, both formal and informal, so that they can get things done and balance the project goals with organizational needs. They should be interested in a broad range of disciplines and be able to handle multidisciplinary issues.

Clearly, as a manager in an R&D organization moves up in the organization, technical skills play a less direct role, while other skills such as human relations and administrative and conceptual skills become increasingly important.

As stated earlier, studies have clearly shown that research groups whose supervisors had high technical skills were the most innovative. On the other hand, those groups that had supervisors who did not possess high technical skills, but were in turn rated highest in administrative skills, were least innovative [Farris, 1982, p. 340]. Thus, the importance of an R&D supervisor's technical skills cannot be overemphasized. This is especially pertinent for organizations interested in productivity and excellence. Experience shows that an individual who does not have the training, the appreciation, and, indeed, the aptitude for science and technology is not likely to provide the necessary visionary leadership in an organization based on science and technology.

Creative people are likely to have good research ideas, but good ideas also come from communication with others. There is considerable research suggesting that communication patterns should be structured so that people can be stimulated by others who do similar work.

In R&D laboratories, only a small percentage (11–18.5%) of all idea generating information comes from the scientific literature [Allen, 1977, p. 63]. However, the scientific literature can be used for purposes other than generating ideas, such as problem definition at different stages associated with the total research process. But even in the problem definition stage, personal contacts provide more than five times the number of messages supplied by written sources (Allen, 1977, p. 65]. Therefore, communication through personal contacts is a crucial aspect of the innovation process.

In the next two sections, two related items—communication networks and the innovation process—will be discussed.

2.5 COMMUNICATION NETWORKS

There is considerable knowledge relevant to R&D activities that is not in books. Many ongoing R&D activities are not documented in the literature for some time. In addition, the written material is static and is a limiting medium for communication, while personal contacts allow one to exchange ideas, analyze data more quickly, and obtain information that is more relevant to the research project concerned.

Because of the complexity of technological problems and the importance of analyzing and synthesizing relevant technical information, verbal communication plays an important role in modern-day R&D activities. Research has consistently demonstrated a linkage between high-performing individuals and projects and an extensive pattern of verbal communication [Tushman, 1988]. Many new ideas are obtained while talking with people who do similar work. Sometimes talking with one person on Monday and another on Tuesday allows two apparently unrelated fields of research to merge in one's mind and leads to a new insight. Personal contacts and verbal communication therefore provide an efficient and effective communication medium within and between research and development communities.

The pattern of communication, however, depends on the nature of research activities. These research activities can be divided into three main areas: research projects, development projects, and technical service projects [Tushman, 1988].

Research Projects. These involve work oriented toward developing new knowledge and concepts.

Development Projects. These are directed toward using existing scientific knowledge to address specific product problems. Generally, these types of projects correspond to technological or experimental development.

Technical Service Projects. These involve solving a specific technical problem using well-known stable technologies.

Based on a study [Tushman, 1988] that compared and contrasted the communication networks of high- and low-performing research, development, and technical service projects, some patterns of communication activities were identified that are associated with high-performing projects. These patterns, as related to project types, are described here.

High-performing research projects showed extensive and decentralized communication patterns. People talked to many others and there were no rules prohibiting exchanges of ideas. Direct contacts and gatekeepers were used to acquire information from professional areas outside the firm. Within the firm, contacts were directed toward individuals who could provide effective feedback and evaluation. Projects were strongly connected to universities and professional societies. In general, there was less reliance on supervisory direction and more on individual initiative and peer decision-making and problem-solving [Tushman, 1988].

High-performing development projects focused on communication patterns directed toward operationally oriented areas (how to get things done; what works, when) both within and outside the firm. Communication outside the firm was moderate and was mediated more often by gatekeepers than in the case of research projects. While there were some direct contacts within the firm, the supervisor mediated much of the communication. There were also widespread and direct communications with the user—for example, marketing and manufacturing [Tushman, 1988].

High-performing technical service projects showed supervisor-dominated communication patterns both within and outside the firm. Communication outside the firm focused on suppliers, vendors, and customers. Communication within the firm related to marketing and manufacturing. In general, the supervisor served as a mediator for all external information sources, and there was more supervisory-dominated decision-making and problem-solving than in research or development projects [Tushman, 1988].

Experience shows that in the development and technical assistance projects of an organization there is an evolution of language, concepts, and values unique to the types of projects undertaken and, at times, unique to the organization itself. This local language and other characteristics make communications with the outside—that is, beyond the organization project boundary—difficult and prone to bias and misunderstanding [Tushman, 1988]. Since communication external to the project (both within and outside the organization) is essential for high-performing projects, the acquisition of information can best be handled via "boundary-spanning individuals" whom Tushman calls "gatekeepers." A gatekeeper then is an individual who links the project to external information sources. Three types of gatekeepers (technology, marketing/manufacturing, and operations) are described in Chapter 11, "Technology Transfer."

Gatekeepers in an organization perform an informal but crucial function. Others working on the project have to feel sufficiently secure and comfortable psychologically to approach gatekeepers with their questions without fear of adverse consequences or personal evaluation [Katz and Tushman, 1981, p. 109].

To encourage gatekeeping, individuals performing this function can be rewarded without being given any formal title or status to their activity. Technology gatekeepers can be easily recognized since they are high technical performers and are able to interact harmoniously with others. For locally oriented projects, first-line supervisors act as gatekeepers for about 50% of the cases [Allen, 1977, p. 163].

A study was conducted to investigate the managerial roles and career paths of gatekeepers [Katz and Tushman, 1981, p. 103]. In a follow-up study 5 years later, it was shown that almost all gatekeeping project leaders had been promoted up the managerial ladder. In contrast, for the nongatekeeping project leaders, only one-half of the promotions were up the managerial ladder. The authors concluded: "This implies that higher managerial levels (in a technology-based R&D organization) demand strong interpersonal as well as technical skills" [Katz and Tushman, 1981, p. 103].

Allen et al. [1979, p. 707] suggest that, contrary to some earlier conclusions, the technology gatekeeping role is important for applied research and development projects where the technology is complex and external sources of information are relevant to the project concerned. For basic research projects and for technical assistance projects, this role is not as critical. In the case of basic research, the problem is universally defined and contacts are best handled directly by the researcher working on the project. In the case of technical service projects, the technologies are well understood and stable; consequently, the organization is capable of providing the needed information internally.

Clearly, the main purpose of the communication network is the organization and processing of information. Also, as discussed above, different R&D activities require different communication networks. R&D managers, recognizing the importance of communication for the innovation process, should facilitate this process. Tushman [1988] suggests that

1. The amount and pattern of communication within the project must match the information processing requirements of the research project.
2. The project must be linked to interdependent areas within the firm.
3. The project must be linked to external sources of information through direct contacts or through the gatekeepers.

To facilitate internal communication (within the work group and interdependent areas within the firm), one must pay attention to the architecture of the workplace and to ways in which socialization takes place. One commonly hears stories of how a scientist thought of an idea while having tea or coffee with a colleague. Americans joke about the sanctity of the British tea breaks. Maybe there is something to their tradition. Sir William Hawthorne of Cambridge University once remarked that institutionalizing (or encouraging) tea breaks or similar social interaction in an R&D organization is quite beneficial; such activities are not common in the United States but they should be fostered.

The effects of office architecture and the nonterritorial office on communication have been investigated by Allen [1977]. In managing an R&D organization, it is important to recognize the need for internal and external communication for the innovation process. A manager should facilitate this process to the degree that resources and organization policies permit. Questions are often raised by upper management as to why it is necessary to have researchers participate in technical conferences and symposia. An R&D manager should be able to justify these activities on the tangible contribution such activities make to the innovation process.

2.6 THE INNOVATION PROCESS

An invention is an idea, a concept, a sketch, or a model for a new or improved product, device, process, or system. Inventing is the creation of new knowledge or new ideas.

The innovation process is the integration of existing technology and inventions to create a new or improved product, process, or system. Innovation in the economic sense is accomplished through the first utilization and commercialization of a new or improved product, process, or system [Freeman, 1982, p. 71].

Various technology-based organizations look at the overall innovation process differently. In a general sense, the innovation process includes (1) identifying the market need or technology opportunity, (2) adopting or adapting existing technology that satisfies this need or opportunity, (3) inventing (when needed), and (4) transferring this technology by commercialization or other institutional means.

The innovation process integrates project need, invention and development, and technology transfer. Ideas and concepts are generated in each of these three major stages; the innovation process is accomplished when these three stages culminate in the utilization and commercialization of a new or improved product, process, or system. Project need and what to research were discussed in Chapter 1. Inventions and development are discussed in this chapter. Technology transfer is one of the key issues in the innovation process. Indeed, the innovation process is never complete without this step. Technology transfer is more fully discussed in Chapter 11.

2.7 FUNDS

While this topic is so obvious that it could be skipped, we have included it for balance. Funds are needed for personnel, equipment, office and laboratory space, libraries, computers, travel, supplies, and so on. This is not the place to discuss research budgets and the like. We wish only to make sure that the reader keeps this element in mind when thinking of the four equally important elements required by an R&D organization. It is important to emphasize the fact that conducting research requires considerable resources. It is indeed an expensive activity. To maintain research excellence, it is necessary to attract talented scientists and have well-equipped laboratory facilities. None of this is probable without sufficient funding support. Organizations that are successful in the technology transfer of their research outputs are more likely to generate customer support for future research. This is particularly true for applied research and development projects. In a way, the steps in the innovation process discussed in the preceding section provide a link to the customer or the sponsor via need identification and technology transfer.

One is always seeking ways to test user acceptance of research output and to determine organization effectiveness. Seeking funds for research can be one way to test the market and user response to the research output, and thus determine organization effectiveness.

Consider the following problems facing two R&D organizations.

Case 1. A premier private university experiences a substantial decline in the number of U.S. citizens applying to its school of engineering. Applications are down 55%. Factors such as high starting salaries for baccalaureate degree holders, rising cost of graduate training, and high opportunity cost of staying in the graduate school contribute to this decline. Increased competition from state universities offering quality graduate training at merely 25% of the private university tuition and a widespread impression that the private university is difficult to get into and lacks the many extramural social activities normally available at large state schools have contributed to that university's problem. Since much of the research at a university is conducted by graduate students with guidance from faculty, lack of graduate students makes it difficult for the university to obtain funding for research and to maintain research facilities.

Case 2. A federally funded environmental research laboratory finds that the national interest in environmental issues is waning. One senior official responsible for supporting such research states that of the many important issues facing the agency, the environment is now at the bottom of the list. In fact the sponsors assert that discovering better ways to protect the environment would only mean the agency would be required to spend yet more resources for environmental projects of little or no value to the main mission of the agency.

Clearly, often external factors beyond the immediate control of an R&D organization affect funding support. Carefully formulated programs and strategies that are user-oriented rather than sales-oriented have to be developed to overcome these problems. In the second case, for example, while the emphasis on environmental issues may have declined, environmental requirements are still there. Research that reduces the cost of complying with existing environmental laws and regulations still remains as vital as ever. Ultimately, the focus for securing research support shifts from solely meeting environmental requirements to including economic considerations as well.

2.8 A CULTURE FOR R&D ORGANIZATIONS

Culture is the human-made part of the environment. It consists of objective elements (e.g., research laboratories, equipment, office buildings, office furnishings, etc.) and subjective elements (rules, laws, values, norms). Among the most important elements of culture are the unstated assumptions concerning "the way things get done in this lab." Some of these assumptions become salient only when they are challenged—for example, is safety more important than production all the time? In some labs it is and in others it is not. It takes something like the January 1986 *Challenger* disaster to find out.

One way of thinking about organizations is to conceptualize them as information processing systems (Daft and Weick, 1984). When the information that must be processed has certain attributes, the structure of the organization must match that type of information. For instance, Keller (1994) studied 98 R&D project groups with a longitudinal design and found that the effective teams showed a match between the structure of the information being processed and the structure of the organization. The more that the tasks the team had to do were nonroutine (i.e., involved radically new technology), the more the successful research teams were capable of processing large amounts of information (e.g., there was much interpersonal contact, permeability of group boundaries, and opportunities for informal, face-to-face communication).

Some organizational cultures are more effective than others. A culture that emphasizes innovation behavior (e.g., where people agree with "creativity is encouraged here" and disagree with "this place seems to be more concerned with the status quo than with change") and has high quality supervisor–subordinate relations (i.e., permits high levels of autonomy and discretion for innovation) is likely to be more effective than a culture that does not encourage innovation (Scott and Bruce, 1994). Furthermore, competitiveness is often not desirable. For example, one experiment compared competitive (the highest producer gets all the reward), individualistic (to each according to contribution), and cooperative (equal share of the reward) conditions for building a tower. Participants randomly assigned to these three conditions had building blocks of different colors, so their contributions could be identified. Dependent variables included number of blocks placed, number of falls of the towers (often due to sabotage), and so on. The major finding: The highest productivity occurred in the cooperative condition. Of course, we do not know whether a team in a research lab working on some project behaves like a group of college students building a tower. Nevertheless, for at least some situations these findings must be applicable [for details see Rosenbaum et al., 1980]. Competitiveness is certainly an aspect of some organizational cultures and this experiment questions its desirability.

Other aspects of organizational culture worth noting are hard work, people emphasis, status emphasis, participative climate, tolerance for disagreement, and frequent rewards. Emphases such as these are best communicated through management actions rather than words [Schneider, Gunnarson, and Niles, 1994]. A laboratory that provides a sense of community, encourages the loyalty of customers, and pays attention to detail is likely to have members who are concerned about the satisfaction of customers (such as the funding agencies that provide contracts to the laboratory) and who will be willing to contribute to every aspect of the laboratory's success.

A word about each of the aspects of organizational culture follows. One can see greater emphasis on hard work in some labs than in others. In some labs, people work hard and very long hours and usually take work home. There is no time for chatting. In other labs, people chat a lot and stop their tasks when it is time to go home. In labs with "people emphasis," the lab comes to a stop if something significant happens to one of its members. Status emphasis is evident when titles, formal dress, or formal language are used. Participation is an important component if people are asked to contribute their ideas and to discuss major decisions, if they have some autonomy in those decisions.

Tolerance for disagreement can be seen when there is frank discussion and when plans are critically evaluated no matter where they come from, whether from top management or from a lowly researcher. In some labs, rewards, recognition, and bonuses are frequent. It seems obvious that these qualities, with the exception of status emphasis, are desirable, but to what extent is every lab at an optimal point on these dimensions?

The point about allowing disagreement deserves special emphasis. When important decisions are made, people often seek others who agree with them. They avoid or reject those who disagree with them. These tendencies result in *groupthink* [Janis, 1972] and in major mistakes. One of the most striking examples cited by Janis concerned a number of the decisions of the National Security Council (NSC) during the Vietnam conflict. Despite their own better judgment, some of the NSC members who personally opposed certain policies contributed to unanimous decisions.

People often feel unjustifiably optimistic about the way their research plan will turn out, they do not give sufficient weight to signals that something is wrong, they reject those who criticize their plans or their accomplishments, they censor themselves when they feel critical about actions of their team members, and they select their critics so as to receive a favorable review of their work. All of these behaviors are aspects of groupthink. Groupthink leads to poor performance.

To avoid groupthink, one needs to bring fresh perspectives into the group. That means tolerating those who disagree—the gadflies. It is even better to appoint a devil's advocate whose role it is to shoot down research designs, to reject drafts of papers, and to warn about disasters that could result from a particular course of action. In the case of very important projects, having several teams tackle the problem from different angles is not a duplication. It is the best way to get a solution.

Finally, in the case of important decisions it is useful to allow a day or two between the decision and the start of the project, as well as to review the decision from other perspectives before committing major resources.

Experience shows that a manager has to watch out more for subservient researchers than for unruly ones. When critical analysis of a manager's proposals is not made early, much is lost. When suggestions made by the manager are taken as commands without discussion and analysis, research excellence is bound to suffer. Related to the groupthink phenomenon is the Not-Invented-Here syndrome, discussed later in this chapter.

The extent to which some R&D units within an organization are considered by the scientists a good place to work, i.e., its internal reputation, defines the unit's culture. A study of 10 science-based organizations employing 1500 scientists was conducted [Jones, 1994]. A climate of innovative work and good working conditions had the highest correlation with good internal reputation. Thus, for effective R&D organizations, a manager needs to foster a climate of innovation and good working conditions.

Jabri [1992] found that scientists who perceived that the tasks assigned to them were appropriate collaborated more with team members, expended more effort on the tasks, and showed more innovation on the job. Job satisfaction and performance were positively correlated when the task allocation was perceived as appropriate, but satisfaction and performance were unrelated when the tasks were seen as low in appropriateness.

2.9 NOT-INVENTED-HERE SYNDROME

The Not-Invented-Here (NIH) syndrome is defined as the tendency of a stable research group to believe it possesses a monopoly of knowledge in its field, thereby rejecting new ideas from the outside [Katz and Allen, 1982]. As discussed previously, communication with the wider scientific community, other researchers within the organization, the user community, and marketing personnel within the organization is crucial for a successful and effective innovation process. The NIH syndrome, then, actually works to the detriment of organizational performance.

As the members of the research group work together longer, the group naturally forms a stable and cohesive project team. Individuals try to organize their work environments in a manner that reduces the amount of stress and uncertainty they must face [Katz and Allen, 1982]. Project members then begin to work comfortably with each other and separate themselves from external sources of technical information and influence by reducing their communication level with the outside community.

It is important to note that it is not a mere reduction in this communication level that causes performance to deteriorate. More importantly, it is the project team's tendency to ignore and become increasingly isolated from sources of information and ideas that makes a crucial difference [Katz and Allen, 1982, p. 16]. As we discussed previously, communication patterns and the need for communicating with outside groups vary according to the nature of the research and development effort. When research and development organization groups are inflicted with NIH, overall performance will suffer. In such cases, research teams will fail to pay sufficient attention to new advances and information within the relevant external scientific community, technical service groups (as opposed to R&D groups) will fail to interact among themselves, and development project members will fail to communicate with individuals from other parts of the organization (for example, the user community, manufacturing, marketing) [Katz and Allen, 1982, p. 16].

To ensure an effective and productive R&D organization, appropriate strategies need to be developed to circumvent and eliminate the NIH syndrome. In a general sense, the strategy needs to focus on keeping the individual research group members from reaching a complacent state—that is, finding ways of destabilizing and energizing the research groups within the organization. Some of the following activities might be helpful:

- Movement of new employees to the research groups.
- Active participation of outside researchers in the research group. This can be attempted by bringing in visiting professors and scientists from other organizations, by establishing a close relationship with premier research universities, and by bringing in graduate students to work with other tenured researchers in the organization.
- Encouragement of the research scientists to interact with the wider scientific community by participating in research seminars, scientific meetings, and professional society meetings.
- Encouragement and facilitation of interaction between development groups (product and process development engineers) and marketing, manufacturing, and user groups.
- Encouragement of interaction of members of technical services groups.
- Establishment of a sabbatical leave program. Though widely practiced in academic institutions, such programs are often looked at with a jaundiced eye by some research and development organizations. Clearly, implementation of a sabbatical leave program in an R&D organization requires considerable resources. Those individuals who are intellectually able to function at the cutting

> edge of technology in research and are able to provide technical leadership for an R&D organization would benefit immensely from the sabbatical leave program. Investing in such programs is well worth the cost.

Sabbatical leaves can be used to develop a new course or to pursue some new line of research. It provides an opportunity for scientists to (a) take part in scholarly activities and research that they would not be able to do during the normal course of activities and (b) interact with a scientific community outside their normal circles. Perhaps, to make the sabbatical program useful to the organization as well as to the individual, the individual should be given new roles and responsibilities to complement the sabbatical leave objectives.

Group members could present "new ideas," technologies, and perspectives acquired outside the laboratory to the group on a regular basis. Experience shows that some of these activities fade away after a while and that a mechanism to stimulate interest in them is necessary. Inviting sponsors of other interested research groups and rewarding such activities may further stimulate participation and interest.

2.10 FIT OF PERSON AND JOB

A word should also be said about the fit of people and environment. A person whose abilities match the demands of the job will be most satisfactory to the organization. If the job makes greater demands than the person's ability, the individual feels unable to cope; if it makes too few demands, the individual becomes restless and bored. A close match between the individual's needs and the job's ability to satisfy these needs leads to job satisfaction. To some extent an individual's needs reflect expectations. People are most satisfied if the job provides what they expect the job to provide. There is empirical research showing that satisfaction is maximal when there is a match between expectation and realization. If one gets more than is expected, it can be dysfunctional. It is a bit like receiving a $100 Christmas gift when one expects a $10 one. Of course, when the expectation is higher than the realization, the person is disappointed or angry, and the effect on job satisfaction is most severe.

It is a good idea to pay attention to the match between personal attributes and organizational cultures. Some people are more competitive than others and feel comfortable in a competitive organizational culture. Similarly, one can analyze each of the dimensions of organizational culture just mentioned to see if the individual would fit that culture.

Some R&D organizations compensate their employees with bonuses and profit sharing rather than with high wages. This is fine for achievement-oriented risk-takers, but more conservative individuals will dislike this method of compensation. Another example is time perspective. If the organization has a long time perspective, requiring individuals to defer gratification for long-term successes will not be appreciated by those with a short time perspective.

2.11 CREATIVE TENSIONS: MANAGING ANTITHESIS AND AMBIGUITY

For a manager in an R&D organization, many questions related to the work environment arise. Answers to these questions are inconsistent and ambiguous. Examples of such questions are as follows:

- In general, what kind of climate in an R&D organization is conducive to technical accomplishment, excellence, and productivity?
- What is the optimum degree of freedom versus control?
- What should be the balance between basic research, applied research, development, and technical assistance?
- Should the scientist be isolated?
- How about the communication network? What is optimum in an R&D organization?
- To what degree is specialization of a researcher important?

Pelz and his colleagues studied 1300 scientists and engineers in 11 research and development laboratories, 5 industrial laboratories, 5 government laboratories, and 7 departments in a major university. Their findings shed light on some of the preceding questions. Based on this study, it was concluded that scientists and engineers were more effective when they experienced a "creative tension" between sources of stability or security on the one hand and sources of disruption or challenge on the other [Pelz and Andrews, 1966b]. This study indicates that achievement often flourishes in the presence of factors that seem antithetical.

Specifically Pelz and Andrews found the following:

1. Effective scientists and engineers in research and development laboratories engaged in both applied and basic research, as well as a wide range of R&D activities (e.g. serving on review panels, providing technical services).
2. Effective scientists were intellectually independent and self-reliant; they pursued their own ideas and valued their freedom, but they also interacted vigorously with their colleagues. They did not avoid other people.
3. In the first decade of their career the effective scientists spent a few years on one main project, but they did not overspecialize. They developed several skills that they used well in the next decade.
4. Mature scientists were interested in both probing deeply and pioneering in new areas.
5. The best work occurred in environments that were not too tightly controlled, provided enough of a challenge as well as adequate security, and did not impose rigid goals of the organization on the scientists. Moderate coordination, allowing individual autonomy, usually resulted in finding the best solu-

tion. But effective scientists were strongly influenced by a variety of internal and external sources, including concerns about the goals of the organization.

6. The most effective scientists were those that influenced key decision-makers of the organization, but whose goals were highly coordinated with the goals of the organization.
7. High performers received personal support and stimulation from their colleagues, but differed from their colleagues in technical style or strategy. In other words, they had complementary talents with their colleagues, and they were well respected and supported by them.
8. R&D teams change over time. As they get "older" they become more and more interested in narrow specialization and less and less interested in broad pioneering. The most useful teams are at a "group age" that has not yet become too interested in narrow specialization but has not yet lost interest in broad pioneering.
9. Effective older teams had members who preferred each other as collaborators, but remained intellectually combative and used different technical strategies.

Thus, in designing organizational cultures for R&D laboratories it is desirable to review some dimensions identified by Pelz and Andrews [1966b]. They emphasize balance between several extremes. It is not desirable to spend all of one's time on applied or on basic research; a mixture of the two is more effective. One should not emphasize extreme self-reliance in a lab; one also needs some interdependence. One should not overemphasize specialization; one needs to be good at many things. Supervisors should not provide too much structure; the subordinate needs some autonomy. Research scientists must find a balance between their personal research goals and those of the organization. Projects should not be too long or too short; a 3-year project is often optimal.

Jobs, particularly in research, should be designed so that they provide opportunities for autonomy, since research personnel is high on this need. They must also provide significance; that is, people should have the sense that what they are doing is important for the organization, for themselves, for the profession, and for society. Finally, jobs should provide feedback.

There is research suggesting that people who are in a good mood are more creative than people who are in a neutral mood [Isen et al., 1985]. Mood can be manipulated by having people think pleasant thoughts for a period of time. This experiment was done with college students and requires replication in a research setting, but it is certainly worth trying to put people in a good mood if possible.

Stimulation by others requires that people be able to communicate easily. In some studies the elimination of physical barriers helped in the operation of R&D laboratories.

A desirable organizational culture allows employees to have a sense of control. In some experiments, those without control became depressed. One increases the sense of control of employees by allowing them to participate in decisions that affect them, such as when to start work, what to study, and when to study it. Management by objectives is desirable since it enables the supervisor and subordinate to sit down periodically and agree on milestones, goals, or values. This review in turn allows for feedback, and it also allows for discussion of why the goals were not reached and for congratulations when they were attained. Cultures that reward frequently are more effective than cultures that do not. That does not mean that one should get rewarded for every success. Rather, there should be uncertainty about getting a reward. However, when something major has been achieved, the reward should be extremely probable.

Goals must be set in such a way that they are specific and difficult, but attainable. There is research showing that this combination of goal attributes results in maximal motivation. This research will be reviewed in Chapter 6, "Motivation in R&D Organizations."

A desirable culture has a family spirit; it uses slogans, myths, and war stories; and it has heroes that transmit the values of the organization. People need to feel that they belong to the organization and that they are important; they also need pride in being members of the organization.

A good R&D culture will accept failure. When all experiments come out as expected, this indicates that the research is too conservative. "If you do not have several failures, you are not doing a good job" should be the way R&D managers talk to their subordinates. Open communication (open doors), acceptance of suggestions, and the assumption that there is always a better way and that the better way does not constitute a criticism of the employee are important values and perspectives for the R&D manager.

The organizational culture should stress a win–win orientation in resolving conflict. This orientation involves looking for a creative solution that will satisfy both sides in an argument. Chapter 9 on conflict will discuss this approach more thoroughly.

2.12 DEVELOP A CLIMATE OF PARTICIPATION

Participation is the right climate for the management of R&D laboratories. Participation makes especially good sense in the case of the management of research. Lawler [1986] argues that it makes sense in any organization, but the kinds of factors that make it desirable are found in particular abundance in R&D laboratories.

Lawler makes the point that participation means moving rewards, knowledge, power, and information flow to the lowest possible levels of an organization. He states: "My prediction is that for participative management to be effective it must put power, rewards, knowledge, and an upward and downward information flow in place at the lower levels of an organization. Limited moves in this direction will, according to this view, produce limited or no results" [p. 43]. Thus, he criticizes many of the proposed panaceas of contemporary U.S. management (such as quality circles, the employee survey feedback, job enrichment, work teams, union management teams, quality-of-work-life programs, gainsharing, and the new-design plants) as doing only some of the job and so achieving only part of the results. In his last chapters, he describes the kind of organization that he sees as optimally participative and successful in the way it deals with organizational change.

The most important point made by Lawler is that there must be congruence between the management actions in the areas of power, reward, knowledge, and information flow. If a management method reduces the level of decision-making in one of these ways but not the others, the effect would be much less desirable than if all four attributes (power, rewards, knowledge, information) were to change together.

The management philosophy that should characterize participative management is as follows:

- People should be treated fairly and with respect.
- People want to participate (this is particularly true in the case of highly educated samples, such as researchers).
- When people participate, they accept change.
- When people participate, they are more committed to the organization.
- People are a valuable resource because they have ideas and knowledge.
- When people have an input in decisions, better solutions are developed.
- Organizations should make a long-term commitment to the development of people because that makes them more valuable to the organization (this is particularly true in R&D organizations).
- People can be trusted to make important decisions about their work activities.
- People can develop the knowledge to make important decisions about the management of their work activities.
- When people make decisions about the management of their work, the results are high satisfaction and organizational effectiveness.

This perspective requires an organizational structure that has very few levels. A structure comprised of a director, a manager level, and researchers organized in functional or disciplinary groups is sufficient. The fundamental grouping should be organizational units that are responsible for a particular product or customer or research area. People should be able to identify with their work group. Each unit should serve some customer and receive feedback about customer satisfaction. Members of the unit should know precisely what the budget is and how it is being spent. They should know what is expected of them from various customers (e.g., funding agencies). They should receive feedback concerning their success in satisfying these customers. Ideally, information about the performance of the unit should be available at frequent intervals. Widespread use of computer networks should improve feedback.

The physical layout of the work group should be egalitarian, safe, and pleasant.

The same informal dress should be worn by all. The physical layout should create team boundaries.

2.13 SUMMARY

The management of R&D organizations is quite challenging. It is difficult to coordinate numerous individuals who are socialized to work autonomously. However, one cannot leave them totally alone, since the organization has goals that its personnel must meet. It is hard to get ideas, funds, and the right climate at the right time and place in order to produce a top-quality research product.

In a real sense the job of the R&D manager is to create the right climate for research. A first-rate researcher, in the right climate, with adequate funding, is likely to come up with important ideas. But providing the right culture is complex. A manager must select people, match them with jobs, match them into teams, do team building, and help develop norms, roles, and standard operating procedures that will result in high levels of innovation. An organization must be developed that will allow people to be maximally creative. Rewards must be provided so that people will be motivated to work hard and to seek excellence. The manager must know how to lead, how to reduce conflict, and how to get maximum advantage of the resources that are available.

In this chapter we examined rather superficially, and at an introductory level, the people, ideas, funds, and culture that are required for excellence in R&D organizations. In the subsequent chapters we will focus in greater depth on the very same topics and will also examine how to evaluate people and how to determine a laboratory's success. Technology transfer and satisfaction of the laboratory's clients are among the outcomes that are measurable and provide clues about the success of the laboratory. We will examine also how a manager can evaluate change in organizations and essentially learn to manage the culture of the R&D organization.

2.14 QUESTIONS FOR CLASS DISCUSSION

1 Discuss the kind of organization (bureaucratic versus organic) that is likely to be the most desirable in many R&D labs.

2 Define the gatekeeper in R&D labs. What are the functions associated with gatekeeping? How can these functions be performed best?

3 What is organizational culture? How can one develop an effective organizational culture for an R&D lab?

4 Discuss participation in decision making in R&D labs. What are the limits (too much; too little)?

5 Develop case studies related to
   - Staffing
   - Communication networks
   - Not-Invented-Here syndrome
   - Creative tensions

2.15 FURTHER READINGS

Allen, T. J. (1977). *Managing the Flow of Technology: Technology Transfer and the Dissemination of Technological Information with the Research and Development Organization.* Cambridge, MA: MIT Press.

Allio, R. J., and D. Sheehan (1984). Allocating R&D resources effectively. *Research Management* **27**(3), 14.

Burgelman, R. A., and L. R. Sayles (1986). *Inside Corporate Innovation.* New York: Free Press.

Matheson, D. (1994). Making excellent R&D decisions. *Research-Technology Management,* **37**(6), 21–24 (November–December).

McGrath, J. E. (1984). *Groups: Interactions and Performance.* Englewood Cliffs, NJ: Prentice-Hall.

Pearson, A. W. (1993). Management development for scientists and engineers. *Research-Technology Management,* **36**(1), 45–48 (January–February).

Pinto, J. K., and D. P. Slevin (1989). Critical success factors in R&D projects. *Research-Technology Management,* **32,** 31–35 (January–February).

Roberts, E. B. (1982). Generating effective corporate innovation. *Innovation/Technology Review,* 3–9.

Root-Bernstein, R. S. (1989). Who discovers and invents. *Research-Technology Management,* **32,** 43–50 (January–February).

Lawler, E. E., III (1986). *High Involvement Management.* San Francisco: Bass.

Tushman, M. L. (1988). Managing communication networks in R&D laboratories. In M. L. Tushman and W. L. Moore (Eds.), *Readings in the Management of Innovation,* 2nd ed. Cambridge, MA: Ballinger, pp. 261–274.

3
CREATING A PRODUCTIVE AND EFFECTIVE R&D ORGANIZATION

In Chapter 2 we introduced the four key elements required for an effective R&D organization. In this chapter we will continue our discussion of these elements, emphasizing in greater depth those aspects that we believe are especially related to organizational effectiveness.

The productivity of an industrial operation usually includes the quantity of its output and its quality. However, in an R&D organization, many units of output are intangible and subjective in nature. Productivity also needs to relate to the objectives and goals of the organization. Consequently, to focus comprehensively on R&D productivity, the concept of "organization effectiveness" is proposed.

Organization effectiveness is a vector that includes quantifiable and nonquantifiable outputs and reflects the quality and the relationship of outputs to broad organizational goals and objectives. Organization effectiveness has a one-to-one correspondence to the general concept of productivity, but it also includes items not always included in productivity—for instance, quality and utility (i.e., relevance to organization objectives). Using this definition, if an organization is very effective, it is very productive, and if it is not very effective, then it is not very productive. Not only should an organization be productive, but it needs to be viable over a considerable period of time. This in turn requires that members be satisfied with the organization.

3.1 ORGANIZATION EFFECTIVENESS

Effectiveness can be determined by a number of different criteria. Table 3.1 lists some criteria that may be used; the reader will think of others. To some extent, the

TABLE 3.1 Criteria of Organizational Effectiveness in R&D Laboratories[a]

Criterion	Measurement Instrument
Quantity of output	Numbers of reports, publications, new products
Quality of the work	Number of patents obtained, number of times publications of lab members are quoted, number of refereed publications per member of lab
Increases in the size of organization	Obtaining more research funds
Absenteeism	Number of persons out of the total work force who are absent without a valid excuse on an average day (counted inversely)
Level of stress	Measured with physiological indexes, number of visits to hospital, frequency of peptic ulcers, etc. (counted inversely)
Level of job satisfaction	Measured with a standardized questionnaire, such as the Job Descriptive Index. Components: Satisfaction with pay, supervisor, organization or company, job, co-workers, working conditions
Pride in the organization	Feelings of pride measured via questionnaires
Congruence of individual and organizational goals	The extent individual goals are consistent with goals as they are reflected in employee and management statements
Profits	Direct profits or return on investment studies where returns are determined from implementation of research products

[a] A good case can be made for each organization developing its own criteria of effectiveness through participation of organization members in a debate that considers (1) different criteria, (2) how they should be measured, and (3) how they should be weighted. Such a debate has the advantage of involving the key members of the organization in the development of its goals. They become committed and ego-involved. The criteria that need to be debated are listed in Table 3.1. Other criteria might be suggested during these debates.

type of R&D organization will determine the criteria. The criteria listed in Table 3.1 are self-evident, but some comments are needed concerning the congruence of individual and organizational goals and the use of profit as a criterion.

First, consider the congruence of individual and organizational goals. If the individual's activities are quite consistent with the activities and goals of the organization, this will result in a better organization than one in which individuals try to do "their own thing" and are not really concerned with what happens to the organization.

Next, consider profit. For a profit-oriented organization, revenues or earnings may provide a good measure of its productivity or effectiveness. However, for a research organization (or a nonprofit organization), other measures are needed.

Nevertheless, a good way to integrate individual and organizational goals is to pay some bonus based on total organizational performance.

In summary, R&D organization output measures can be subjective or objective, discrete or scalar, and quantitative or nonquantitative, and there can also be qualitative aspects associated with them. The relationship of output measures to organizational goals must also be included. An interesting categorization of output measures in terms of result, process, and social indicators has been proposed by Anthony and Herzlinger [1984].

Different organizations (governmental, commercial, educational) will weigh the available criteria differently. It may be a useful exercise for the key teams of a lab to devote some time to a discussion of how the various criteria should be weighted. Agreement on how to do that is likely to increase the congruence of individual and organizational goals, and possibly reduce role conflict within the organization. Of course, in pure research the publication criterion is weighted more heavily, and in applied research the product that has been invented or developed is the key output that must meet certain specifications. These specifications themselves can be stated as criteria (e.g., product should cost less than a certain amount, should weigh less than a certain amount, should have certain performance characteristics, and so on). Sessions that are devoted to goal clarification and how specific criteria will be used to determine their attainment by the individual or the organization will take time, but will be of great value in creating a good climate of cooperation within the organization.

Blake [1978, p. 260], commenting on organization effectiveness, suggests that the criterion for evaluating the effectiveness of an R&D organization should be the record of its success or failure in meeting its objectives. He recommends a set of questions that would form a basis for determining R&D organization effectiveness:

- Are project cost schedules met?
- Are project time schedules met?
- Are time schedules kept that show both original estimated costs and actual costs of the projects?
- Are records kept that show both the estimated completion time and actual completion time for the projects?
- Is there clear delineation between overruns and cost increases caused by change in the scope of the projects or other proper causes?
- Is there significant scientific fallout?

Szakonyi [1994] proposes ten R&D activities and six operating levels to measure organizational effectiveness. The proposed activities are as follows:

1. Selecting R&D
2. Planning and managing projects
3. Generating new product ideas
4. Maintaining the quality of the R&D process and methods
5. Motivating people

6. Establishing cross-disciplinary teams
7. Coordinating R&D and marketing
8. Transferring technology
9. Fostering collaboration between R&D and finance, and
10. Linking R&D to business planning

These activities are evaluated in terms of their level of operation.
The six proposed levels are as follows:

1. The issue is not recognized.
2. The initial efforts are made toward addressing issue.
3. The right skills are in place.
4. Appropriate methods are used.
5. Responsibilities are clarified.
6. Continuous improvement is underway.

The proposed activities and levels of operation can be modified to suit organizational needs. This approach can be used as a diagnostic tool to identify deficiencies and ways to improve organization effectiveness.

Clearly, there are a number of ways of looking at organization effectiveness. Viewing organization effectiveness, and thus productivity, as a vector, the following relationship is proposed:

$$\text{Productivity} = \text{Effectiveness} = \text{Output} \times \text{Quality}$$

Output Measures

Output has three categories: *process measures, result measures,* and *strategic indicators.* A quantitative or a qualitative measure can be assigned where possible; and these measures, where appropriate, relate to organizational objectives. Further description of these proposed measures follows.

Process Measures. These measures are process-oriented and relate to activities carried out by an organization or its subunit; they also relate to short-term, day-to-day activities of the organization. Some examples in a research organization may be

- Number of times technical assistance was provided to an operational unit
- Number of responses sent to enquiries from outside scientific or internal unit
- Number of visitors to the organization
- Number of administrative types of actions handled

Result Measures. These would be tangible, measurable outputs expressed in terms of an organization's objectives and goals. Some examples are

- Number of technical reports published
- Number of refereed papers published
- Number of patents generated
- Number of major innovations developed and adapted for commercialization
- Dollar amount of external research grants obtained
- Return on R&D investment

Strategic Indicators. These indicators would focus on long-term and strategic aspects of the organization. Examples include

- Reputation of the research organization
- Ability to attract highly qualified scientists
- The degree of customer (sponsor organization) satisfaction with research output
- Stability of research funding
- Ability to attract research support for new high-risk research projects
- Job satisfaction level of the employees

It is clear that there are many ways in which one can assess the effectiveness of an organization. Thus, when thinking of an effective organization we should include all the variables because any one of them is likely to be biased or contaminated by extraneous factors. Any one criterion could be biased because of the way it is confounded with other variables of which we are not aware, or because the way it is measured may not be as accurate as possible. On the other hand, if we utilize many criteria and there is some degree of convergence across these criteria, then we are reasonably sure that we are dealing with a meaningful overall criterion (e.g., the weighted sum of the above-mentioned criteria), which can be used as a means of assessing the effectiveness of an organization.

3.2 WHO ARE THE INVENTORS AND INNOVATORS?

Individual capabilities, availability of resources to pursue research and development, and the ethos of a scientific community are all relevant to understanding inventors and innovators and the milieu in which they are likely to create and invent. The following are some ideas related to creativity and the characteristics of inventors.

Creativity

To invent or to innovate requires creativity. A very good account of creativity is given by Barron [1969], who discusses the majority of the tests that are available that purport to measure creativity. These tests do not specifically focus on R&D

personnel; however; the attributes are those that Barron and others have discovered in their research with creative people. In doing the study, Barron and others obtained nominations of creative people from different professions and studied these individuals. They found a number of attributes frequently associated with creativity. Of the more than 30 attributes, the significant ones are

- Conceptual fluency (that is, being able to express ideas well and to reformulate the ideas as one proceeds)
- The ability to produce a large number of ideas quickly
- The ability to generate original and unusual ideas
- The ability to separate source (who said it) from content (what was said) in evaluating information
- The ability to stand out and be a little deviant from others
- Interest in the problem one faces
- Perseverance in following problems wherever they lead
- Suspension of judgment and no early commitment
- The willingness to spend time analyzing and exploring
- Genuinely valuing intellectual and cognitive matters

Research on creativity [Freiberg, 1995] indicates that the most creative scientists are those who had creative mentors. For example, most Nobel Prize winners have studied under previous laureates or had people around them that inspired them. Native intelligence is important, but it must match the scientist's social world.

Psychologists argue that the larger the talent pool from which a person comes, the higher the intelligence. Thus, we can expect that the most intelligent people would be found in China (population 1.2 billion) and India (with 0.8 billion). But these countries have relatively few Nobel Prize winners because the social environment is not optimal. To be creative, one must be surrounded by creative people. Those who work in highly creative laboratories are more creative than those who work in less innovative environments.

A study of creativity in molecular biology laboratories found that social interaction was a major factor. Psychologist Kevin Dunbar, who did that study, put it this way: "One person might provide one premise, another might provide another premise, and the third person might draw a conclusion from the two premises" [Freiberg, 1995, p. 21]. An additional factor was the extent that people in the laboratory used analogies and metaphors. Laboratories that use these devices a great deal had the most creative scientists.

Finally, research finds that to be creative one must think differently from the way others think. That means that the more one uses past knowledge, for example spending a lot of time examining the literature, the less likely one is to be creative. One could even say, without too much exaggeration, "Expose yourself to the literature only superficially, to make sure that you are not reinventing the wheel. Then work on the problem first, try to solve it without looking too much at what others have written, and after you solve it read the literature very thoroughly to place your solution into a scholarly context."

Sternberg and Lubart [1995, p. 2] ask the question, What exactly is a creative person? They argue that it is a person who generates many ideas that are relatively novel, appropriate, and of high quality. Creativity is domain-specific. Research has shown that children who were unusually good at one domain (writing, art, music, crafts, science, performing arts, public presentations) were usually not especially good in others. In a study of 2400 eminent Western creators in various fields between the years A.D. 850 and 1935, there were few examples (17%) of eminence in more than one field (e.g., painting and sculpture) and only 2% of examples of eminence in unrelated fields (painting and philosophy).

Age is related to creativity rather differently. For example, mathematicians make their major contributions early in their careers; social scientists do so much later, and philosophers are usually quite late in reaching their peak.

Older people are not as creative as younger scientists, but that is not because they are no longer able to generate new ideas, but because they are distracted by competing activities—administration, writing letters of recommendation, attending ceremonies where they get awards, giving large public lectures, and so on. They also are not inclined to take risks to the same extent as younger scientists.

What is valued as creative in one historical period or in one culture is not seen as creative in other historical periods or cultures. For example, Confucius defined creativity as taking what is already known and improving it a little. In the West, on the contrary, we value revolutions and major shifts in perspective and would not call Confucius' recommendations creative.

The Creative Person and Creative Work

The attributes of the creative person include: does what others think is impossible; is nonconformist; is unorthodox; questions societal norms, truisms, assumptions; is willing to take a stand. The creative person questions basic principles to which everyone subscribes.

Sternberg and Lubart [1995] argue that there are six personal resources that are necessary for creative work:

1. *Intelligence.* An I.Q. of less than 120 (average of American university students) is not helpful in generating new ideas, but one does not need a huge I.Q. in order to be creative.
2. *Knowledge.* One needs to know what others have done on the subject, but one does *not* have to know everything. In fact, if one tries to know everything, one is likely to spend all available time reading the literature rather than coming up with new insights.
3. *Thinking Style.* Questioning what others think is true, thinking about what is unusual, profound, and important. "Creativity is 99% perspiration and 1% inspiration" means that the person thinks a lot.

4. *Personality.* To be creative, one must be willing to take chances. One must also be willing to be ridiculed, because many creative ideas strike ordinary people as stupid or crazy. They also shake the status quo. For example, before the theory that bacteria causes infections was developed, Ignaz Semmelweiss, a Hungarian physician-researcher, suggested that patients in maternity wards might be dying because of germs on the unwashed hands of physicians. This idea was ridiculed even after he had demonstrated that by his strict hygiene measures he had reduced maternity ward mortality rate dramatically. It also helps not to be arrogant, because arrogance reduces the chances that others will accept the idea. Another attribute is that creative people like to create their own rules for getting things done.
5. *Motivation.* Creative people have high energy and produce a lot. They are task-focused and absorbed in what they do. They are in love with what they do. They are not 8-to-5 employees; they work night and day on their problem. Since they produce a lot, some of what they produce has some chance of being valuable. People who produce little have a very low chance of producing something valuable. It is only by trying a lot of experiments that it is possible to determine how they will come out.
6. *Be in the Right Environment.* We already made that point above. Creative people need supervisors who leave them alone or who help them find the best work conditions. For example, some people are more creative when they are with others, but the majority are most creative when they are alone. A wise supervisor can detect what conditions are best for the particular scientist and will arrange for those conditions to occur most of the time.

These six attributes have to have a minimal value for a person to be creative, but every one of them does not have to be at an extremely high level. People can compensate for being low on one by being high on another aspect.

Can Creativity Be Developed?

Sternberg and Lubart [1995] argue that creativity can be developed. They recommend teaching people a few principles:

1. Learn to redefine problems. Don't just accept what you're told about how to think or act [p. 285].
2. Look for what others do not see. Put things together in ways that others don't; and think about how past experiences, even ones that may initially seem irrelevant, can play a part in your creative endeavors [p. 286].
3. Learn to distinguish your good from your poor ideas, and pay attention to their potential contribution [p. 286].
4. Don't feel that you have to know everything about a domain in which you work before you are able to make a creative contribution [p. 286].
5. Look at the large picture.

6. Persevere in the face of obstacles, take sensible risks, and be willing to grow [p. 287].
7. Do what you really love.
8. Find environments that reward you for what you love to do [p. 288].
9. You can compensate for what you do not have. For instance, a person who does not have all the knowledge might be able to compensate by being extra-intelligent or by being in an extra-helpful environment.
10. Learn that the major obstacle to creativity is not the environment, but the way you look at the environment.

Creative Phases

Creativity has phases [Sternberg and Davidson, 1995]: first *preparation* (the problem is defined and goals are diffuse, but there is a general direction), then *incubation* (different ideas float in one's subconscious until they join together in a new way of thinking), and finally *production* (one generates a lot of ideas, does a lot of experiments, and thinks of many different ways of looking at a problem). An insight that suddenly provides a new perspective may come after the incubation. At some point in this sequence, one has to switch from generating ideas and insights to criticizing them. One has to select the idea that is to be sold to the scientific community. That means that criticizing, improving, extending, condensing, sifting, combining, correcting, and testing are all important ingredients of creativity. It is just that they must come at the right time, after the ideas have been generated. During the criticism phase, it is important to talk with others to see how acceptable the insight or ideas will be to different audiences.

As mentioned briefly in the previous chapter, an interesting study by Isen et al. [1985] suggests that people can become more creative if they are in a "good mood." Subjects were randomly assigned to experimental groups (where manipulations of good mood were provided) and to control groups (where a neutral or negative mood was created). The mood manipulation was achieved by asking the subject to give associations to specific words. For the positive mood, the words were positive; for the negative mood, they were negative. The experimenter then scored the extent to which the associations were "unusual." Previous work has shown that creativity is higher when people make unusual associations. Standardized norms were available to obtain an objective measure of "unusualness" (i.e., if a response is frequently given by similar samples, it is not unusual). Thus, if the study were generalized to R&D laboratories, it would seem that managers in such laboratories must be particularly concerned when their subordinates suffer from poor morale. Such a condition may be particularly detrimental to creativity.

Characteristics of Inventors and Innovators

Views on characteristics of inventors and innovators naturally vary. Commenting on inventors and innovators, McCain [1969, p. 60] has suggested that the amateur scientist is nearly extinct and that formal training is almost a prerequisite for inventions and new scientific concepts. The individual normally has spent a very substantial amount of time and effort absorbing existing knowledge in his/her particular discipline while obtaining a graduate degree. These individuals tend to be above average in intelligence, as measured by ordinary testing methods.

However, even these highly trained, intelligent individuals with backgrounds in a scientific discipline are not very likely to make any substantial new contributions. In some fields, the average number of publications during the lifetime of a person with a doctorate is just a bit over one! Of those who make some contributions, only about 10% contribute more than half the scientific publications in a given field [McCain, 1969, p. 60].

Education and aptitude then form the first signaling mechanism in identifying inventors and innovators. Among this select group, however, only a small proportion will produce many inventions and innovations. Empirical studies [Charpie, 1970, pp. 7 and 17] further suggest other characteristics that contribute to innovation. Focusing on the individual, some characteristics of a successful innovator and inventor are as follows [Charpie, 1970, p. 7]:

- Strong technical background
- Able to deal with *things* rather than people
- Fluency in discussing ideas rather than handling processes in a formal organization
- More at home with technical products than with marketing problems
- Inclined to be disdainful of the professional judgments of others
- Committed to innovative concepts and product notions

Recall, as discussed earlier, the distinction between inventors (basic research) and innovators (applied research). At the University of Sussex, United Kingdom, 29 pairs of similar innovation projects were examined. In each pair, one project was successful and the other less so. There were clear differences within pairs that fell into a consistent pattern of successes. The characteristics of a successful innovator and the related organization structure that was implied can be summarized as follows [Twiss, 1992]:

- Successful innovators are seen to have a much better understanding of *user* needs.
- Successful innovators pay much more attention to *marketing*.
- Successful innovators perform development work more *efficiently* than those who failed, but not necessarily more quickly.
- Successful innovators make more *effective use of outside technology and outside advice* (even though they perform much of the work in-house).
- The responsible individuals in their successful attempts are usually more senior and have *greater authority* than their counterparts who fail. This could be because of the successful individual's previous record.

None of these factors can be taken in isolation, but clearly, individuals working together on a research and development project who exhibit these characteristics are likely to show a higher degree of success.

Characteristics of inventors and innovators discussed here cannot be measured precisely, and they represent a rather subjective evaluation of human capabilities, behavior, and accomplishments. Consequently, any rigid formalization of these characteristics is likely to lead to their misuse and thus be counterproductive. For example, many individuals without extensive formal education or other characteristics mentioned here have made important contributions to inventions and innovations in the past and they will continue to do so in the future.

Wainer and Rubin [1969] studied 51 technical entrepreneurs who founded and operated R&D companies. They checked if the personality of these entrepreneurs predicted company success—that is, company growth. They found that those who were most successful were high in need for achievement. Such people like to compete with a standard of excellence, and they have many fantasies in which they accomplish something especially important. They usually take moderate risks, they are high in taking individual responsibility, they are interested in the outcomes of what they do, and they are able to delay immediate rewards so as to get larger rewards later. In addition, the researchers found that the most successful of the entrepreneurs in this sample were moderate in need for power (wanting to be the boss). Those who were moderate in need for achievement but high in need for affiliation (wanting to be with people) were also relatively successful. This suggests that an entrepreneur who is good in getting help from others can also succeed, even if his need for achievement is not extremely high.

3.3 ODD CHARACTERISTICS OF INVENTORS AND INNOVATORS

Rosovsky [1987] raised the question, Is there any substance to the theory that a significant proportion of scholars possess difficult and childish personalities—that is, The Amadeus Problem? While it is not prudent to generalize on personality traits, many R&D managers would agree with this assertion. In the film *Amadeus,* Mozart was characterized as infantile and fundamentally boorish. While his behavior was atrocious, his musical gifts were divine. A manager may have to deal with the Amadeus Problem or Complex when a researcher displays exceptional scientific talent but is a difficult, inconsiderate, and unpleasant person to work with.

Some inventors and innovators—men and women of science—are the essence of modesty and kindness. Many of them, however, are not likely to be so characterized. Few have the fine human qualities of Einstein. Many inventors and innovators have well-developed egos and incredible hubris. Some cases described below will give the reader a better understanding of the problem.

Let us take the case of Wolfgang Pauli at the Institute for Advanced Study, Princeton University. Pauli, of course, was a brilliant physicist, discoverer of the "Pauli Particle," and inventor of the "Exclusion Principle," which is one of the pillars of new physics. Pauli used to put people down at physics conferences whenever presenters were not being clear or correct according to his own thinking. This once happened to Robert Oppenheimer at a seminar in Ann Arbor, Michigan [Regis, 1987, p. 196]. While Oppenheimer was lecturing he covered the blackboard with equations. All of a sudden, Pauli jumped up, grabbed an eraser, and cleaned the whole blackboard off, saying it was all nonsense!

Pauli's uncontrolled behavior continued 20 years later when Frank Yang was lecturing at the Institute for Advanced Study on the topic of gauge invariance [Regis, 1987, p. 196]. Yang, a Nobel laureate, had barely started when Pauli interrupted him with a question: "What is the mass of this particle?" Yang replied that it was a complicated problem and that he had not come up with a definite answer yet. Pauli retorted that this was not a sufficient excuse. Yang, who was a model of politeness and reserve, was so stunned that he had to sit down and collect himself [Regis, 1987, p. 196]. Pauli did not feel that he had done anything wrong; instead he thought that it was Yang who was not responding appropriately. Pauli left a note in Yang's mailbox suggesting that Yang had made it almost impossible for Pauli to talk to him after the seminar.

Pauli was not a modest person. He often complained to his colleagues that he was having a hard time finding new physics problems to work on, because he knew too much [Regis, 1987, p. 196].

There is also the case of Kurt Godel, the brilliant logician, probably the greatest since Aristotle, who was also at the Institute for Advanced Study. Godel published his work on general relativity in 1949 and at the Institute he was regarded as utterly profound and inexpressively deep [Regis, 1987, pp. 47 and 63]. This great logician and mathematician, however, believed that his food was being poisoned and that his doctors were trying to kill him. He died of malnutrition.

Edward Wilson, a world renowned evolutionary biologist, wrote about his experiences with James Watson, the co-discoverer of the structure of DNA. Wilson stated that when Watson was a young man in the 1960s, ". . . I found him the most unpleasant human being I had ever met." And at the biology department meetings at Harvard, ". . . Watson radiated contempt in all directions." [Wilson, 1995, p. 42]. Wilson further commented that Watson did not acknowledge his presence as they passed in the hallways and Watson was ". . . supremely self-possessed and theatrically condescending" [Wilson, 1995, p. 43]. Wilson praised Watson for his great discovery and his brilliance and credited Watson's unpleasant and hostile behavior toward him as responsible for redoubling his own energies in the evolutionary area.

3.4 RESEARCHER'S RELATIONSHIP WITH MANAGEMENT AND PEERS

Many researchers have a rather negative view of managers or directors of research organizations. As an example, at the Institute for Advanced Study, Oswald Veblen, the famous physicist, suggested that the Institute did not need a director. Instead, he proposed that the Institute should have a rector, who would not have any power or authority to hatch forward-looking plans or schemes or develop any new institutional policy [Regis, 1987, p. 128]. Some scientists at the Institute joked that a good director should be "a little stupid" so that he would not come up with new ideas that might change the status quo of the institution. When Oppenheimer was being considered for the position of the Director of the Institute, some scientists advanced the truly shocking notion that perhaps the faculty could manage quite well on its own with only an administrator to manage business affairs. Later on when Oppenheimer became the Director of the Institute, he quite seriously suggested that the Institute could probably run quite well without a faculty [Regis, 1987, p. 129]!

One of the directors of the Institute for Advanced Study stated that many of the scientists claimed that they wanted to be free of routine administrative matters so they could focus their energies on research and scholarship. The director suggested that they did not mean a word of it. Although they want opportunities for research and scholarship, they also want managerial and executive powers [Regis, 1987, p. 38].

Power sharing is essential in an R&D organization. Yes, researchers do, in fact, want to share the managerial and executive authority with the administrative structure. Researchers, in particular, want to share those aspects of managerial and executive authority that effect their research activities. Researchers on the other hand need to understand that accountability and some administrative duties also go with such sharing. This simply cannot be avoided. In sharing these powers, researchers will have to do some administrative work, meet scheduled deadlines, listen to the views of others, compromise where there are differing views, and not engage in guerrilla warfare after the decisions are taken.

Some of the negative views that researchers have about management are based on their experiences with organizations. There are cases at universities and research organizations where the administrative structure seems to grow at a proportionally higher rate than the size of the research group or faculty. As an example, at the Institute of Advanced Study, when Oppenheimer was the director he did not spend all of his time running the place. He even used to do some actual research. He had only one secretary, a business manager, and another administrative person. A few years later after he left, the Institute had an Associate Director and several assistants and secretaries [Regis, 1987, p. 285]. The size of the administrative staff seems to grow geometrically and thus many scientists in R&D organizations and faculty members at universities feel that the administration lives for itself while research and teaching activities become side issues. Managers would be well advised to look periodically at the administrative structures in their own offices and question the necessity for every job.

3.5 FORMATION OF TEAMS

There are many attributes that need to be considered when forming a team. The work of Pelz and Andrews (1966b) has already indicated that effective teams are characterized by the support of members for each other's work, great respect for other members, and complementary skills, strategies, and approaches. Respect requires similarity in level of ability, as well as similarity in basic attitudes and values. Yet team members must have varied skills and specific attitudes that are different and complementary. For example, they may all be top-level scientists with similar values toward autonomy, yet differ in their disciplines and in their attitudes toward specific methods of data collection. Empirical support for this point has been obtained in laboratory studies [Triandis et al., 1965], indicating that creativity was highest when team members were similar in their abilities but different in their specific attitudes.

Similarity in ability is important because it is undesirable for people to look at their co-workers as being intellectually too different. A person feels uncomfortable with co-workers who are either "dummies" or "geniuses," and thus cooperation will suffer. At the same time, complementarity is desirable for many personality attributes. For example, a person who would like to dominate a team can get along much better with people who like to be dominated than with others who also want to be in charge. A person who is talkative gets along better with those who like to listen. A diversity of viewpoints on how to approach a research project and different skills concerning its actual operation may result in a better research project [Janis, 1972]. If members of the team are too different, cooperation will suffer; if they are too similar in perspective, *groupthink* may result (see Chapter 2).

What is the optimal team number? There is research suggesting that five is the best number for a discussion group. In larger groups, people feel they do not have enough time to present their ideas, the leader becomes more autocratic and monopolizes the available time, and competing subteams may develop. Smaller groups often lack a definite leader, and they may not develop clear goals or may not have enough different perspectives to avoid groupthink.

The effectiveness of a team depends on the quality of the people in it and the coordination of its activities. However, how these variables impact on team effectiveness depends on the task. There are three types of tasks that need consideration:

Divisible Versus Unitary Tasks. Divisible tasks can be done by different people—for example, checking the references in a reference list (one could divide the checking across as many clerks as there are pages of references to be checked). Unitary tasks cannot be divided—for example, understanding this paragraph.

Maximizing Versus Optimizing Tasks. Maximizing tasks have a criterion with no limit—for example, find as many references as you can. Optimizing tasks have a criterion with an optimal level—for example, determine how much space you need for this project (too much space is wasteful and too little results in an ineffective project).

Disjunctive Versus Conjunctive Tasks. In disjunctive tasks, if one member has the correct solution, the others will necessarily agree—for example, the root of a quadratic equation. In conjunctive tasks, every member must agree—for example, in a jury or in a committee in which everyone has veto power.

In disjunctive tasks, if the probability that one member has the correct solution is P, and the probability that no one can solve the problem is Q, where $Q = 1 - P$, the theoretical probability that the group will solve the problem is $1 - Q^n$, where n

is the number of members. Clearly, the more members you have, the better the group's chances of being successful. In the case of conjunctive tasks, the opposite is true. A solution is more likely if the group is small than if the group is large.

Incidentally, empirically, groups do less well than the theory ($P_g = 1 - Q^n$) predicts, primarily because there are losses in efficiency that occur when people discuss various solutions. Similarly, in divisible tasks, one would expect n people to produce n times the output of one person, but empirically this does not happen. The group often produces less. This lack of individual responsibility, called "social loafing," results when there is no clear identification of each person's output. Usually when one can identify the output of each individual, social loafing is small or nonexistent. Finally, unitary tasks are better done by individuals than by groups.

Osborn [1957] advocated the use of "brainstorming" to increase the creativity of small groups. According to his theory, when people generate ideas in the nonevaluative climate of a group and get stimulated from each other, they produce more ideas. However, the evaluation of this proposal has not supported this idea. For example, Dunnette et al. [1963] arranged for 48 research scientists and 48 advertising personnel to either do brainstorming or work individually. In 23 of the 24 groups a greater number of different ideas was produced by the individuals than by the groups. Not only did individuals produce more ideas when working alone, but also the quality of these ideas was not inferior.

An important aspect of good group problem-solving is the development of a wide range of alternatives. This can best be done when people think of a variety of courses of action independently of others. However, when it comes to evaluating these alternatives there is an advantage in having many different critics who examine the solution from as many points of view as possible. This critique also can be done individually, but groups can be more effective when the evaluation requires memory for previous events. Groups are better in remembering complex material than are individuals, who often forget some fact that others remember and can supply during the evaluation.

In forming effective R&D teams, leadership, individual problem-solving style, and work group relations affect innovative behavior directly and indirectly through their influence on perception of the climate for innovation [Scott, 1994]. The results of this study also provide evidence that innovative behavior is related to the quality of the supervisor-subordinate relationship.

3.6 GENERATING NEW IDEAS

Precise definition of how new ideas are generated and how results from R&D are converted into innovation are hard to articulate. In an article entitled "Serendipity or Sound Science?" Sutton [1986] points out that the Nobel Prize to Burton Richter and Samuel Ting resulted from discoveries that, though unexpected, were nonetheless the result of a lifetime of careful research. Sutton shows that the way to unexpected results lies not in accidents, but in excellence and in thorough scientific investigation with the best possible intellectual and laboratory resources.

An article entitled "The Acid Test of Innovation" [Bell et al., 1986, p. 32] provides several examples of innovation; it concludes as follows:

> Serendipity, self-interest, concentrated and coordinated work: all these are characteristics of successful innovation. But the single most important lesson of these examples is the importance of keeping in touch. Academics need to know what their local industries can do and what they might be interested in making and marketing in the future. Companies, large and small, benefit by having key personnel who can keep up with scientific literature and what is going on in university departments to which they have easy access.

Many innovations occur when facts from previously unrelated fields of knowledge are brought together in a creative solution. For example, a process used in the chemical industry may prove useful in the textile industry. Many other innovations require the redefinition of the problem. For example, for thousands of years horses were used to draw chariots, until it occurred to some warriors that they could ride them, giving them speed not available until that time.

Since such redefinitions are inhibited by conventional wisdom, people without such wisdom (e.g., the newcomer) are often more creative than those steeped in conventional ideas.

MacKinnon (1962) has summarized research on creativity that suggests that creative people are more open to their feelings, have a better understanding of themselves, have a wide range of interests, and have many interests that U.S. culture classifies as feminine (e.g., an interest in the arts). They tend to be uninterested in small details and more interested in the broad picture and its implications. They possess cognitive flexibility and verbal skills and are good communicators and intellectually curious, but they are not interested in policing their own impulses.

Intelligence and creativity are not correlated in the case of IQs in the 120 (ability to do college level work) to genius range. In other words, there are many examples of modestly intelligent individuals who are extremely creative and extraordinarily intelligent individuals who are not.

Creativity depends on both people and environment. A creative environment allows the creative scientists to feel free to work in areas of their greatest interests, provides them with many rewards and recognition, allows them to have broad contacts with stimulating colleagues, encourages them to take moderate risks, and tolerates some failures and nonconformity.

Numerous techniques have been suggested that supposedly improve creativity. Among them we will mention teaching the scientist to ask questions such as "Should I adapt, modify, reduce, substitute, rearrange, reverse, or combine the processes under consideration?" Brainstorming [Osborn, 1963], synectics [Gordon, 1961], lateral thinking, need assessment [Holt et al., 1984], and combinations of the above [Carson and Rickards, 1979] have been suggested. There are also analytic techniques [for a review see Twiss, 1992] which follow logical analysis (the analysis of the attributes of the products that need to be developed) and morphological analysis (the study of the needs of customers, technologists, and marketers, as well as the systematic monitoring of technological developments).

Each of these techniques has enthusiastic proponents, but systematic evaluations of the effectiveness of the techniques are lacking. In the few cases where careful evaluation was done, it did not support the claims of the proponents. However, one could argue that careful evaluation in laboratory settings may not generalize to the field.

The general idea of most of these techniques is to involve many people in the creative process, to use a number of different ways of generating ideas, to find ways to systematically eliminate those ideas that are unlikely to be workable, and to keep eliminating until one idea survives. This new product would be one of a myriad of potential ones, but many perspectives will converge in support of it.

In many of these techniques, one is supposed to "suspend criticism" during the idea-generation phase. Thus, in brainstorming one is allowed to suggest any idea, no matter how unworkable. In synectics, one is supposed to link apparently irrelevant elements and be free from the constraints of critical judgment and the boundaries of orthodox ideas. One states and restates the problem, makes analogies, uses fantasy, and is encouraged to produce paradoxical ideas, such as "dependable unreliability" or "living death." In lateral thinking, one is supposed to challenge assumptions, focus attention on different aspects of the problem, generate many solutions, and introduce irrelevant ideas and even discontinuities in thinking about the problem. In need assessment, one examines the existing, future, emotional, and rational needs of customers, technologists, and marketers with respect to a particular product.

Since systematic evaluation of these techniques is not available, we offer this advice. Speak to several "expert" trainers who advocate each of these techniques. Select two or three of these techniques on the basis of how promising they may be in relation to your particular products or problems. Randomly assign problems to techniques and have experts lead your groups through the creativity phase utilizing each technique. Evaluate results by examining which technique produced the best set of results.

This approach seems wise because it is likely that *your* products or problems can be better solved with one approach than with another. In short, it is not clear that a particular technique will prove effective for all problems. Your problems may well have some industry-specific characteristics. Thus, by experimenting, you should be able to identify the unique combination of techniques that is most helpful in solving your particular kind of problem.

Successful innovators pay close attention to their users' needs and desires [Quinn, 1985], avoid detailed early technical or marketing plans, and allow entrepreneurial teams to pursue competing alternatives within a clearly conceived framework of goals and limits. A number of important patterns contribute to innovation [Quinn, 1985, p. 77], as discussed below:

Atmosphere and Vision. This includes providing the proper environment, value system, and atmosphere (perhaps culture of the organization) to support the innovation process. Perhaps an executive vision (having goals that move the organization toward societally valued achievements) is more important than a particular management background. Managers with executive ability project clear long-term goals for their organizations that go beyond simple economic measures. Such vision, combined with a creative organizational culture, is likely to lead to an innovative organization.

Orientation to the Market. Since innovation involves doing research and development activities that are commercially useful, innovative companies inevitably have to tie their activities to the realities of the marketplace. This means keeping in touch with the user.

Small, Flat Organizations. This means an organization with two or three levels and project teams that are small (fewer than seven).

Multiple Approaches. Since many positive results might come from unexpected approaches, it is important not to narrow the investigation too early. Thus, management should not overdirect the approach used for research and development activities, at least not early on.

Developmental Shootouts. It may be desirable to use parallel and competing developments for an activity. While the cost of such an activity may seem high, this duplication may provide the most efficient and effective output. These developmental competitions or shootouts among competing approaches perhaps can be handled best when the project reaches a certain prototype stage. Quinn [1985] points out that one of the problems associated with such an approach is the issue of managing the reintegration of the members of the losing team.

Skunkworks. According to Quinn [1985, p. 79], for every highly innovative enterprise in the research sample he studied, a small-company environment was emulated by using groups that functioned in a "skunkworks style." In this approach, small teams of engineers, technicians, designers, and others were placed together with no other intervening organizational or visible physical barriers to developing a new product from the idea stage to the final commercialization stage. This approach has been used successfully in many Japanese companies. Quinn [1985, p.79] gives the example of Soichiro Honda, who was known for working directly on technical problems and who emphasized his technical points by personally working with other members of the "skunkworks team."

Interactive Learning. While "skunkworks" emulate the highly interactive and motivating learning environment that characterizes many successful ventures, there is also the need for interactive learning achieved by close contact with the wider scientific community. Even the largest research organization represents only a small fraction of the total research investment internationally and, in turn, only a small fraction of the enormous intellectual and technological resources available and necessary for generating new ideas and innovation.

3.7 EMPHASES ON ASPECTS OF ORGANIZATIONAL CULTURE

Culture includes values—that is, conceptions of the desirable. For example, values specify how much freedom people should have to determine the kinds of research problems they will tackle or what level of equality there should be within the organization (i.e., whether people should feel "great respect" and distance from their supervisors or be able to walk into their offices and feel close to them).

Related to values are norms, which are ideas about desirable behaviors for members of the organization. Organizations have norms about many things (for example, how much to produce). A norm many research organizations have is that its members should try to publish at least one paper in a refereed journal every year. Members of cultures also have ideas about rewards. Who should get rewarded, when, and under what conditions? What kinds of activities should get rewarded and what activities should not be rewarded? What kinds of schedules for these rewards should be adopted and used? How should people be socialized in the organizations so that they become "good" members? How should they keep up to date? What sorts of norms should one have about attendance at scientific conferences? How should one develop new skills? Should one take courses? Should one go back to a university?

An organizational culture that fosters excellence in R&D work and that is internally coherent and dominates a laboratory is highly desirable. No specific or comprehensive list of cultural elements is possible. In what follows, we propose some crucial areas a manager may want to consider:

- Nurturing the ethos of modern science as an important aspect of organizational culture as postulated by Merton [1973] and discussed in detail in the section entitled "Who Are the Inventors and Innovators?" When the ideal ethos (universalism, sharing of scientific knowledge, disinterestedness in terms of commercial or financial benefits, organized skepticism involving detached scrutiny of scientific discoveries) becomes part of the culture of the organization, science and innovation can flourish.
- Tolerating an innovator who may not always work well within the existing administrative procedures.
- Providing meaning to every person's contribution in terms of organization goals—personal goal congruence.
- Recognizing the importance of interaction with the wider scientific community and users (who can play an important role in the innovation process) and encouraging such interaction.
- Encouraging the various gatekeeper roles, as further described in the chapter entitled "Technology Transfer."
- Recognizing and rewarding excellence—both technical and managerial.
- Finally, developing a culture with a "can-do" attitude directed toward the needs of the customer.

3.8 ETHOS OF A SCIENTIFIC COMMUNITY

Robert Merton [1973, p. 270], commenting on the normative structure of science, stated that "the institutional goal of science is the extension of certified knowledge. The technical methods employed towards this end provide the relevant definition of knowledge: empirically confirmed and logically consistent statements of regularities." He further stated that the ethos of modern science includes universalism, communalism, disinterestedness, and organized skepticism [Merton, 1973, p. 270].

Universalism. This expression suggests that scientific discoveries, whatever their source, are subjected to preestablished impersonal criteria. The truth and scientific value of a scientific discovery is independent of the personal or social background attributes of the individuals. Merton further stated that "universalism finds further expression in the demand that careers be open to talents . . . to restrict scientific careers on grounds other than lack of competence is to prejudice the furtherance of knowledge. Free access to scientific pursuits is a functional imperative" [Merton, 1973, p. 272].

Communalism. Communalism implies that scientific findings should be shared equally among all members of the scientific community. Since findings of science are a product of social collaboration within the scientific community, the ownership of such discoveries is a property of the commons and the rights are assigned to the wider scientific community. Merton [1973, p. 273] suggests that scientific discoveries

> . . . constitute a common heritage in which the equity of the individual producer is severely limited . . . Property rights in science are whittled down to a bare minimum by the rationale of the scientific ethic. The scientist's claim to "his" intellectual "property" is limited to that of recognition and esteem which, if the institution functions with a modicum of efficiency, is roughly commensurate with the significance of the increments brought to the common fund of knowledge.

Disinterestedness. Merton [1973, p. 276] has stated:

> A passion for knowledge, idle curiosity, altruistic concern with the benefit to humanity, and a host of other special motives have been attributed to the scientist. The question for distinctive motives appears to have been misdirected. It is rather a distinctive pattern of institutional control of a wide range of motives which characterizes the behavior of scientists. . . . The demand for disinterestedness has a firm basis in the public and testable character of science and this circumstance, it may be supposed, has contributed to the integrity of men of science.

Scientists make few spurious claims, and, in fact, they would be ineffective since much of what is produced by scientists undergoes scrutiny, replication, and review by fellow scientists. Since scientists do not personally benefit from many of these discoveries, disinterestedness (in terms of commercial or financial benefits) forms an important characteristic of the scientific community.

Organized Skepticism. Organized skepticism is interrelated with other characteristics of a scientific community. It is both a methodological and institutional mandate. This involves a temporary suspension of judgment and the detached scrutiny of scientific discoveries. Numerous scientific conferences and meetings held throughout the world embody many of the characteristics described here.

In many ways, in a modern society, it has become possible to further strengthen these characteristics due to the ease of communication among scientists. Fostering this ethos is essential for productivity and excellence in an R&D organization.

The ethos of the scientific community, as is the case with such principles, can never be universally achieved in practice. As Bok [1982, p. 151] has stated, the force of competition, lure of prizes, and fame, among other things, can have some negative effects on this ethos. Over the years, nonetheless, the peer review and refereed publication process and the universal nature of science has helped to make these aspects of ethos the norms of the scientific community.

Discussing the rationality of science, Newton-Smith [1981, p. 44] states: "According to Popper, truth is the aim of science. But the scientific condition is one of ignorance. . . . Popper's thesis of the utter inaccessibility of truth leads him to reconstrue the goals of science as that of achieving a better approximation to the truth, or as he calls it, a higher degree of verisimilitude." For Popper, when an experiment turns out as the hypothesis predicted, this only means that the hypothesis has not been refuted. Popper asserts that positive evidence favoring a hypothesis does not necessarily constitute evidence in its favor [Newton-Smith, 1981, p. 45].

Some science historians would argue that science does not and cannot get hold of the "truth" in any objective and impersonal sense [Regis, 1987, p. 217]. Ed Witten, one of the inventors of the superstring theory, suggests that whether we get to the "truth" or not, we learn new things as we develop new theories. He suggests that as we learn new things, "We develop more powerful laws, laws that unify principles and give us more accurate descriptions of more and more phenomena. It doesn't mean that the old stuff was wrong. It just wasn't complete" [Regis, 1987, p. 211]. Basically, new discoveries mean we know more than we did before, and it is a step forward.

3.9 SUMMARY

In summary, this chapter has comprehensively defined organization productivity in terms of organization effectiveness and has presented suggestions regarding what might constitute organization effectiveness. Creating such an organization, its team structure, and its culture have been discussed. Further information on creative tensions and staffing was included in Chapter 2, "Elements Needed for an R&D Organization." We have tried to focus on such interesting questions as the way in which new ideas are generated and who are the inventors and innovators. Ideas and empirical data available on such issues cannot be precise. Thus, rigid formulation of information about the characteristics of an innovator or inventor is treacherous ground and should be avoided. Instead, in managing a productive and effective R&D organization, a manager should foster the environment so it will be conducive to the characteristics of successful innovators and inventors.

3.10 QUESTIONS FOR CLASS DISCUSSION

1. What are the criteria of R&D lab organizational effectiveness?
2. What kinds of people are most creative?
3. What can be done to increase lab creativity?
4. How should unique characteristics of inventors and innovators be handled by research managers? Develop a case study to focus on the issues.

3.11 FURTHER READINGS

Lander, L. (1995). Improving the R&D decision process. *Research-Technology Management,* **38**(1), 40–43 (January–February).

McGrath, M. E. (1994). From experience—the R&D effectiveness index: a metric for product development performance. *Journal of Product Innovation Management,* **1**(3), 213–220 (June).

Ransley, D. L. (1994). A consensus on best R&D practices, *Research-Technology Management,* **37**(2), 19–26 (March–April).

4

JOB DESIGN AND ORGANIZATIONAL EFFECTIVENESS

A major consideration in designing jobs is the match between the requirements of the organization and the requirements of individuals. To design a job optimally, one needs to consider the abilities, interests, and personality of the individuals as well as the needs of the organization. For example, the organization may find it best to have people follow precisely the rules and regulations that the organization develops, but individuals often find it much more satisfying if they have considerable freedom in deciding how to behave within the organization. The individual's freedom, on the other hand, cannot be unlimited, so jobs have to be designed in such a way that a balance is achieved between the needs of the organization and the needs of the individuals. Some individuals, because of their personality, have an especially strong need for autonomy and will require a job that is designed with far more freedom than is necessary for the majority of individuals.

The organization must be concerned with the compatibility of individual goals with those of the organization in order to maximize the motivation of individuals and to minimize friction among them. However, individual needs can be satisfied in a number of ways, and they tend to change with experience, maturity, and the individual's stage in life. For example, security may be less important to an unmarried 20-year-old than to a married 45-year-old. Security can also be satisfied in different ways (e.g., a social security system that operates over a long period or a high salary for a short period of time). In matching individual and organizational goals there is necessarily some give-and-take in both directions, probably with the individual giving more than the organization, simply because the organization cannot bend as easily as can an individual.

Managers can accomplish the melding of individual and organizational goals by (a) selecting people whose personal goals are already compatible with those of the organization and (b) using participative management, which sets goals in such a way that individuals can accept them and the organization can attain its objectives. Some "negotiations" of such goals is healthy, since the organization can reach its goals by a large number of paths. For example, commercial organizations can make a profit in many ways, and it is more likely that they will do so if they use goals that are compatible with goals of their employees.

In dealing with job design, we need to consider the match between the person and the job in greater detail. On the individual's side, we have the person's abilities as well as needs. On the job's side, we have requirements that can be conceived both in terms of ability requirements and job attributes that satisfy needs. When there is a good match between the job ability requirements and the individual's ability, the individual is more likely to be satisfactory from the standpoint of the organization. In that case, the individual is more likely to be promoted and achieve his goals within the organization. When there is a match between the needs of the individual and the job's ability to satisfy those needs, the individual is more likely to be satisfied with the organization, to find the job enjoyable, and, as a result, to stay with the organization and participate in its activities more frequently. For example, low absenteeism is likely to be associated with such satisfaction.

Jobs should be designed so that people can define the functions. It is a mistake to think in terms of the traditional bureaucratic structure, in which the organization writes the job description. Those who are members of a team should write their own job description, and the team must do some of the job defining. Of course, the team leader has to provide some guidelines and ensure that the jobs that are so defined are consistent with corporate goals. But the details can be left largely to the job holders. Once a talented scientist is given a broad objective compatible with the goals of the organization, a lot of job designing can be done by the researcher.

An organic organization allows job incumbents to develop job definitions and allows the research team to define the jobs of its members. Competence is used as the main determinant of status in the hierarchy of the organization. Peer review of projects and job definitions proposed by job incumbents can shape the job definition better than management review.

4.1 JOB ATTRIBUTES

There is considerable literature that is consistent with the Hackman and Oldham [1980] analysis of job attributes and how they are related to satisfaction. The general argument in this literature is that jobs that provide sufficient variety, autonomy, task identity, and feedback are more satisfying than jobs that do not have these attributes. *Variety* refers to the ability to do different jobs at the particular job site. *Autonomy* refers to the ability to decide for oneself what should be done. *Task identity* refers to having a job that can be identified as a distinct unit. For example, completing a particular research project, or saying that this particular discovery is associated with this particular individual, gives the specific activity more task identity because that activity can be associated with the name of the investigator.

On the other hand, in a research job involving many investigators, where the complete job is done at different organizational levels and locations, the situation reduces the identity of the job. *Feedback* refers to knowledge of how well one is doing the job.

Hackman and Oldman's theory argues that a job is high in meaningfulness when it has task variety, task identity, and task significance, so that people think that what they are doing is important. High meaningfulness in turn creates satisfaction. Similarly, they argue that autonomy, which allows a person to be more responsible for the job activities and outcomes, also leads to satisfaction. Finally, feedback is very important. People should know how well they are doing and whether what they have been doing is having an impact. A person who works for years on something without receiving any attention (for example, nobody quotes his/her scientific work) is not receiving feedback.

So, in research, as in other kinds of work, the best possible job is one that is designed so that it is variable, has significance and identity, allows the investigator considerable autonomy, and provides feedback.

There is a substantial literature that deals with these issues [Loher et al., 1985]. Thus, in designing research and development jobs, it is well to remember that maximizing variety, autonomy, and feedback, as well as the meaningfulness of the work, is likely to be associated with satisfaction.

While the R&D field has its own frustrations, nearly 75% of more than 4000 respondents to R&D Magazine's Career Satisfaction Survey indicated that they would choose the same profession again [Dougherty, 1990]. A majority of the respondents said that they were satisfied with their jobs, and this was attributed to the opportunity to address challenging problems that the work setting provided.

4.2 PHYSICAL LOCATION AND COMMUNICATION

Another aspect of job design concerns the way jobs are physically located in relation to each other. There is considerable literature that indicates that arranging jobs in such a way that communication is either increased or decreased can result in improved performance in research and development jobs. For example, Allen [1977] and Szilagyi and Holland [1980] have indicated that the probability of communication increases rather sharply with a decrease in the distance between co-workers. Moreover, Szilagyi and Holland [1980] have reported that increased social density (a situation in which there were many employees per square foot who were able to talk to each other quickly) resulted in such employees experiencing task facilitation. Co-workers exchange information much more easily; as a result, they were more effective and also more satisfied with their jobs. There is also a study by Morton [1971] that found that removing spatial barriers improved communication between innovators and their customers.

On the other hand, one should not oversell these ideas because there are some studies, such as Thompson's [1967], that indicate that there are situations when barriers are desirable. For example, if one has a rather creative group in a bureaucratic organization, physical barriers between that group and the bureaucratic organization can be extremely desirable.

There are differences also between scientists and engineers in the way they communicate [Allen, 1977]. Scientists communicate through the literature and through professional meetings. They spend a good deal of time talking with outsiders and exchanging experiences. On the other hand, engineers spend most of their time communicating with customers and vendors and doing actual experiments within the laboratory. The motivation of the two groups is to some extent different because the scientists are much more individualistic and require recognition as individuals by the scientific community. Publishing is vital for them. In contrast, the primary concern of engineers is the success of the organization. For the scientist, the number of colleagues in the same discipline who are working on the same problem is associated with success; for the engineer the number of colleagues in the same laboratory is related to success. In other words, we need to distinguish between scientists and engineers. The motivational and organizational courses of their success are not identical. This has implications for the effective management of an R&D organization, as well as for job design.

In designing jobs, we also ought to consider the flow of information. Research and development jobs are particularly susceptible to this factor because there is so much task uncertainty and novelty associated with such jobs. There is also the need to communicate with the wider scientific community. We must consider the conditions under which jobs will be better performed because they are designed so that people can exchange information that clarifies and improves the task. To some extent this has to do with the way we organize the job in relation to other jobs. For example, how we place one job next to or away from another, or the arrangement of the desks of particular scientists or engineers, may be crucial in determining how effective they are going to be. Some of the research literature, such as Cheng [1984] and Katz and Tushman [1979], discuss these matters and point out that on jobs for which there is no organized body of literature to which technologists and engineers can turn when they face a new problem, they will obtain information from others who have similar experiences. Thus oral communication is extremely helpful.

Oral communication permits rapid feedback, decoding, and synthesis of complex information, and it fits especially well into scientific settings where most of the research ideas are as yet unformed and difficult to articulate [Katz and Tushman, 1979].

However, the literature suggests that too much communication may be dysfunctional if the actors who exchange information do not share a common language. In one study [Katz and Tushman, 1979] that involved 350 academic research units located in six countries, the researchers found that in most situations in which there was low uncertainty about how to do the job and high ambiguity about the long-term goals of the job, there was a positive relationship between communication and the success of the unit, but a negative relationship between communication outside the unit and performance. If people started talking with others outside their units and lacked a common language, the greater the amount of communication frequency the lower the performance. In short, people spent so much time clarifying what they meant that they were not effective in performing their tasks.

These findings are not consistent with the points made by Pelz and Andrews [1966b], who found that among 1130 scientists in academic and industrial laboratories *both* intra- and interunit communication was highly desirable. However, an explanation of the discrepancy between the findings by Cheng and Katz–Tushman on the one hand, and those by Pelz–Andrews on the other, may be that Pelz–Andrews dealt primarily with applied scientists. In their case, perhaps the difficulties in not having a common language are not as great as in the case of those doing more basic research, who are trying to develop new concepts and a new language at the same time that they are communicating.

A comment about the physical work environment of the R&D personnel might be in order. As discussed previously, scientists have a tremendous need for autonomy. In addition to providing a formal and informal supportive management environment, R&D personnel should be given considerable freedom to modify and personalize their office environment within the resources available for such improvements. In the case of R&D personnel, more than in other groups, one may find some office modifications and arrangements unusual and bordering on grotesque. Our experience indicates that some of these problems may seem trivial, but in practice they are much more difficult to resolve. Exhorting a researcher to change his office arrangement or keep his office better organized simply does not work. Unless the physical situation in the office is clearly adversely affecting job performance or safety (for example, creating a fire hazard), it is best to tolerate such idiosyncrasies.

4.3 CAREER PATHS

In designing jobs, it is also desirable to think of the total career path of the scientists. One of the problems in a research and development setting is that there are not sufficient opportunities to reach the top of the organization in terms of pay and prestige for those engaged in purely technical activities. The jobs are designed in such a way that administrators receive higher compensation, but this is inconsistent with many of the values and needs of technically gifted individuals. To the extent that the engineer or the scientist is dissatisfied with his job definition, he is more likely to turn to other sources of satisfaction—for example, family or civic matters. There is evidence [Bailyn and Lynch, 1983] that indicates that this is exactly what happened to those engineers who were less satisfied with their careers. Keenan [1980] reports a study by Gorstel and Hutton that indicates that many engineers who became part of the management team did so rather reluctantly and because of the absence of sufficient opportunities for technical careers, rather than because they enjoyed the management position.

In other words, we have a situation in which the role a person really wants (a top-level technical job) is not available in the organization. Furthermore, only the managerial jobs pay well. This situation results in low morale and inefficiency. It is a problem that most organizations must face squarely and try to solve. It is often necessary to design jobs to fit the needs of employees to some extent.

When thinking about job design we should also realize that engineers go through various stages in the development of their careers. For example, Thompson and Dalton [1976] interviewed over 200 scientists, engineers, and managers and identified several stages for the development of an engineer's career. They argue that in order to perform well an engineer should go through four different stages.

1. The engineer should work with a mentor, who can teach him/her how to design and carry out projects and how to be successful in relating to clients and upper management. During this stage, the mentor obtains the projects, designs the broad outline of the project, and fits the project into the activities of the organization. The apprentice does the detailed work, makes sure that everything is accurate, and follows up on all details. It is obvious that during this stage the definition of the job makes the apprentices appendages of their mentors, and so they must be physically located very close to each other in order to develop the proper interpersonal relationships.

2. The engineer assumes responsibility for a definable portion of a project or process, works independently, and produces results that are significantly identifiable with him/her. The professional begins to develop credibility and a reputation as a person who knows a great deal about a particular area. The professional now manages more of his/her own time and accepts more responsibility for the outcomes. Relationships with peers and fellow professionals now become very important, while the relationships with the supervisor or mentor are less significant. This requires a different kind of job definition and a different kind of physical arrangement. For example, the professional, in this case, could be located far away from the supervisor.

In general, organizations consider Stage 2 valuable, but not especially desirable. If a person stays in Stage 2 for a very long time, the chances are that he will be fired or moved to some other job that is not very important. In other words, the expectation is that the engineer will get out of Stage 2 and into Stage 3 if he is going to be considered "successful."

3. This stage is somewhat different. Here the engineers apply their technical skills to several areas rather than to a specific project. They get involved in external relationships with suppliers, with clients, and with new business ventures, and they begin to do things that benefit others and the organization in general. They become involved in the development of other people. Many engineers stop at this point, and are considered very successful.

4. In Stage 4 the manager exercises a significant influence over the future direction of a major portion of the organization. He/she tends to engage in wide and varied interactions both outside and inside the organization; he/she is also involved in sponsoring and developing promising people who might fill future key roles in the organization. Generally, people in Stage 4 spend their time in three ways:

(1) They are innovators who contribute to the future of the organization by supplying innovative and original ideas that might shape the organization; (2) they are internal entrepreneurs who bring together resources, money, people, and ideas in order to pursue new developments (for example, new research projects), and (3) they are upper-level managers who form policy, initiate programs, and monitor the progress of the organization.

It appears clear from this discussion that there is an increasing managerial component in the activities of the engineer as he/she moves from Stage 1 to Stage 4. However, it is important to remember that the technical side of the activities can remain a very substantial component of the engineer's total activity. It seems appropriate, then, to reward people who do purely technical work in spite of the fact that they are not supervising a large number of people. In other words, the managerial career and the technical career should not be viewed as inconsistent. On the contrary, sometimes professional employees in a managerial position can take better care of their interests as professionals by increasing their decision-making latitude concerning research and by their control over resources relevant to their scientific work.

Roberts [1978, p. 6] asserts that "even with academic scientists, some of our own studies here at MIT show that faculty are more productive when they have a mix of work activities. There's even a finding that mixing research with administrative work helps increase creativity and idea generation!"

It would seem, therefore, that technical competence and management responsibilities are not inconsistent. Nonetheless, in job design and in structuring an organization, the problem still remains of providing promotion opportunities for technical personnel, who may not want management responsibilities at higher levels where they can no longer contribute directly to R&D. A mechanism—using multiple hierarchies—for addressing this problem is discussed in the next section.

4.4 DUAL AND TRIPLE HIERARCHIES

One of the ways of dealing with the problems outlined above is to develop a dual or even a triple hierarchy within the organization. In the dual structure, organizations develop an additional hierarchy with technical positions that parallel the positions in the management hierarchy. These technical positions comprise a professional hierarchy that has the same degree of control, authority, and compensation as the corresponding positions in the management hierarchy. However, Schriesheim et al. [1977] reviewed the literature and concluded that dual hierarchies have been generally unsuccessful at resolving conflicts between professionals and their employee organizations and at providing alternative career opportunities and reward systems.

Apparently, the most important reason that these organizational structures failed is that promotion in the professional hierarchy was, by definition, "a movement away from power." In addition, there were signs of failure among those who followed the professional hierarchy because they felt they lacked parity with the managerial hierarchy and because the evaluative criteria used were inequitable. For this reason, Schriesheim et al. [1977] suggested a triple hierarchy as an alternative structure for managing professional organizational conflict.

The triple hierarchy provides three different advancement opportunities. The managerial hierarchy is available to those who desire advancement to managerial positions. For those professionals who desire only professional duties (here "professional" implies scientific, research, and technical duties) the professional hierarchy remains a viable option.

The third hierarchy is occupied by professionals who have key administrative jobs, as well as regular professional duties. They have hierarchical authority in those areas where professional values and organizational requirements are most likely to diverge.

This type of organization is similar to the organization of major research universities. In such universities the administrators are usually individuals with a good research record. Thus, they are able to relate to the faculty. At the same time, they have other attributes that enable them to interact successfully with government officials, trustees, alumni, major donors, and so on. Some administrators are closer to the research process, while others are closer to the political activities needed to run a university. A successful university has the right mix of administrators. Similarly, R&D organizations need some people who will deal with the politics, some with the technical aspects, and some with both. The right mix results in the best R&D organization.

We will use the terms *technical, professional-liaison,* and *management hierarchy* to discuss the advantages of the triple hierarchy. In the university setting, these terms might correspond to the head of a department, a dean, and a university president. The head of the department is able to evaluate a faculty member's professional qualifications; a dean is able to evaluate a head of department; a president is able to evaluate a dean. Usually, the head of the department is technically very competent and even does research. The dean is also technically competent and in some cases does research. The president is often a generalist and rarely does research.

The argument is that the triple hierarchy can deal with all three of the problems: the domination of the organization by the management hierarchy, miscommunication, and inadequate evaluation procedures. In the triple hierarchy, managers have less power, since much of it has been taken away by those in the professional-liaison hierarchy. While managers and professionals in the dual hierarchy have different perspectives and often miscommunicate, in the triple hierarchy the technical people interact mostly with those who are in the professional-liaison hierarchy rather than with those in the management hierarchy. As a result, they are able to communicate and get along better because they share the same values. In the dual hierarchy the managers judge the technical people, whereas in the triple hierarchy the professional-liaison hierarchy individuals, rather than the managerial structure, evaluate the professionals. The result is that those who best understand their perspective are the ones who are judging how well the technical people are doing. There is some evidence that the triple hierarchy is effective. For example, Baumgartel [1957] and Pelz [1956] found that when the positions of research director or administrator were held by individuals with professional or scientific backgrounds, researchers felt more protected and work units had higher productivity and morale. Additional support came from studies by Lawrence and Lorsch [1967], Likert [1967], Marcson [1960], and Mintzberg [1973].

Can this triple hierarchical approach be successfully implemented in an organization, and is it really practical? In formal organizations, by necessity, both the organizational structure and individual responsibilities are rigidly defined, while the triple hierarchical organizational structure requires more flexibility and cross or parallel communications. When this triple hierarchical approach is successfully implemented, many unintended and secondary benefits have occurred, such as retaining highly qualified technical staff and minimizing the complacency process of the organization. (For the purpose of this discussion, *complacency* is defined as a state of being uncreative, stale, and aging.)

4.5 CENTRALIZATION AND DECENTRALIZATION

Another broad issue of job design is whether to use centralized or decentralized structures. According to Allen [1977], in a centralized project or project structure a majority of the people working on a particular activity report to one project manager, who makes the assignments. They receive management reviews from this particular individual and are also physically located near that person. On the other hand, if fewer than 50% of the personnel are reporting to this one person, then the project is considered decentralized.

Allen [1977] and Marquis and Straight [1965] suggest that both the decentralized and the centralized structures can be effective, but under different conditions. In decentralized structures all the information needed to do the job is available to most members. In centralized structures the information must be obtained from one or two specific individuals who have most of the relevant information. The decentralized structure is effective when the flow of knowledge has to be relatively fast and the projects are of long duration. On the other hand, the centralized structure is more effective when the projects are short-term and the flow of knowledge is not especially rapid. The decentralized structure is better when a lot of new information comes in and out of the project area, requiring a flexible system of organizing as well as a great deal of communication and cooperation among people participating in the project.

According to Peters and Waterman [1982, p. 15], excellent companies combined centralized and decentralized structures. These companies have given autonomy to many aspects of the organization down to the lowest level. However, when it comes to the core values that are dear to the company, they are fanatics about being centralists. Peters and Waterman [1982, p. 314] suggest that the hybrid alternative for organization structure responds to three crucial needs: a need for efficiency, a need for innovation, and a need to avoid rigidity.

Many other aspects of centralization and decentralization for large organizations are beyond the scope of this book. As organizations become larger, the hybrid (centralized and decentralized) approach is quite appropriate at the lower levels. There still is a need to provide some further division of the large workforce. Divisions based on product line, geographic location, or specific project are among the possibilities.

4.6 KEEPING THE RESEARCHER AT THE INNOVATION STAGE

An individual goes through a number of stages in a given position. One first goes through a socialization stage, then an innovation stage, and, finally, the stabilization stage. In the stabilization stage the individual is less creative, less risk-taking, and less productive. While individual differences exist, the stabilization stage is normally reached by an individual after 6–8 years of being in a given position. How a person is socialized into an organization makes a lasting impression. Thompson and Dalton [1976] have presented a suggestion that might be helpful in keeping a researcher at the innovation stage and minimizing the stabilization process.

The idea is to limit the tenure of supervisory personnel to approximately 5 years. In such a case, after the 5-year period the individual will have to return to technical function. The individual will thus have a strong incentive to remain current on the technical side of activities. In addition, there will be an opening so that another person can move into that particular managerial activity. There should be a budget that allows the retraining of personnel, so that senior people who are not aware of the new technologies can have sabbaticals that allow them to catch up with new developments in their fields.

Lateral transfers can also be used to motivate technical personnel. The idea of working on a new project every now and then can be quite refreshing. It provides an opportunity for the person to become motivated and avoid becoming stale. Many managers are against such transfers because they do not want to lose good people, but if these transfers become the norm for the laboratory, then managers should be able to get their share of good people in the course of the total change occurring in the laboratory.

Some organizations use a matrix structure in which the employee has two bosses: one who is responsible for a particular project and another who is responsible for a particular function. For example, there might be a project manager to whom the individual reports, plus another person in the same technical field (functional manager) as the individual, who looks after the development of his career and makes sure that high standards of professional activity are maintained. In such a case, the functional manager may undertake to transfer a person from one project to another in order to increase his skill. It is important to note that Peters and Waterman [1982, p. 307] state that "virtually none of the excellent companies spoke of itself as having formal matrix structures, except for the project management companies like Boeing." They further point out that even at Boeing, where many of the matrix management ideas originated, what is meant by "matrix management" is not what is generally thought of. In an organization like Boeing, people operate in a binary fashion.

Individuals are either (a) part of the project team and responsible to that team for getting tasks accomplished or (b) part of a technical discipline almost all of the time. There is no day-in and day-out confusion as to which team they belong to. One responsibility has clear primacy. A matrix organization where a person reports to different bosses complicates the organization and should be avoided.

Another approach has been to institute manpower reviews designed to improve the skills of the employee. The idea is that every 6 months the employee discusses with second- and third-level supervisors his or her activities and reviews career progress. Decisions are then made concerning transfers and current assignments so as to maximize the development of skills.

Another idea is "career monitoring." This system assigns an engineer or scientist to a new project not less than every 4 years. Anyone in the same job for more than 4 years has his or her name sent to the chief engineer or the technical director of the laboratory, who checks to see what is happening and then makes a judgment concerning whether the particular individual should continue to remain in that job. The general idea is to make sure that assignments are such that repetitiousness and overspecialization do not occur.

Bailyn [1984] suggests that research and development organizations should develop career programs based on relevant information about the employees working in these organizations. This information would take into account the orientation of each scientist. If, for example, his/her orientation is quite academic, there would be more opportunity for the scientist to undertake activities that maximize scientific growth. On the other hand, if the engineer or scientist is not that concerned with academic work, a career that emphasizes support of ongoing research might be more appropriate.

Company or college courses or exposure to professional journals can be used to promote the career of a scientist or engineer. However, Thompson and Dalton [1976] report that they do not find any correlation between high and low performance on research and development jobs and the level of education of the individual. This may be due to the fact that to begin with, most R&D personnel are highly trained and are high achievers. In a modern R&D organization, a doctoral-level education is the norm for R&D personnel. Undoubtedly, some unique individuals without this advanced training have made, and should be given opportunities to make, significant contributions. Thompson and Dalton found that complexity and the challenge of the job were much more strongly linked to job performance than the educational level of the researcher. Such findings would suggest that an individual's success is related to how the organization structures the job rather than the educational preparation of the individual.

Providing a sabbatical leave, though routinely practiced in many university environments, has not been widely used by R&D organizations. Those individuals who have the motivation and the ability to take advantage of such opportunities should be encouraged to do so. If the managers really believe and want to practice the dictum that "people are our most important resource," then the cost of providing sabbatical leaves should be worth the investment. There should be no confusion about the extensive resources and organizational flexibility needed to allow such leaves. Excellent research organizations can and have successfully implemented these sabbatical leave programs.

4.7 JOB DESIGN AND CONFLICT

Sometimes the design of jobs increases the chances of conflict in organizations. This is particularly true because the goals of scientists are sometimes not the same as those of the organization. The goals of scientists are more likely to reflect scientific values, while the goals of the organization often reflect concern for profitability [Keenan, 1980; Souder and Chakrabarti, 1980].

An important attribute of job design is the degree of autonomy that is allowed on the job. There are two kinds of autonomy [Bailyn, 1984]: *strategic autonomy,* the freedom to set one's own research agenda, and *operational autonomy,* the freedom to implement the agenda in different ways. Bailyn discovered through empirical studies that at the beginning of a technical career operational autonomy is more important than strategic autonomy. As the employee becomes more and more experienced, it is essential for the employee to have a certain amount of strategic autonomy. Thus, in the design of jobs we ought to assume that the employee will have a certain amount of autonomy. However, the kind of autonomy that we design for the job may differ depending on the level of the incumbent in the particular job.

There is a considerable literature that discusses the conflict between professional values and organizational goals. Professional values reflect concerns for scientific development. To the extent that scientists regulate their behavior to reflect such professional values, they are likely to remain up-to-date. Thompson and Dalton [1976], however, point out that this emphasis on keeping up with professional development often clashes with basic organizational goals.

Scientists tend to identify more with their profession than with the institution in which they practice. Research administrators, as they move up the ladder, increasingly tend to identify more with their organizations than with their profession. These different tendencies may provide a clue to some conflicts that may emerge between scientists and research administrators. Since the relationship between research administrators and research scientists in an organization is an important variable in determining the success of the organization's research activities, the research administrators need to find ways to minimize this apparent conflict. Ross (1990) has suggested four ways to optimize the administrative role. These are:

1. Maintaining a service orientation toward the organization
2. Remaining flexible
3. Promoting free and complete communication
4. Ensuring that the main goal is the furtherance of research objectives

LaPorte [1967] describes the main sources of conflict between professional and organizational goals:

1. There is often a clash between profit and technological innovation, since an interesting technological development may not necessarily be profitable.
2. The expression of professional desires and goals is often different from management because the individual wishes to be autonomous, while management wishes to integrate the organization.
3. Professionals seek freedom from procedural rules, while managers emphasize them.
4. Professionals seek authority relations based on professional status, whereas managers rely on bureaucratic position and power.
5. Professionals seek rewards contingent on professional status, while managers emphasize rewards that match the organization in strength and status.

Pelz [1956] identified four types of conflict that occur in technical organizations. Type I is mostly technical conflict that occurs with peers and is related to such things as technical goals, milestones, means of achieving particular goals, and interpretation of data. Type II is interpersonal conflict with peers (for example, likes and dislikes, trust and apprehension about the goals of peers.) Type III is conflict that occurs between supervisor and subordinates about technical or administrative matters such as technical approach, milestones, and schedules, while Type IV is conflict that occurs between the supervisor and the subordinate about interpersonal matters such as power, authority, rules, and procedures. Evan's [1965a,b] empirical research discovered that technical conflict is twice as prevalent as interpersonal conflict in governmental and industrial laboratories.

A rather common type of conflict occurs when different parts of the organization have different mandates and, as a result, try to maximize different criteria. An example is the conflict between marketing and research and development groups. The research and development groups are likely to wish to develop products that meet certain technical criteria, while the marketing groups are likely to wish to develop products that will sell well. While the two sets of goals are not incompatible, there are situations when the coordination of the activities of these two groups is complicated and requires very careful attention by management.

The problem has been analyzed by Souder [1975] and Souder and Chakrabarti [1980], who have identified three possible approaches or mechanisms for the coordination of the research and development groups and the marketing groups. The three approaches, which Souder has named "stage-dominant," "process-dominant," and "task-dominant," are described below.

In the *stage-dominant approach* the research and development groups, as well as the marketing groups, create formal structures that have specialized functions; these groups have narrowly defined responsibilities and specifically limited activities. People in an organization have responsibilities that are directly tied to their functional specialties. For example, the engineers limit their activities and responsibilities to the technical side, while marketing people, who are concerned about the way the public reacts to products, have specific responsibilities that deal with the way the product is likely to be received by the public. These formal structures are also reflected in the way the particular job is transferred from one group to another. There are formal and institutional transfer points where the research and development people hand the job to the marketing people, or the marketing people hand it back to the research and development people, with a special ceremony.

In the *process-dominant approach* there are no apparent transfer points, and the parties hand the job to each other, back and forth, without any kind of ceremony. There are no apparent cases where one group builds up and another builds down in order to get extra manpower to complete a particular job during a particular phase. Rather, people move in and out of the activities and the groups as needed. The interaction between the technical side and the marketing side is almost continuous and people do not have points at which they say "that is your job" and "I am not going to deal with that," but rather they are continuously involved in the development of the product. The incumbents in this process are specialists, of course. However, engineers, for example, understand a fair amount about the marketing situation, and the marketing people understand quite a bit about the technical aspects of the project. There is no paperwork that is filed to indicate transition points where the job goes from the research group to the marketing group, or from the marketing group to the research group. The products are expected to and do oscillate from one group to the other.

The *task-dominant approach* is characterized by even greater flexibility. Here the incumbents have a strong orientation and focus toward the task and the end product, and they talk in terms of "our" product rather than "our" and "their" functions. It is not "I am the technical person" or "I am the marketing person," but rather "I am the person who is interested in that product." There are no transfers of authority in this case, and personnel do not go on and off the team as the product is developed. People are specialists, of course, but they do not function as specialists. They are part of the team that is developing the product, and the fact that they happen to be scientists or engineers, economists, or public opinion experts is irrelevant. All they are doing here is participating in a project-oriented team. Thus, in the task-dominant approach the individuals are in continuous contact with each other as a team. This is very different from the situation in which there are formal structures where the research group meets with the marketing group and where people have identities as a "research person" or as a "marketing person."

Souder argues that each of these approaches has some advantages under some conditions, so that one should not generalize and say that one of these three approaches is better than the other under all conditions. Obviously, when you have an organizational structure that allows people to be specialists, they can practice their special skill and become very good at the particular functional activity.

Unfortunately, while the specialists can be especially good in their activities, they may not have very much understanding of what is going on in the other group. The advantage of high specialization is that specialists can do some tasks very fast and extremely well. At the same time, a disadvantage is that there is little coordination with other activities.

Souder has outlined a number of factors that are relevant to maximizing the effectiveness of one or the other of these three organizational structures. Those who are facing the specific question of how to organize teams should consult the original publication [Souder and Chakrabarti, 1980], which shows that the criteria that one wishes to maximize will influence the type of organizational structure likely to be optimal. These criteria include environmental factors (e.g., environmental uncertainty, dynamics), task factors (type of technology, the type of innovation), and organizational factors (the nature of the organization, the complexity of the organization, the kinds of communication patterns, and the division of responsibilities).

4.8 SUMMARY

In this chapter we have examined job design, broadly defined, as it contributes to organizational effectiveness. We have considered how managers can interrelate individual and organizational goals for effectiveness, how job attributes suggest that certain people would be more or less effective doing some jobs, and how physical layout can help communication. Jobs should be designed to promote the interrelationship of personal and organizational goals, match available personnel, and be consistent with the physical layouts that are good for communication. We then turned to career paths, which inevitably have implications for the organizational structures that are likely to be effective in an R&D laboratory, and that discussion resulted in an examination of the advantages and disadvantages of various organizational structures. Keeping the researcher on the path of innovation and designing the job so as to take advantage of productive conflict and avoid destructive conflict were also discussed.

4.9 QUESTIONS FOR CLASS DISCUSSION

1 What kinds of hierarchy are ideal for an R&D lab?

2 What are the advantages and disadvantages of a matrix form of organization?

3 How should jobs be designed in an R&D lab?

4 How can a lab minimize destructive conflict and maximize productive conflict?

4.10 FURTHER READINGS

Perry, T. S. (1995) How small firms innovate: designing a culture for creativity. *Research-Technology Management,* **38**(2), 14–17 (March–April).

Rosenbaum, B. L. (1990). Techies really aren't like the rest of us. *Business Month,* **135**(2), 74–75 (February).

Schriesheim, J., M. A. von Glinow, and S. Kerr (1977). Professionals in bureaucracies: A structural alternative. In P. C. Nystrom and W. H. Starbuck (Eds.), *Prescriptive Models of Organizations.* New York: North-Holland, pp. 55–69.

Souder, W. E. and A. K. Chakrabarti (1980). Managing the coordination of marketing and R&D in the innovation process. In B. V. Dean and J. L. Goldhar (Eds.), *Management of Research and Innovation. TIMS Studies in the Management Sciences,* Vol. 15, New York: North-Holland, pp. 135–150.

Turpin, T. (1995). Occupational roles and expectations of research scientists and research managers in scientific research institutions. *R&D Management,* **25**(2), 141–157 (April).

5
INFLUENCING PEOPLE

An important aspect of a manager's job is influencing others. Top management often needs to influence the rank and file, lower management needs to influence middle management, and so on. In this chapter we will consider how attitudes are formed and changed, and how a person can influence others.

"Influence" may sound like manipulation, and some people may find it objectionable on moral grounds. Yet the evidence is clear from several studies that managers who are influential with their bosses are better managers and can be more helpful to their own subordinates. For example, some studies show that those managers who are influential with their supervisors have subordinates who are more satisfied with their jobs.

Furthermore, an important aspect of lower or middle management is to get resources from top management. Such resources in the form of budget approvals, space allocation, and the like are vital for the unit the manager is administering. Since research output can never be completely predicted and since considerable time lapse exists between resource inputs and research outputs, R&D managers' ability to influence upper management and sponsors could be crucial in acquiring needed resources for research. This ability of influencing people could be important for an R&D manager for internal purposes as well. For example, many research projects require the collaboration of researchers in the manager's own group as well as periodic assistance from external groups. The manager's ability to influence people should give him or her the necessary tools to (a) provide order and purpose and (b) integrate the contributions of different participants to the research effort.

A principal investigator, on the other hand, often has to deal with researchers within his or her own team and also has to work with the immediate supervisor, the sponsor, and many individuals in the support offices of an R&D organization. The ability of a principal investigator to understand the attitudes and motivation of different people and to influence these people could make a crucial difference in getting the job done.

In influencing people, some important aspects deal with attitude and attitude change, as well as with communication alternatives and outcomes. These two major topics along with a case study and its analysis are discussed in this chapter.

5.1 ATTITUDE, ATTITUDE CHANGE

A central construct for understanding influence is the construct of attitude. An attitude is an idea, charged with affect, predisposing action. For example, Mrs. Top Manager's attitude toward Department X getting additional space is based on several beliefs about the space problems of Department X, how well Department X is doing, the future of Department X research, and so on. Each of these ideas has some emotion, positive or negative, attached to it. For instance, if Mrs. Top Manager thinks that Department X is doing well, she is also likely to feel more positively about Department X getting additional space. Such feelings become intentions (e.g., self-instructions to approve the extra space), which often lead to action (approval of the space).

To change an attitude, one needs to consider the phases of the attitude change process. These are *attention, comprehension, yielding,* and *remembering,* followed by *action.* For example, if top management wants to change the attitude of the rank-and-file toward a new policy, it must produce a communication that will be attended. If the employees place the memo describing the new policy in the waste paper basket, unread, it will obviously have no effect. But even if it is read, it must be understood. Comprehension is the phase when the attitude change message is understood. However, understanding does not necessarily mean acceptance. The reader might understand it and reject it. Thus, *yielding* refers to the phase in which the reader not only understands, but goes along with, the suggestion or the message. While yielding is very important in attitude change, if one is interested in action, one needs two more phases: First the person must remember yielding, and then the factors involved in the action (norms, roles, previous habits, affect toward the behavior, perceived consequences, and so on, as discussed in Chapter 6 on motivation) should not cancel out the intention to act. In other words, the analysis of attitude change shows that it is a complex process in which the attempt to change attitudes and behavior can be "derailed" by many factors. In thinking of the best means to change someone's attitude, it is useful to be analytic and to consider and anticipate all those factors and, if possible, make them favorable to the change.

The analysis of attitude change also requires thinking of the factors that are important in this change. There are four factors to be considered: the source of the attitude change, the message, the medium, and the audience. The source is the person or group that produces the attitude change message. For example, if Department X wants to get more space it could send its manager to talk to the top management, or it can send a departmental resolution to the top management or ask some allies to suggest to the top management that more space is needed. In each of these examples both the source (manager, department, allies) and the message are different. One must also consider the medium. For example, one might attempt to influence either face-to-face, through a written document, or by using a video presentation. Finally, one must consider the characteristics of the audience. Who are we trying to influence? Depending on our analysis of audience attributes, we can develop different strategies for attitude change. If the audience is very intelligent, research shows that we must produce a message that presents the position we advocate and that also convincingly deals with any objections to that position. This has special implications for an R&D organization whose participants are especially intelligent.

5.2 FINDINGS FROM ATTITUDE RESEARCH

There are many experimental findings about attitude change (for further reading see Triandis [1971] and Cialdini [1985]). We will summarize a few that may be of special interest to managers.

Sources. What kinds of sources are most effective? Those sources that the audience trusts most and finds most attractive and most similar to itself. It is good to use a source that has high credibility with the particular audience.

Message. What kind of a message is best? Here we need to consider whether we are going to have the greatest difficulties changing the audience at the level of attention, comprehension, yielding, remembering, or action. If the level of attention is the weak spot, we need a message that is dramatic and includes a simple slogan. If the level of comprehension is the weak spot, we need a message that is clear and draws a definite conclusion. If the message needs mostly to influence the yielding stage, it must link the goals of the audience with acceptance of central ideas of the message. If remembering is likely to be the weak spot, then repetition is useful (as in advertising). If action is the weak spot, the message needs to address issues that may derail the intention to act such as incompatible norms, roles, or perceived consequences.

A good message starts with the good news, since people open up when they hear good news. The amount of change advocated must not be too great or too small. That is a very tricky point. If one asks for too much change, one loses credibility; if one asks for too little, one will not get as much as is feasible. Thus, it is important to know something about the audience's "range of noncommitment."

An example about buying a car may be helpful in explaining this idea. When the average manager goes to a car dealer, the chances are that he has a range of prices in mind—for example, $15,000 to $20,000. If the dealer were to show him a $14,000 car that the buyer likes, the salesman would not make as much profit. If he were to show him a $30,000 car, he would not make a sale. So the challenge is to know the range and to present a car that is just a little higher than the range, say $22,000. There is a phenomenon in perception known as "assimilation and contrast." Stimuli that are similar to a given category are assimilated into that category, because they appear to be more similar than they really are, while stimuli that are different from the category are contrasted—that is, they are seen as more different than they really are. Thus, the $22,000 car will appear not too different from the $20,000 car, while the $25,000 car may well appear as different as the $30,000 one. The $22,000 car is thus assimilated into the range of noncommitment, while the $25,000 car is contrasted from it.

There is another way to change the range of noncommitment, and that is to change the audience's "level of adaptation." The level of adaptation is a kind of neutral point. It separates "expensive cars" from "inexpensive cars." Obviously what is expensive for one buyer might well be inexpensive for another. This neutral point turns out to be the geometric mean of all the stimuli that are salient at the time of the particular judgment.* Some of these stimuli will influence the judgment more because they are most recent, and some will influence it more because they have occurred very frequently. If we take all these stimuli into account and take the geometric mean, we get the level of adaptation for a particular judgment. Therefore, if 10 stimuli are salient in somebody's head, we multiply their values and take the 10th root, and the result of that calculation is the level of adaptation.

Now, obviously, one can change the level of adaptation by exposing a person to stimuli. For example, a salesman may say, "I know you do not want to spend that much money, but let me show you this beauty of a car here. Is it not a work of art?" The car turns out to be worth $25,000. The customer does not buy it, but it goes into his calculation of the level of adaptation. If the customer, at the start, has cars in mind worth $15,000 and $20,000, his level of adaptation would be $17,320. Now if in addition, he has been exposed to a $23,000 and a $35,000 car, the level would be $22,168. So, now a $22,000 car would appear "slightly inexpensive"!

A good message links the proposal to the rewards that will be received by the audience if the proposal is accepted. For example, if the manager asks for a new building, a good message might mention that the new building eventually will be named the (Mrs.) Top Manager Research Center!

An effective message anticipates objections and shows that they can be overcome. As an exercise, the reader might think of objections based on various components of the model discussed in Chapter 6 on motivation and think how they might be dealt with. Perhaps the best way to do this is to read the case that follows.

5.3 BEHAVIORAL SCIENCE DIVISION CASE†

The Laboratory for Behavioral Sciences at Government R&D Lab (GRDL) has been under attack by the Scientific Director, Dr. Brown, because it has experienced substantial reductions in grants and contracts. Much of this reduction is attributable

*The geometric mean is the *n*th root of the product of the values of the *N* salient stimuli, e.g. $(S_1 S_2 \cdots S_N)^{1/N}$.

†From the files of H. Triandis and David Day. The events are true, but all the names of people and laboratories are fictional.

to an unfavorable climate for support of such research by the federal government. However, a team of managers from other divisions of GRDL who inspected the laboratory identified several internal problems as well.

The laboratory used to have a sizable research program consisting of $2–3 million a year, but in recent years grants and contracts amounted to only about $750,000. The director, Dr. Park, formed the laboratory 15 years ago and has been highly instrumental in obtaining outside grants and contracts. Four researchers from various behavioral sciences are also connected with the Laboratory. Dr. Link, a psychologist, deals with artificial intelligence. Dr. Henson, a sociologist, deals with public opinion sampling. Dr. Duff, a geographer, employs aerial photography to develop information about economic geography. Dr. Barron, an economist, conducts market surveys.

In addition, three other researchers also work at GRDL: Dr. Clay, a political scientist; Dr. Goa, a sociologist; and Dr. Harden, a geographer. The laboratory provides substantial space for the offices of these associates and their research assistants, for computer outlets that link the laboratory with a main computer of GRDL, and for a library that has a full-time librarian. The central administration of GRDL feels that the current rate of activity does not justify the space that it occupies, and it has announced that the laboratory should confine itself to about one-half of its present space so that the other half might be used by another laboratory that is expanding.

The visiting inspection team, which consists of division managers, has identified numerous problems. For example, there is conflict between Dr. Park, the director of the laboratory, and Dr. Link, who is working on a problem that has occupied him for 7 years but has not yet produced any publications. However, Dr. Link is very optimistic that the problem will eventually lead to a major publication of high theoretical value. The inspection committee estimates that the probability is around 0.3 that such a publication might be forthcoming. Dr. Park has identified an interest by the Office of Naval Research (ONR) in work related to the research that Dr. Link has been doing, and he is pressuring Link to drop his current activity and try to obtain a contract from ONR. The work that the ONR contract would pay for is estimated to have a high probability (around 0.8) of resulting in several publications. However, because this work is going to be applied, these publications may not make an impact on the writer's whole scientific field and may, instead, only solve a specific problem. Current income in the form of overhead received by the laboratory from the work that Link does is very small, whereas the work that would be done under the ONR contract is expected to bring in considerable amounts. Although Link's current traveling, which causes him some concern, could be reduced significantly if he were to work on the ONR project (the new project would employ a research associate to do most of the traveling), he is resisting pressure from Park.

The visiting committee identified many other cases of conflict between Park and his co-workers. Park was critical of Harden for spending too much of his time repairing antique cars instead of doing research. Harden, however, says that his publication record is at least as good as the upper third of the members of the laboratory, and therefore it is none of Park's business how he spends his time. The laboratory has no clear standards concerning either the number or the quality of the publications of its associates. For example, Park criticized Henson's publication record for 1985. Henson, however, feels that his record is satisfactory and improving and that in his field it is pretty good. In any case, 1986 was better than 1985. Park was also critical of Henson and Duff because they failed to collaborate in the bidding for a contract with the National Institute of Education (NIE), which required the skills of these two researchers. Because Henson and Duff do not get along and prefer to work independently rather than collaboratively, the opportunity of an NIE contract was lost.

In talking with Goa, the inspection team learned that he felt isolated and found his colleagues uncooperative. He tended to attribute this to the fact that he was Portuguese and that, in his view at least, there was some discrimination toward foreigners in the laboratory.

The morale of many of the researchers was low because their pay was generally low. This was particularly the case with Clay, who received his doctorate from a prestigious Ivy League university and consequently felt that he should be receiving a higher salary than the other researchers.

Many of the researchers at the laboratory were critical of Park because they felt that he was not sufficiently involved in the activities of their own research groups, was unaware of specific difficulties experienced by these research groups, and did not seem to appreciate that many of these difficulties had been overcome and that a number of projects had come to a successful completion. Park's view, however, was that he was much too busy to keep detailed track of such matters because he often had to be in Washington to gauge the interest of several federal agencies in various kinds of research in order to submit appropriate research proposals and to increase the funds from grants and contracts currently administered by the laboratory.

According to the inspection team, one of the problems of the laboratory was that Park made most of the decisions independently. Others had very little opportunity to register their views, and, in fact, Henson, Duff, and Barron almost never argued or raised issues with him. The majority of those associated with the laboratory felt that Park provided a setting within which they could carry on their work without having to go to Washington to find money to support their research. Thus, they were willing to put up with Park's pressures, but at the same time they resented them.

People did not seem to identify with GRDL. The majority of the researchers felt that it was a convenient setting for their research, but if they could work elsewhere they would just as soon do so.

The major current problem facing the laboratory is that the central administration of GRDL wants to take away half of its space. Park has called everyone to his office to discuss what might be done to prevent this from happening. What recommendations would you have concerning attempts by the laboratory to change the attitude of the administration? What changes in working procedures may be instituted within the laboratory to improve its internal functioning and internal relationships?

5.4 CASE ANALYSIS

In thinking through this case try to consider who the best source of attitude change might be (e.g., Dr. Park, a friend of the top administrator, a committee?) What would be the best message (exactly what would it say so as to shift the level of adaptation, present the good news first, etc., as discussed earlier)? How should it be presented (face-to-face, in writing, etc.)?

There are a few additional points about attitude change that can be made. One technique that can work well is to have several sources. One of these might take an extreme position and thus change the audience's level of adaptation. Then the other source could make a "modest proposal" that would be immediately accepted.

Another approach that sometimes works is the "foot in the door" procedure. With this tactic one may be able to get the audience to accept a *very* modest proposal and then, in time, ask for more. There is also research on the way fear and threats of punishment can be mixed into the message. The general finding is that only modest levels of fear or threat can be effective. If one puts too much fear into the message, the audience rejects the message.

This "foot in the door" analogy can be particularly relevant to research activities. For example, many research project outcomes are uncertain, while the resources required for completion of the entire research effort may be substantial. Consequently, asking for all the resources necessary for the research project at the outset may not be a prudent course of action. One approach that works quite well is to ask for sufficient research funding to do initial, or pilot, programs to ascertain the feasibility of the research approach. This way one has a "foot in the door" and the sponsor is more likely to continue with the funding than if this initial step had never been taken.

Sources that are extremely credible and knowledgeable can have more of an effect on attitude change because they are likely to be listened to very carefully and because they can get away with advocating more change. For example, to go back to our car buying case, if a credible source such as the editor of a national magazine on cars were to argue with our buyer that a $25,000 car was a better investment than any of the other cars he had in mind, there is a good chance the message would be effective, even though the gap between the buyer's upper limit and the advocated position is now very large.

Face-to-face communication is generally more effective than communication in writing or through other media. However, in certain situations, written communication can be more effective—for example, if the argument is very complex and requires the audience to think about it step by step.

When an audience is highly involved with an issue it is difficult to change its attitude. In general, audiences that are intelligent, complex in their thinking, and self-confident are also very difficult to change. Such audiences usually attend and comprehend, but they do not yield. On the other hand, audiences that are less intelligent and neither confident nor involved in the issue may not even pay attention to the message. In other words, the communicator has different problems with different audiences. As a result, the relationship between variables such as involvement, intelligence, and self-confidence and attitude change is an inverted U. For very low or very high levels of these variables there is little attitude change; for moderate levels there is a fair amount of change.

5.5 COMMUNICATION ALTERNATIVES AND OUTCOMES

There are three kinds of communication that take place in organizations: interpersonal communication (e.g., among colleagues), group communication (e.g., a principal investigator talking to his team), and organizational communication (e.g., a top manager attempting to change attitudes of the rank and file). In interpersonal communication the major weak point is yielding; in organizational communication the major weak point is attention. So, for different kinds of communication the communicator has to develop different strategies in order to overcome different problems.

In reading the discussions of various communication alternatives presented here, one may feel that there is an attempt to manipulate others unfairly. This is not the purpose of this chapter. We are dealing with very intelligent people; consequently, any attempt to manipulate them is likely to backfire. In addition, one's ability to influence others regarding a project or an issue is fundamentally limited by the merits of the project or the issue. At times, managers and even researchers are frozen in a position or a mindset that does not allow them to review the project or an issue without preconceived ideas. In such cases, it is helpful to unfreeze a person so that he or she is able to see your viewpoint. Some of these communication alternatives might provide you with a winning edge.

Take Advantage of the Other's Cognitive Habits. It is possible to influence others by taking advantage of the cognitive habits they already have. For example, most people think that "expensive" is equivalent to "high quality" when it comes to purchasing goods. This can easily generalize to research, so that they may think that elaborate and expensive research is better than research done on a shoestring.

Another habit deeply ingrained in human thinking is reciprocity. We believe that we must return a favor; or if someone makes a concession, we feel obligated to make one in return. One can take advantage of this tendency. Conversely, someone who is aware of the tendency can learn to resist manipulation. For example, suppose your boss wants you to do a job you really do not want to do. He or she might be able to influence you by asking you to do something that is even worse, thus shifting your level of adaptation in the negative direction, and then provide a concession by asking you to do the first job. Since the boss has made a concession, you have the tendency to make a concession also. This leads to the boss manipulating you and to your agreeing to do the job. If you can analyze the situation and identify the manipulation, you can resist it better.

Use Information Optimally. When making judgments we use the information that is most readily available. Information that is used frequently is more available than information that is used rarely. It is the same principle as putting the things you use often in the front of your refrigerator near the door, and the things you do not use so frequently in the back. You can use this principle to influence others. For example, suppose you want to do research on X and your colleagues or supervisor are not interested in this kind of research. You mention the names of people doing research on X, you describe studies that have utilized that kind of research, and so on, thus shifting the cognitive availability of the information. When the time comes to decide on a new direction for research, the fact that information on this kind of research is more available in the cognitions of your colleagues can make a difference.

Use Analogies. We think by analogy. One way to convince your colleagues is to find good analogies when a particular kind of research you want to do has been successful. The analogy does not even have to be in the same area of research, as long as there is some congruence. The analogous research may be in another field but has similar attributes, such as when and where it was done and the configuration of people who participated. Incidentally, the difficulty most people have in accepting a truly original idea can be traced to the same phenomenon. In that case they lack a ready-made analogy. As we discussed earlier, laboratories that use analogies and metaphors are more effective than laboratories in which the norms do not call for the use of such devices.

Use Repetition. Repetition can be helpful in influencing people. It has the tendency to make some ideas more available, and it also shifts the level of adaptation. For example, if your level of adaptation is at 10 units and you get exposed to three events at 11 units, the new level of adaptation would be 10.74. So, while 11 units was "high" before the repetition, now it is almost "neutral."

Use Prior Imagination. Another well-known approach is to use prior imagination. "Imagine what would happen if we do this research." You describe exciting possible outcomes. That kind of line can prove quite convincing. Of course, there has to be some substance to the argument. When someone is using this approach, be on your toes!

Use Positive Experience. Finally, one tactic is to resurrect a positive prior experience. For example, you might describe to your colleagues a past event that was very positive to them, and link that event to your proposal.

Get People to Commit Themselves. Once your colleagues are committed to a course of action, they are more likely to change their subjective probabilities of success. For example, until a proposal has been committed to paper and sent to a sponsor, there may be a 0.5 subjective probability that it will succeed. Once work has started on it, the subjective probability is likely to rise to 0.7 or 0.8. The general tendency is to make our feelings line up with our current behavior. Thus, this is something that one can take advantage of in influencing others.

Choose the Right Source of Influence. The choice of who is the source of influence is critical. A source who is physically attractive and who is similar in many dimensions (particularly values and general goals) to the audience, and with whom the audience is quite familiar, is a good one. Thus, if you are a biologist and you want to convince your supervisor, who is a physicist, arranging for a very attractive physicist to talk to your supervisor may be a better approach than talking to him directly. If such a person is also one of your supervisor's old cronies, that is even better. One might think that this approach sounds a bit naive. How can an intelligent manager be influenced by this? Since many judgments related to research funding are subjective and many research outcomes are unpredictable, some of the approaches presented here may make the crucial difference in getting a positive response. Try it and you may be pleasantly surprised.

Get Help from Others. Many judgments we make in life do not have objective bases. When we deal with ambiguous situations and we do not have a good way to test them objectively by using data, we rely on the opinions of others to validate our judgments. These others are usually chosen because they are similar to us and because they have comparable levels of knowledge, prior experience, and the like. Thus, in many cases one might be able to convince another by first convincing that person's peers. The central person is usually more ego-involved with the issue than are his friends. So, while he may pay attention and understand the issue, he may not yield as readily as his friends who, as third parties, are less ego-involved and thus more likely to yield when presented with a reasonable argument. Once the peer group has yielded and is convinced, it is easier to persuade the central person to yield and become convinced. Again, you can learn to resist such manipulation by becoming aware of how such a system works; and if you see your friends shifting in a given direction, identify the influences on them.

Stress Rare Events. Finally, events appear more desirable than they really are if they seem rare. For example, a scientific breakthrough can be seen as even more desirable if you can convince others that such breakthroughs seldom occur. Naturally, you may also be able to convince them about other issues if you convince them that the event is of great value.

Up and Down Communication. In most U.S. organizations there is a lot of downward communication and very little upward communication. There are many channels of upward communication, however, that are not immediately apparent. For example, suggestion systems, quality circles, and management by objectives are different ways to increase upward communication. It is important for managers to support such channels in order to correct the imbalance in the communication flow that is typical in most organizations.

Another channel of upward communication that is particularly effective is briefings given to visitors where management is present, or briefings given to management in order to seek its input. During such briefings it is not prudent to ask for more funds or more staff. These briefings should be designed to provide information succinctly and to seek comments from management. Critical comments should be encouraged, and management concerns should be brought out. Management would, of course, understand that, to the degree possible, its concerns will be dealt with. Our experience indicates that this strategy for upward communication, though requiring some preparation and planning, works quite well. This also gives management a genuine feeling of participation in the project. A caution, however, is in order. In cases where management is likely to be autocratic and dictate a new direction for the project or place other poorly thought-out restrictions or demands, such an approach will naturally not work.

Sideways Communication. There are studies of communication channels that show that people who are in central positions where communication flows through them are more satisfied than people who are on the periphery of the organization. Another problem is that the accuracy of the communication is not always high. Communications are often distorted as they move from person to person. "Telephone," a well-known party game, is unfortunately repeated in everyday life in some organizations. The distortions that usually occur are of three kinds: (1) some detail is dropped out, (2) some aspects become more salient than other aspects, and (3) the values of the communicators distort the message in the direction of wishful thinking. Of course, the more relay points there are between the originator and the target of a message, the greater the distortion.

There are some differences in the kinds of communications that are most helpful to scientists, particularly those doing basic research, and to engineers, particularly those doing development work. The former get more of their information from journals; the latter get it mostly from face-to-face communications. An effective lab needs both kinds of communication, and management will do well to ensure that both types are widely available. Highly effective labs encourage publication, since that is the ticket that allows the scientist access to the prepublication work of other scientists. Participation at professional meetings is expensive, but managers would be foolish to save on travel to such meetings since their people are usually stimulated by such meetings.

Resistance to Change. Resistance to change is very common in most organizations and is one of the problems a communicator must overcome. It is often helpful to analyze a problem of resistance to change in terms of "forces opposed" and "forces supporting" the change. Such an analysis can often suggest where communications should be sent, and what the message should be in order to overcome the resistance. Research has shown that major change requires some people who can become "opinion leaders" and influence others. Such people are often secure in their positions, influential, and of high status. They have extra resources that allow them to take some risks. Once they adopt the innovation, it is likely that some others will follow, and the trickle becomes a flood.

Ingratiation and Impression Management. There is also quite a bit of research on the way people can influence others by using ingratiation, or *impression management,* by being vigilant about what aspects of themselves they present to others. Ingratiation, for instance, can be effective in getting a boss to like a subordinate more.

There are several tactics that work in ingratiation. One is "other enhancement," which involves the subtle use of flattery. To be most effective, this approach requires that you flatter the boss behind his/her back, with comments about qualities the boss would like to have but about which he/she feels uncertain. "Opinion conformity" is agreeing with the boss, particularly about pet projects that others are not too enthusiastic about. "Rendering favor" is most effective if the donor is seen as really intending to do it, and at a cost. It is least effective if the donor is seen as enjoying rendering the favor.

There are problems with "other enhancement" with which one must be careful. For example, if the boss has a poor opinion of himself, praising him can backfire. Conversely, if the boss has too good an opinion of himself, praise can also backfire: "What? You did not expect me to do this well?" You can also make the boss anxious by praising him after a good performance that he is not sure he can pull off again.

"Opinion conformity" also has its tricky side. While agreeing with a rare and not commonly held belief can be effective, agreeing with a belief that most people hold can appear unoriginal and might seem to be an attempt to manipulate the boss. One approach is to disagree on trivial topics, thus lowering the boss's level of adaptation for your agreement and then suddenly agreeing on something that matters to the boss. This appears to the boss as a "real coup," since you have created a reputation for being "difficult to sell."

Impression management involves making claims (e.g., I can do this difficult job) that the other might be able to accept. The boss will usually challenge the claim. A successful claim, therefore, is one that is not challenged or that can stand up to the challenge. If you can make the claim stick, you have won. But if the claim proves illegitimate, you lose, and not only are you likely to feel shame, guilt, or embarrassment, but you may even be fired. Thus, in making a claim, you have to analyze in advance the possible consequences and their value. Basically what you have to do is to figure out the risks and the benefits from making a claim. If you can assess the probability and the value of each consequence and can come up with a positive outcome, it may be worth making the claim.

If you make a claim that is challenged and invalidated, you must do something to account for your behavior. You can, for instance, provide a *justification* (e.g., I wanted to test my limits, to see if I could develop this new system), you can claim to be *innocent* of the claim (e.g., I did not really mean the complete system but just wanted to develop a concept for the system), or you can make an *excuse* (e.g., my colleagues thought I could do this). The more severe the predicament that results from a successfully challenged claim, the more you will have to give an account of why you made that claim in the first place. All of this can be computed ahead of time, taking into account probable consequences and the value of these consequences and trying to maximize outcomes.

If you really goof, you need to develop a good apology. There are a number of possibilities to choose from. For example, you might indicate that your previous action was an aberration that was not at all typical of you. Ideally an apology has the following components: (1) admission of guilt ("the boss was right, I was wrong"), (2) a description of what should have been the correct behavior ("so the boss knows you know"), (3) a disparagement of self for misbehaving (e.g., "it was stupid of me"), (4) a promise of appropriate behavior in the future, and (5) an offer of compensation (if possible).

On the other hand, if a desirable event occurs, impressing management often involves making sure that others know about it and requires an effort to increase the perceived desirability of the event (e.g., it resulted in a new contract). Boasting can be counterproductive, but there are tactics that can get the same effect (e.g., "I know you expected 100 units and I produced 150; if I had been on the ball and done such and such, I would have produced 200"). Such an approach can be particularly effective in an R&D organization where researchers are expected to be high performers and are naturally optimistic about what they can accomplish. Some people are more skilled than others in presenting those aspects of themselves that will impress management. A scale exists to measure "self-monitoring" [Snyder, 1979]. Those high in self-monitoring present themselves better, because they stress those aspects of themselves that are likely to make a good impression. Those low in self-monitoring present themselves the way they are, and do not change their presentation tactics from audience to audience. High self-monitors are good actors, they remember the traits of other people better, they are guided by the situation, they define who they are according to the situation, and they know more about their audience than low self-monitors. The low self-monitors know more about themselves, and they do not try to impress others.

Discussion

Which of these various tactics is most likely to work under what conditions? The greater the difference in the status between two people, the more indirect the ingratiation must be in order to be effective. If a first-level manager meets the lab director, it would look foolish to compliment her about how good a job she is doing. However, an indirect compliment is another matter (e.g., "I heard about your new research project at the professional meeting I attended last week"). Similarly, the presentation of positive information about the self must be more indirect if there is a large status gap (e.g., I had such difficulties publishing three refereed papers this year), and doing a favor is best avoided unless the other explicitly asks for it. Opinion conformity is the most effective tactic for low-status persons, while compliments and doing favors are the best tactics for high-status persons. High-status persons can improve their self-presentation by linking successful events with trivial mistakes (e.g., spilling coffee).

5.6 SUMMARY

In summary, there are a number of ways to influence others, and consideration of the various tactics outlined in this chapter can improve one's strategies and chances of successful influence.

In reading this chapter some may feel that the information may be used to manipulate the behavior of others. This is not the purpose of the material presented here. Opinion and views vary. At times, people get attached to hopeless causes or to views based on erroneous information. The methods of influence presented here might provide tools to overcome some of these difficulties. Manipulative or unethical behavior is neither proposed nor condoned. Furthermore, we chose to include this material here because some people do use these methods to exert influence, and their effectiveness can be reduced when these tactics are known to the public.

5.7 QUESTIONS FOR CLASS DISCUSSION

1 Which of the techniques of interpersonal influence is likely to backfire? Which of the techniques of interpersonal influence is likely to prove especially effective in R&D labs?

2 Now that you have had a chance to read more about influence, how would you change your plans about the Behavioral Science Division case? Go back, reread it, and think of Dr. Park's predicament. How can he approach the problem, taking advantage of some of the ideas you have just read?

5.8 FURTHER READINGS

Cialdini, R. B. (1985). *Influence*. Glenview, IL: Scott, Foresman.

Ffalbe, C. M. (1992). Consequences for managers of using single influence tactics and combinations of tactics. *Academy of Management Journal,* **35**(3), 638–652 (August).

Gilbey, J. (1995). I promise not to call you a moron again: how scientists and administrators can win friends and influence people. *New Scientist,* **146,** 53–54 (April).

Triandis, H. C. (1977). *Interpersonal Behavior.* Monterey, CA: Brooks/Cole.

6

MOTIVATION IN R&D ORGANIZATIONS

Goals determine a substantial amount of human behavior [Locke et al., 1981]. Motivation to achieve these goals is a major factor in researcher performance and in organizational effectiveness. For these reasons we devote a full chapter to this topic. Individuals have goals and organizations have goals. For maximal organizational effectiveness it is important to make these two sets of goals compatible. In fact, that is the major role of management. The R&D manager must have a clear understanding of both sets of goals and find ways to make them similar, overlapping, and at least noncontradictory.

Organizational effectiveness depends on (1) individual motivation for organizational effectiveness (i.e., individual goals that are compatible with the goals of the organization), (2) individual performance (just because one has the right goals does not automatically result in effective performance), and (3) adequate coordination of individual performances.

Performance depends on more than motivation. One must have adequate skills and abilities and proper training, and there must be a good match between the individual and the organization goals. Coordination depends on adequate communication, and it can be improved when there is participation by employees in decisions that affect them and when organizational goals overlap with personal ones.

In order to understand performance better, it is useful to focus on a model that links the probability of an act to particular determinants.

6.1 A MODEL OF HUMAN BEHAVIOR

For our purposes here an *act* is a short sequence of behaviors that eventually results in some outcome, such as the publication of a paper or the development of a good research design. In other words, we are using the word "act" in a very specific way. Hundreds of these acts are necessary to produce a publication or to develop a product. What we are trying to understand is what makes these small acts more or less probable.

There are two variables that are important in this case: previous habits and self-instruction. For example, when a person says, "I should look up these references," that is a self-instruction or behavioral intention. Research has shown that behavioral intentions predict behaviors quite well [Triandis, 1977, 1980].

The model thus states that the probability of an act is dependent on two kinds of variables: habits and behavioral intentions. However, even when people have the proper habits and intentions to carry out a particular act, they may fail to do so because external conditions may not be favorable. We utilize the concept of *facilitating conditions* in order to explain the phenomenon that even though the individual may have all that is required, the act may not occur. Reasons beyond the intentions of the individual may not allow it. For example, there may be a lack of proper equipment or there may be distractions in the environment. Facilitating conditions can be measured both with data obtained "outside the individual" (e.g., by asking objective observers, who know the conditions of work well, to judge if the act can occur) and with data obtained from "inside the individual," by measuring the individual's sense of "self-efficacy." This can be measured by asking the individual, "Can you do that?" A scale can be constructed that measures the individual's beliefs that the behavior can take place under different kinds of circumstances. The circumstances described in the scale can be more and more difficult. Those who think that they can do the behavior under the most difficult circumstances are highest in self-efficacy. Thus, a high sense of self-efficacy is an especially important facilitating condition. For instance, we can ask, "Can you solve this equation?" A person who says "No" is very low in self-efficacy. Those who answer "Yes" are higher in self-efficacy. A person who says "Yes" when the question is "Can you solve this equation when you are waiting to board a plane in a noisy airport?" is very high in self-efficacy.

Consider a more specific example. If a person said, "I will look up this reference," but the book that contains the particular reference is not around, the probability that the act will occur decreases. Facilitating conditions modify the probability that habit and intention in themselves will result in the act. They reflect the situation within which behavior may occur.

For those who enjoy the precision that mathematical statements provide, the first equation of the model is

$$P_a = (W_H \cdot H + W_I \cdot I)F \tag{1}$$

where P_a is the probability of an act, W_H and W_I are weights that are positive numbers between 0 and 1.00 and sum to 1.00, H is a measure of habit, I is a measure of intention, and F is a measure of the facilitating conditions.

The weights depend on the novelty of the act for the individual. When the individual is faced with a new situation, the weight for intention is 1.00 and the weight for habit is zero. However, as the person performs the act over and over again the weight for habit keeps increasing until it becomes 1.00 and then the weight for intention is zero. For instance, when one learns a new skill (e.g., riding a bicycle) in the early phases, one's behavior is under the control of intentions, but at the end it is entirely under the control of habits. Once behavior is under the control of habits, it is difficult to "explain" it to others without actually carrying out the act and observing one's own behavior.

Another variable that shifts behavior to habit control instead of intention control is stress. When people are under stress, as in an emergency or under time pressure, their behavior is under habit control. That is why there is so much drilling of emergency procedures in the military or on ships. In an emergency one cannot depend on an intellectual analysis of the situation. One must have the right habits.

To make a behavior automatic under habit control, one needs several hundred trials (Schneider, 1993). But for some jobs the advantages of having a behavior under habit control are immense. Consider these examples. Suppose the job requires a two-category judgment or a four-category judgment (e.g., is this an airplane, a missile, a bird, a shadow?) To match a new stimulus with the correct response on a two-category task takes 0.7 seconds; on a four-category task it takes about 2 seconds if the response has not yet become automatic. If it has become automatic (after several hundred trials) the two-category task or the four-category task take the same time: 0.002 seconds! If the workload changes, the responses that are not automatic require more time, but the responses that are automatic do not require more time. Of course, to get the response to be automatic requires about 10 hours of practice. Thus, it is only in some critical jobs, such as air controller, that this type of training is economically justified.

Another advantage of responses under habit control is that they are not forgotten. Once you learn to ride a bicycle, you can ride one 10 years later. In short, automatic responses are faster, do not degrade with time, and stay in long-term memory. But only for some jobs it is justified to spend the time to make responses automatic. While in most research work the development of automatic responses is unnecessary, there are components of jobs, such as learning the value of the integral of a particular equation, that may speed up the work of scientists.

Determinants of Habits

What are some of the variables that determine the habit? Habits build up as a result of previous rewards. We call such rewards "reinforcements" because they reinforce the link between stimulus conditions and behavior. Behavior is a function of its consequences. As people engage in a particular behavior in the presence of a certain configuration of stimuli, and when desirable events follow the behavior, the probability increases that the configuration of stimuli will in the future produce the same behavior. The behavior eventually becomes automatic, without thinking. When this happens, we say that the act has become "overlearned" and occurs under the control of habits. In that case, behavioral intentions are not relevant as explanations of the behavior.

Determinants of Intentions

Let us now examine what determines behavioral intentions. There are three classes of variables that are relevant for the determination of behavioral intentions. They are social factors, act satisfaction, and perceived consequences.

Social Factors. Social factors include roles, norms, self-concept of the person, and interpersonal agreements.

1. *Norms.* Ideas about correct behavior for all members of the organization. They emerge in discussions among members of the organization. For instance, arriving at 8 a.m. would be a norm since it applies to all members of the group.
2. *Roles.* Ideas about correct behavior for the specific position that a member of the organization holds. These are evident when a person says to himself or herself, "I am supposed to be doing this because it is my job." In short, the role has become embedded in the person's thinking and has certain activities associated with it. The probability of these activities (acts) increases when the person thinks that he is doing the job. If the researcher feels it is his job to keep the supervisor informed, he is more likely to do it. For instance, what behaviors are expected of a "principal investigator?" In some cases, these expectations are quantitative, such as "producing three papers a year." In other cases, they are qualitative—for example, the expectation of an important scientific contribution, or the development of a new product that will benefit the company.
3. *Self-Concept of the Person.* This includes the ideas a person has about the types of activities that are appropriate for him or her; for example, if a researcher feels it is appropriate for him to present his views, even though they differ from others, he is likely to participate actively in discussions and meetings.
4. *Interpersonal Agreements.* These are similar to management by objectives. The supervisor and subordinate agree that the subordinate will try to reach a particular goal. Interpersonal agreements increase the probability that the goal will be reached through behavioral intention (self-instruction). Some research projects use milestones that are really interpersonal agreements as conceptualized here.

Act Satisfaction. The second class of variables that determines behavioral intentions is satisfaction associated with the act itself. Many acts are enjoyable in themselves, such as eating certain types of foods, playing the piano, or working on computer problems. Often such acts associated with pleasure have been formed through classical conditioning. In other words, the activity itself was associated with pleasant events in the past and is pleasant to think about, so this factor involves affect (emotion) toward the behavior itself. This affect motivates the person to self-instruct to do the act, and this in turn becomes the behavioral intention that causes the behavior. Working on a challenging research project or working with a noted scientist could fall in this category.

Perceived Consequences. Finally, the *perceived consequences* of the act are also important. When we do something, such as publish a paper, we perceive certain consequences. For example, when we publish a paper, we might have the perception that this could lead to a promotion, to recognition, or to a particular reward. It is obvious that each of these consequences is probabilistically associated with the act since there is no certainty that the behavior will have the particular consequence. For example, if the scientist publishes a paper, the probability of promotion may be 0.60; the probability that there will be some recognition associated with the paper may be 0.90. Thus, each act has associated with it a probability between zero and one. So the person says to himself, "If I do such and such, then there is a high probability (or low probability) that x will happen." In this case, x is a consequence. Each consequence also has some value to the person. For example, some people would see a promotion as very desirable, but others might not. Obviously, if the consequence has a positive value attached to it, it will increase the probability of the behavioral intention. If the consequence is perceived as negative, it will decrease the probability that the corresponding behavioral intention will be activated. For each of the acts the person may consider, there is a whole string of consequences, each of which has some probability and some value attached to it. To obtain the total effect of these perceived consequences, each person must multiply the probability and the values for every consequence and then sum these products. Intelligent people will make better estimates of these probabilities and values than others.

Thus, we can say that behavioral intentions are a function of (1) social factors such as roles, norms, the self-concept, and interpersonal agreements, (2) the affect toward the act itself, and (3) the total value of the perceived consequences. Because there are some people who are susceptible to social factors, and others who are susceptible to perceived consequences of the act, each of the three factors can now be given a weight. For example, people who have been socialized to be very sensitive to the views of others, and who have received lots of rewards and punishments in their interactions with others, develop great sensitivity to social norms. Their behavioral intentions are much more influenced by the social factors than by the other two sets of factors. On the other hand, people who have been socialized to be quite independent often give attention to how much pleasure they can get out of a particular situation. Thus, they are likely to pay a great deal of attention to the affect that is attached to the act. Still others are quite interested in the future and in the way the act is going to bring "good outcomes." Such people look at the consequences of the act and are likely to give weight to those consequences.

The consequences of the act can include job autonomy, vacations, fringe benefits, and the opportunity to use time flexibly. For example, to work at home when you otherwise are expected to be at the office can be highly rewarding. Setting difficult but reachable goals with feedback is one of the ways in which a supervisor can motivate a subordinate. In addition, it has been found that interesting work that provides both challenge and variety can be rewarding. Deadlines are like an interpersonal agreement that can be rewarding and that can function as a goal. Recognition, promotions, and the opportunity to grow, to receive more pay, or to have a more secure job can all be motivators.

It is also useful to consider situations that are demotivating. One such situation occurs when the employee feels that the organization discriminates against him or her. Other causes for demotivation are poor interpersonal relationships with a supervisor or with peers, low pay, indifference by the organization, lack of promotion or recognition, and having to work for an incompetent supervisor.

Again, for the sake of those who like mathematical formulations, what we have said above can be summarized by the following equations:

$$I = W_S \cdot S + W_A \cdot A + W_C \cdot C \tag{2}$$

$$S = R + N + S_C + I_A \tag{3}$$

$$C = \sum_{c=1}^{n} P_C V_C \tag{4}$$

where I is a measure in intentions
A is a measure of the affect toward the behavior itself
C is a measure of the value of the consequences
S is the social factor that reflects roles (R), norms (N), the self-concept (S_C), and interpersonal agreements (I_A)
Pc is the probability of a consequence
V_C is the value of the consequence
W_S, W_A, W_C are weights that are positive numbers between zero and one and that sum to 1.00
c stands for consequences 1 to n

If a supervisor wants to change the behavior of a subordinate, every one of these variables may be influenced, and of course combinations of these variables may be optimal. For example, the supervisor can associate pleasant events with the desired behavior, so that even a minimal quantity of the desired behavior may elicit the pleasant event (a nod, a smile, a pat on the back, etc.). A discussion of roles, norms, and the resultant interpersonal agreements can influence the S-factor. A discussion of the probable consequences of the particular behavior can influence the P_c. The association of important values of the subordinate with the desired behaviors can lead to higher C. Goals are most effective if they are specific, difficult, and attainable. Such goals can become interpersonal agreements.

Facilitating Conditions

There are a number of factors that facilitate the performance of a behavior. Most of them are situational, such as helpful conditions, the right setting, or access to the resources needed to carry out the behavior. However, there are also internal conditions over which the individual does not have much control, such as the person's physiological state (e.g., hormonal balance), beliefs that the behavior is possible and likely to lead to the successful reaching of goals (sense of self-efficacy), and the level of difficulty of the task relative to the person's ability. For instance, no matter how intensive a researcher's intention to invent a new product, and how brilliant the past record of inventions (habits), there are situations in which no invention will be possible because the person is feeling depressed, or he believes that he is not able to have a new idea, or the task is much too difficult relative to the available talent. Some of these conditions can be measured objectively, and others may be estimated by objective observers of the total situation. The point about the F component of Equation 1 is that when it is zero, it can bring the probability of the act to zero, no matter how high the levels of habits or intentions.

Links between desirable behavior and challenges, variety on the job, recognition, promotions, growth, extra pay, extra security, and so on, are too obvious to mention in detail. One can also motivate people by providing deadlines.

Of special importance in R&D labs is whether the organization rewards reasonable risk-taking, innovation, and creativity. Does the organization provide feedback and rewards for good work? What kind of facilitating conditions and environment is the organization providing for motivation of this unique group of talented individuals—the researchers?

Common sense suggests that job satisfaction results in high productivity. However, the empirical evidence is not supportive of this expectation. This happens in part because one can get high production without satisfaction (e.g., in coercive situations, such as among slaves and in prisons, where people may not be able to eat if they do not produce enough), and one might also be able to have high satisfaction without much production (as in situations in which the management lets workers do whatever they like). The evidence suggests that those who receive high pay and have supportive supervisors are high in "extrinsic" satisfaction, which in turn leads to high performance. On the other hand, high performance leads to intrinsic satisfaction—that is, to people enjoying the work itself. If we perceive the work situation as equitable—that is, if our effort is rewarded about as equally as that of others—we are more likely to be satisfied than if we feel that others are getting more for their effort.

Scientists crave visibility and recognition. That is certainly true in the West. However, some caution is required in East Asia. People in some East Asian cultures feel distinctly uncomfortable when they are made to "stick out" from the group, even when that means that they are being complimented. A supervisor in that part of the world should first "test the waters" by privately talking to the subordinate. "I think your accomplishments require public recognition. You did extremely well and I want everyone to know that. I am planning . . . ." See how the subordinate reacts. If the subordinate is clearly embarrassed and does not want public recognition, do an about face: "I respect your wishes and will not do what I had planned." But this last step is complicated. It could be compared to such East Asian occasions as Chinese banquets or the Japanese tea ceremony, where one is frequently offered a special dish or tea. The correct response in such situations is to refuse it politely. The host then insists, and after a few refusals the guest agrees to accept the tea or delicacy. Thus, offer the opportunity for a special recognition again, at another time, and see what happens. The reluctant East Asian scientist may change his mind.

Nevertheless, in general, managers should provide opportunities for visibility (e.g., invitations to give a lecture or to make presentations to important customers) as a reward. There are many studies of compensation, and this is a topic we will not discuss in detail, except to point out the desirability of linking the scientist's behavior to the goals of the laboratory by offering personal rewards, prizes, or recognition for actions that promote such goals.

At different stages in their careers, people need a different mixture of rewards. Young scientists and engineers need to increase their skills in order to learn more. So training, growth, and transferring to different jobs can be seen as rewards. In middle career (age 35–50), recognition, esteem, and visibility are the most important rewards. In late career (50–70 years), security, health and pension benefits, recognition, and visibility are the important rewards.

Support for the Model

There are numerous studies that support this formulation [for a review see Triandis, 1980]. We will mention only two as examples. In one study, foremen were instructed by the experimenters to behave as SOBs. In a control group, foremen were instructed to behave normally. Half the workers were doing a new job; the other half did a job that they had been doing for a long time. The instructions to the foremen influenced the productivity of the workers *only* when the workers were doing a new job. In other words, when the job was under habit control, the supervisor's behavior was irrelevant, but when it was under intention control a supervisor who treated a subordinate badly depressed the subordinate's performance. In other studies [Fiedler, 1986a] the effect of the leader's intelligence (which is relevant to the utilization of intentions, as we will see below) and years of experience (which is relevant to the extent the leader's behavior is under the control of habits) were related to the effectiveness of the team under the supervision of the particular leader. Under conditions of time pressure, stress, or emergency (when habits are likely to control the behavior), the experience of the leader correlated with group effectiveness. Under those conditions the leader's IQ was unrelated to effectiveness. However, under conditions of low stress the opposite pattern of correlations was obtained. In short, when people are under stress they use their habits more than their intentions, and so they do not utilize their intelligence as much.

The example with the foremen who were SOBs for experimental purposes makes another important point: It is possible for people to be very dissatisfied and yet be highly productive. For example, an employee who sees high productivity as a means to a promotion out of a boring job may be low in job satisfaction but extremely high in productivity.

In fact, the factors that determine productivity are *not* the same as the factors that determine job satisfaction. Productivity depends on how many high-effort, high-quality behaviors are attempted by the person. The model we just described indicates the factors that will lead to high productivity: beliefs that others expect high productivity; the person's beliefs that high productivity is appropriate and that he/she is the type of person who is highly productive; instructions from supervisors that point to high but attainable goals; specific goals; the availability of clear procedures for reaching goals; feedback from supervisors concerning goal attainment; enjoyment of high-effort, high-quality behaviors; beliefs that such behaviors will have desirable consequences (e.g., promotions); and the conversion of the intention to produce high-quality outputs into habits—that is, automatic behaviors that the person carries out without thinking.

Job satisfaction, on the other hand, depends on how much one gets (resources such as status, training, money, goods, services) relative to what one expects. If one gets slightly more than one expects, this will boost productivity, but the effect is likely to be short-lived. One soon rationalizes that the extra resources obtained are "well deserved." If one gets less than expected, one is dissatisfied. Expectations depend on our *perception* of what we bring to the job, relative to what others bring, and what we get out of the job, relative to what others get. So, a researcher who believes that her international reputation is much greater than that of her colleagues is likely to be dissatisfied with the same pay as her colleagues. Note that we are discussing *perceptions,* not reality. It is perceptions that determine expectations.

In other words, the manager of an R&D laboratory who wants subordinates to be productive must ensure that the norms of the laboratory (perception of what people are expected to do, what is "proper" behavior) call for high effort and high quality and must present difficult but attainable, specific goals to subordinates. Furthermore, the environment should be structured in such a way that there are clear procedures for reaching such goals, and feedback is provided when the goal is reached. Rewards should be given liberally, for both minor accomplishments (a nod, a pat on the back) and major accomplishments (special prizes and awards). This will link the high productivity behaviors to enjoyable situations and to beliefs that such behaviors result in benefits.

A manager who wants subordinates to be satisfied should provide as many rewards as feasible (see below for varieties of rewards) and also realistic expectations concerning such rewards. Publishing wage surveys that indicate that the laboratory pays better than average, for instance, would be helpful. The fact that a famous scientist at another laboratory is underpaid is worth mentioning to one's subordinates. Discussing how much value is placed on various factors that one brings to the laboratory—advanced degrees, years of experience, publication record, editorships, listings on the masthead of specific journals, honorary degrees, elections to high-status positions in scientific societies, and so on—can be helpful. This is true because a subordinate may think that one of these factors is worth much more than does a supervisor, thus creating discrepancy between what the subordinate expects and what she is likely to get. For example, does the laboratory *really* care if a scientist gets an honorary degree? Usually such events promote the individual but not the lab and may not improve the lab's productivity. Or is a book summarizing a program of research worth as much as *N* refereed publications? Again, a major discrepancy between subordinate and supervisor *perceptions* can occur, and clear discussion of such issues can be most beneficial.

6.2 CHANGING THE REWARD SYSTEM TO SUPPORT TECHNICAL CAREERS

Thompson and Dalton [1976] suggest that there are a number of things organizations can do to improve the motivation of technical personnel. For example, they can pay for performance and not position. In other words, people who are doing first-rate work that is important to the organization should receive the same pay, regardless of their title or position level. But there are objections to this idea. The implication of such change is that a person whose performance deteriorates gets less pay, and this is something that many employees find very difficult to accept.

Another approach is to increase the visibility of those who are extremely effective. Most organizations do this by promotion, but usually that means promotion to management and many employees want to continue their technical activities rather than become members of management. Organizations have to find ways to provide recognition that do not require promoting the employee. There are a number of ways to do this. One is to structure the organization with a triple hierarchy. Other activities that provide recognition may be as simple as having the successful performer make presentations for the top management, inviting them to meet important customers, or giving them nonmonetary honors and awards. These honors and awards, of course, have to be meaningful and reflect genuine recognition to be effective.

At one research lab a researcher is annually selected by a vigorous peer review process to be the "Researcher of the Year." The Researcher of the Year is given a reserved parking space for the year, and he or she is invited to all senior management conferences. The Researcher of the Year serves on the peer review panel for the selection of the next year's honoree.

Another way is to equalize the status symbols that are used in managerial positions for those professionals who are exceptionally effective. For example, there is no reason why the office of a researcher should be smaller than the office of a manager. Another possibility is to increase office privacy or to provide attractive furniture for the researcher.

Commenting on motivating creative employees, Cuadron [1994] suggests that scientists and inventors are not always attracted by traditional incentives like titles and promotion. They seek opportunities to create, the freedom to innovate, and recognition for their scientific breakthroughs. One of the most tangible recognitions, of course, is the opportunity to participate in the commercial success of the products of their innovations. By sharing the profits with key contributors, R&D organizations can provide incentives for innovation and successful product commercialization [Cuadron, 1994].

As mentioned earlier, different mechanisms should be used to reward individuals during the various periods of their career [see Hall and Mansfield, 1975]. For the young engineer or scientist, during the first period of their career, the most important reward is self-fulfillment and growth. Satisfaction and the sense of accomplishment that goes with it are maximized when such a person feels that the job provides such opportunities. For the person in mid-career, organizational recognition and esteem derived from the organization are the most important elements of motivation. For the person in late career, security and a good system of health and pension benefits, as well as organizational recognition, are important rewards. Such persons should have the sense that they have contributed to the organization and that the organization is grateful for the contributions they have made.

6.3 STRUCTURING THE ORGANIZATION FOR OPTIMAL COMMUNICATION

People are more motivated if they have clear goals and know how their job fits the goals of the organization than if they do not have this information. Thus, structuring the organization for optimal communication can help individual motivation.

There has been a good deal of literature on the question of how to expose members of research and development laboratories to the information they need to have to do their jobs well. One concern has been the accessibility of technical literature to the members of the laboratory [Fisher, 1980]. To ensure that people will become acquainted with other activities of the laboratory and with new technical developments, there has been a greater focus on increasing interdependence among projects within a laboratory.

It has been argued that people should become aware of activities in other parts of the laboratory because they can often pick up ideas from seeing what others are doing [Allen, 1970; Fisher, 1980]. One idea has been to increase the sharing of such facilities as coffee pots, restrooms, and computer equipment in order to increase interaction and the likelihood that people in the laboratory will get to know one another well.

It is obvious that members of the laboratory should be encouraged to participate in national meetings and professional societies, to hold offices in professional associations, and to serve on the editorial boards of journals, since all these activities increase communication and are likely to bring new information to the laboratory. Establishing contacts with academic scientists who are working in the same general area as the laboratory can also be very helpful [Fisher, 1980]. Finally, the architecture of the laboratory can have some beneficial effects for the flow of information.

Allen [1970] has described the effects of a so-called nonterritorial office that was built by the research and development section of a small chemical firm. In this case, all of the office walls were removed and an individual could choose to work anywhere that suited him in the area, depending on what was convenient. The effect of this change appeared to increase communication, both in the number of communications per person and the number of individuals with whom the average engineer communicated. This, of course, would not work well in situations where projects are of longer durations and considerable uninterrupted time is required for research activities.

6.4 TYPES OF REWARDS

A variety of factors can be used to motivate an individual. Foa and Foa [1974] have analyzed the motivation that is provided by different resources. They have identified six resources, with money, of course, as an obvious one. However, a person can be motivated by the services the organization provides, such as (a) a good benefit package and (b) assistance in finding housing or day care. There are a number of other activities (some of which may sound paternalistic) that could be included under "services."

Still another factor is status. People often make very fine distinctions about status. For example, in a study of a restaurant [Whyte, 1948], it was found that different kinds of cooks had radically different statuses! Foa and Foa [1974] also mentioned love as a motivator. An individual can be motivated by having a very good relationship with a supervisor who provides emotional support and help in solving personal problems. This kind of motivator is used much more in Japan than in the United States and is consistent with other aspects of Japanese culture. Still another reward is information. For instance, training or the opportunity to grow can be a very important reward. Goods are important motivators in some organizations. For instance, special discounts for particular products that are produced by the organization or gifts given on certain occasions can be motivating, at least for some people.

The variety of rewards is not exhausted by the ones just mentioned. For example, giving time to an employee can also be a reward. The superior can accomplish this by paying attention to an employee's problems or by granting time off when there is a family crisis or when the employee needs to get away from it all. Allowing the employee to work at home is yet another form of reward.

An analysis of the way these various rewards function suggests that there are hierarchies of such rewards. Maslow [1992], for example, has argued that there are some basic physiological needs—for example, for food, water, and sleep—that have to be fulfilled before the next higher needs can be activated. After these basic needs are met relatively well, the next higher level needs—protection from danger, threat, and deprivation—become important. These are followed by social needs that include the need for love and for acceptance. The next level includes ego needs, which, according to Maslow, involve the need for self-confidence, for achievement, for competence, and for knowledge. Finally, the highest need, self-actualization, can be activated. This is the need to develop one's own potential and to maximize self-development.

Maslow has conceived of these needs as hierarchically structured. Although the evidence for a multilevel hierarchy of needs is very weak, physiological needs are the basis for all others. If physiological needs are not satisfied, then other needs do not become activated. The evidence for this point comes from studies of hunger that were done during the Second World War [Guetzkow and Bowman, 1946]. In these studies, volunteers agreed to live on 900 calories a day. This starvation diet resulted in an extremely disturbing experience for the participants. They stopped functioning as normal adults; they no longer had any interest in development, in sex, or in interpersonal relationships. Their only concern was obtaining food. Food dominated their thoughts, their dreams, their everyday life. This example supports Maslow's thesis and suggests that at least a two-level hierarchy is valid. Some other theorists, such as Alderfer [1972], have argued that a three-level hierarchy can be identified. He called them existence, relatedness, and growth needs (the ERG theory). Existence includes physiological and safety needs, relatedness includes membership and self-esteem needs, and growth includes self-actualization needs.

The presence of a goal can create a need. As indicated above, there is evidence that individuals who have been given specific, difficult, but attainable goals are much more motivated to work hard to obtain them than those individuals who have not been given any goals or who have been given goals that are too easy or to difficult [Locke, 1968]. Thus, management by objectives—where specific, difficult, but attainable goals are established for a research project and where the manager and researcher discuss and agree on specific goals and later review the extent to which these goals have been reached—is an approach to motivation.

Another factor that is very helpful in motivating people is receiving feedback. Feedback takes many forms. It can include evaluation by the supervisor, formal recognition by the organization, receiving data of various kinds about how well one is doing, and comparisons with others or comparisons with the self at different points in time. Here information regarding scheduled versus actual specific goals' completion rate can be used as one mechanism for such feedback.

The research shows that at least in some organizational settings, certain kinds of feedback are more effective than others. For example, Herold and Parsons [1985] have found that formal recognition is the most powerful form of feedback. Other forms of effective feedback are positive supervisory behavior, positive comparisons with the self at another point in time, and positive comparisons with internalized standards.

Some feedback can be ineffective because it makes people defensive. For example, negative comparisons with others, along with negative evaluations by co-workers, may do more harm than good.

The evidence generally indicates that managers need to provide both goals and feedback. One without the other is unlikely to be effective [Becker, 1978]. To be most effective, feedback should be given by the person who knows the most about the employee, with the least delay possible. It should be positive and relevant to the job, referring specifically to the goals, and be frequent enough to be noticed [Brickman et al., 1976; Ilgen et al., 1979]. If individuals are allowed to set their own goals when the assigned goals are easy, they are likely to set difficult ones; and when the assigned goals are difficult, individuals are likely to set easier ones [Locke et al., 1984; Murphy et al., 1985]. Goal setting is more effective than instructions such as "do the best you can." It becomes even more effective when the person is aware that he/she will be evaluated and when the person receives positive cues (e.g., when somebody says, "this is a fun job, it is a challenging job"). The combination of these variables is most effective [White et al., 1977]. Providing a challenge can be particularly effective with a researcher.

6.5 REWARD SYSTEM DISCUSSION

We asked earlier, What acts does the organization reward? For example, does the organization reward innovation? Does the organization reward having original ideas that are not very popular? One can analyze the way the organization gives rewards for specific kinds of behaviors and come up with a profile which suggests why an organization is not very creative or not very successful. The reason usually is found in the nature and frequency of the rewards that are being distributed. You can have rewards that are given every month, such as salary, but this is not nearly as motivating as rewards that occur with a variable schedule. There is evidence that a variable schedule of rewards is much more motivating than one that occurs on a regular basis [Saari and Latham, 1982]. Receiving recognition after each publication is less effective than getting a major recognition following a series of publications.

While motivation is an important aspect of individual performance, we must not neglect to mention that the availability of proper skills and adequate training is also crucial to good performance. Furthermore, the rewards that the person receives from the organization should be tied to organizational performance. Otherwise, the person may function extremely effectively, but his or her performance may have no impact on the organization. Consider, for example, the case of an employee who is inspired on the job to invent something that could make a million dollars. However, the organization has neither the need for such a product nor the resources to take advantage of the invention. Such a person is performing well at the individual level, but not at the organizational level!

The most important principles of compensation are (1) equity, (2) competitiveness, and (3) link to performance. Equity is achieved by making sure that employees are rewarded according to their education and merit. Competitiveness requires salary surveys. Links to performance are difficult to establish but are important. Specifically, if salary is the major means of compensation, it does not correlate sufficiently to performance. Bonuses do so much more. Systems of compensation that review the employee's achievements every few months and that provide a raise according to the outcomes of these reviews link compensation and performance even more effectively.

In such evaluations an important issue is what attributions the evaluator uses to account for the behavior of the subordinate. In other words, is the behavior attributed to ability, task difficulty, effort, or luck? After a failure it is important for the effort attribution to be made, both by the employee and by the supervisor. An attribution to ability is likely to result in giving up. Effort is by far the best attribution for supervisors to make, and that is particularly so in the case of failure. In short, they should say, "try a bit harder and you will succeed."

Which exact mixture of rewards such as profit sharing, salary, fringe benefits, vacations, working at home, and so on, is likely to be most effective will vary depending on the attributes of the employee, as well as on the organization and its environment. Based on data from 33 high-tech and 72 non-high-tech firms with research and development units, Balkin and Gomez-Mejia [1984] concluded that high-tech firms place greater emphasis on profit sharing than traditional firms. Firms that are just getting established usually cannot afford to pay high salaries, but are able to offer a share of the profits. Of course, exactly what share depends on many factors, including profitability, sales volume, the stage of the product's life cycle, and attrition rates. However, profit sharing seems to be desirable in the case of commercial R&D organizations because such firms depend so much on innovation, and new products introduced by their scientific personnel can easily be associated with particular improvements in profit. Furthermore, those scientists who receive large shares of the profits do not have to try to become managers. They already have an excellent income doing technical work. Thus, some of the problems discussed in the previous chapters, concerning the need for dual and triple hierarchies as a means of motivating professionals to stay in technical work, can be solved if profit sharing is used wisely.

Atchison and French [1967] examined which of three systems of pay for scientists and engineers is perceived to be most equitable. One system was based on job evaluation, one was based on years of professional experience and judged quality of performance, and one was based on Jaques' [1961] notion of time span of discretion. According to Jaques' idea, the more important the job, the longer it takes to obtain feedback on how well one is doing. Also, the more important the job, the longer it takes for a higher authority to review one's performance. Hence, the time span of discretion (how long it takes before one's work is reviewed) can be used as a measure of the importance of the job. Atchison and French [1967] found that the traditional job evaluation method of determining pay and the time span of discretion method were considered more equitable than the method of using years of professional experience and quality of performance. However, the results may be due to the fact that people object to having others evaluate their performance. There are studies on performance appraisal that show that more than 80% of those appraised think that they perform above average (which, of course, is a statistical impossibility).

Since a large number of researchers work for not-for-profit organizations, government agencies, and universities, pay can hardly be tied to the profits of the employing organization. No one really knows for sure how pay for researchers in such organizations is determined or should be determined.

Individual researchers are at a distinct disadvantage in negotiating an equitable salary. Bargaining units where union representatives negotiate wages exist in few universities and research organizations. The leverage for obtaining equitable salaries through individual negotiations comes primarily from the threat of losing the researcher. However, the individual faces considerable uncertainty when starting a new position, and financial and social costs for relocating can be high for both the individual and the family. This, therefore, works to the advantage of the organization. Good starting salaries, with low potential for salary increases as the researcher gains experience, become normal management practices. This, in turn, creates dissatisfaction among experienced researchers.

Comparability salary surveys provide one of the best leverages. When such surveys continually show that salaries in certain research organizations, universities, or governmental labs are markedly lower than in comparable organizations, recruitment for new employees, especially for inducing quality researchers, suffers.

Take the case of the U.S. government research organization in which one of the managers stated that "due to a salary lag of 20 to 25 percent, he has not been able to recruit a single researcher holding a Ph.D. in engineering from one of the premier research universities during the last five years." Such pay policies may not affect work quality immediately because researchers hired prior to the period when the pay disparity developed are not likely to leave at once. But the long-term consequences of such policies on the quality of the research staff are predictable and inevitable.

Some may argue that market mechanisms properly determine salaries. However, President Derek Bok, in his commencement address to the Harvard Class of '88, provided comprehensive and compelling arguments to disprove this myth. There are few salaries that are determined by such mechanisms. For instance, salaries of chief executive officers of corporations are not really determined by a free market mechanism. If that were the case, why would the president of Chrysler get $16 million a year more than the president of Toyota?

Organization policies that attempt to provide competitive and equitable salary compensation to the researchers, in the long run, are likely to attract and retain highly talented research staff. In an R&D organization, nothing is more crucial than the quality of the research staff.

6.6 SENSE OF CONTROL AND COMMUNITY

Organizational effectiveness does depend on individual motivation and individual effectiveness. It obviously depends on individual performance, but it also depends on communication and coordination among individuals and also between the individual and the organization. Certain techniques, such as gain sharing and profit sharing, are techniques for bringing the goals of the individual and the organization into line with each other and are also methods for motivating the individual. One of the most successful plans to motivate employees has been developed by Lincoln [1951] and involves profit sharing.

In addition, openness of information is necessary so that the individual knows what the organization expects and hopes to get from him. Job rotation can help the individual get a better feel for what the organization is trying to achieve, and intrinsic rewards (getting a kick out of doing the job) that are tied to individual performance can help the individual line up his own rewards and goals with the goals of the organization.

Finally, designing jobs in such a way that individuals have a sense of control over their activities is very important. Individuals must feel that a lot of what they do is consistent with their goals. Thus, the individual should have a certain amount of choice. Individuals who see themselves as having some alternatives are much more satisfied and have a greater sense of control than individuals who are told "this is it." Having only one alternative is demotivating.

Furthermore, self-esteem is linked to a sense of control. In other words, people who see themselves *in control* have higher self-esteem; conversely, having high self-esteem often means the individual is in control. There are also studies showing that people who do not feel that they are in control feel depressed [Langer, 1983]. The need for control is so strong in humans that in certain experiments [Thomson, 1983] people have been found to prefer receiving a punishment in the form of a loud noise that they could control, rather than a reward in the form of chips which they could later use to buy something, but over which they had no control. In other words, something as fundamental as reward and punishment can change meaning when matched with control and no control.

Thomson [1983] has also suggested that job design is relevant to developing a sense of control because there are jobs in which the individual experiences so much role conflict and role ambiguity that the individual does not feel in control. Such jobs have been associated with dissatisfaction and high turnover.

To achieve a high level of satisfaction, good morale, and productive R&D activity requires that scientists and engineers be given as much a sense of control over their job environment as is feasible. It is also important to give employees a sense that they are being appreciated by the organization. Good management should coordinate the goals of the individual and the organization by providing rewards that shape the goals of the individual. Specific activities that increase the sense of community within the laboratory can also be helpful in linking the individual to the organization. For instance, allowing the individual to be "in-the-know" can be very helpful and very motivating; however, too much information can confuse. A good principle is that a person should be given as much information as is desired without being overloaded with facts by management.

A true story, from our file, that occurred in a government R&D laboratory, with the names changed, of course, illustrates the way a supervisor can demotivate a subordinate. Section 6.7 describes the case, as developed by Harry Triandis and David Day.

6.7 A FEDERAL R&D LABORATORY CASE

Some people believe that there is considerable inconsistency in the way vacation time, flextime, being allowed to work at home, and other personnel matters are handled by the various department managers. Some managers are very liberal, while others are not at all. Some employees see their friends in other sections of the organization behave in ways in which they are not allowed to behave.

While the personnel department allows considerable flexibility in such matters, it interferes when performance evaluations are done. A case in point is Dr. Blank, who is a researcher and has provided the following account. He was asked by his supervisor, Dr. Ablex, to do some work that was rather unskilled, simply because the work had high priority. Dr. Ablex assured Dr. Blank, at the time the work was assigned, that this would not affect his income, but when the personnel department did an evaluation of Dr. Blank they found he was doing less skilled work and demoted him. Dr. Ablex claimed that he could do nothing about the personnel department's action. Contributing to Dr. Blank's demotion were, however, several other factors. First, it appears that the high priority work that Dr. Ablex wanted done precluded publication of papers. Yet, the personnel department considered publication as one of the criteria for performance evaluation. Second, the job assigned to Dr. Blank required his undivided attention and was such that one person could perform it without subordinates. So, there were no technicians, graduate students, research assistants, or associates under Dr. Blank. The personnel department, on the other hand, considered the extent a person supervised others as a criterion for evaluation. Finally, Dr. Blank had been encouraged by directors of the R&D laboratory to participate in national and international associations and committees of his discipline. However, Dr. Ablex did not want Dr. Blank to spend his time in such activities instead of doing the high-priority study, and he provided no funds for travel. Rather than paying for such travel out of his own pocket, Dr. Blank simply did not participate in committees, but the personnel department considered participation in national and international committees as a factor in the evaluation of Dr. Blank's performance. The result is that Dr. Ablex's job assignments and behavior created conditions that made Dr. Blank look professionally isolated and without influence—hence the recommendation that he be demoted. Dr. Ablex's behavior has been a problem in other ways and to other people as well.

In this example Dr. Ablex is getting his high-priority job done, but the results, in terms of motivation, are devastating to Dr. Blank. The situation is demotivating in a number of ways: There is little overlap between the goals of the organization and the goals of Dr. Blank; the consequences of pursuing organizational goals have been disastrous for Dr. Blank; it is unlikely that Dr. Blank is enjoying this high-priority job; and while such jobs may have to be done occasionally, to have them become so central to a person's career is undesirable.

The case also indicates that the situation in Dr. Ablex's division is unsatisfactory since some people get privileges that others do not. Such inequities are bound to be demotivating. One can defend the principle that some inequities are unavoidable, but when they have to happen it is good to evoke Rawls' principle that equality is desirable, except when inequality is to the advantage of the least powerful. In other words, if the young scientists are given some extra privileges so they can finish their dissertations, that is fine; but if those with power are given the privileges, that is undesirable. In any case, Ablex should have discussed the rules he uses in granting such privileges, and his subordinates should have an opportunity to argue and to participate in the formulation of the rules, and once these rules are set they should have been followed.

6.8 SUMMARY

Behavior depends on habits and self-instructions (behavioral intentions). The former are shaped by patterns of rewards. The latter depend on social, affective, and cognitive factors. The social factors reflect norms, roles, self-definitions, and interpersonal agreements. Facilitating conditions, such as self-efficacy, are also most important in understanding whether behavior will or will not occur.

6.9 QUESTIONS FOR CLASS DISCUSSION

1 Go over the model presented in the chapter and analyze it from the point of view of how to develop specific procedures in an R&D lab that will favor increased productivity.

2 Go over the model again, and analyze it from the point of view of how to develop specific procedures in an R&D lab that will favor job satisfaction.

6.10 FURTHER READINGS

Cuadron, S. (1994). Motivating creative employees calls for new strategies. *Personnel Journal,* **73**(5), 103–106 (May).

Lawler, E. E., III (1986). *High Involvement Management.* San Francisco, CA: Bass.

Riggs, W. (1994). Incentives to innovate and the sources of innovation: The case of scientific instruments. *Research Policy,* **23**(4), 459–469 (June).

Scott, S. G. (1994). Determinants of innovative behavior: A path model of individual innovation in the workplace. *Academy of Management Journal,* **37**(3), 580–607 (June).

Triandis, H. C. (1980). Values, attitudes, and interpersonal behavior. *Nebraska Symposium on Motivation, 1979.* Lincoln, NE: University of Nebraska Press.

7

DEALING WITH DIVERSITY IN R&D ORGANIZATIONS

As organizations are becoming more culturally diverse, there is a greater need for dealing effectively with this diversity. Research teams with one scientist in Asia, another in Europe, and a third one in North America are now much more common than they were a few years ago.

With electronic mail, extensive and inexpensive international communications are possible. Talent can be found in many places, and the very top people from every continent often like to work together. Joint research projects with scientists who are different in culture, gender, age, sexual orientation, discipline, organizational level, and function are becoming more common.

In addition, many of the world's problems cannot be solved by people from a single discipline. Interdisciplinary work is essential if such problems are to be solved.

Many graduate departments in the United States have equal numbers of students of Asian as of North American cultural backgrounds. Many of these professionals will collaborate in the future, in any number of different places.

The Space Station is such an expensive research project that it requires funding from the United States, Russia, Japan, and the European Community. Scientists from all these regions will work on it. Antarctic expeditions often have members from half a dozen countries. Women are increasingly becoming engineers and scientists and work side by side with men in many R&D laboratories. Young scientists often work with older and more experienced mentors. Sexual orientation and physical disability have nothing to do with talent. In fact, if Freud's theory of sublimation is correct, people who repress their homosexual tendencies, to avoid "coming out of the closet," may be especially hardworking and creative.

7.1 ASSIMILATION AND MULTICULTURALISM

Until the 1960s the emphasis was on assimilation in much of the Western world. People from a minority group were expected to change and become like the members of the rest of the work environment. Even today in many laboratories in Asia, and even Europe, assimilation is still emphasized.

But the world is changing. Multiculturalism is emerging as the typical pattern in many places. In multiculturalism each group is allowed to retain those of its own attributes that it finds most important. People are allowed to satisfy their special needs and rights to express their cultural identity. People are treated equally, regardless of ethnicity, sex, sexual orientation, or physical disability.

Multiculturalism is increasingly recognized as desirable, as people realize that talent is not linked to specific demographic factors, and the optimal utilization of personnel resources requires that all talent be used.

Yet some of the old attitudes still persist. All humans are ethnocentric [Triandis, 1994]. That is, they use their own culture as the standard to judge other cultures. The more another culture is like their own culture, the "better" it is.

People are attracted by similarity. In one experiment, psychologists measured how much people liked a large number of first names. They found that the more the letters of a name were the same as the letters of a respondent's first name, the more the respondent liked that first name! In short, we are "wired" to like similarity. As a result, many of us find that "otherness" is a deficiency. We feel that having an organization that is too diverse is undesirable. We think that people who express discomfort with our values are a bit strange (to say the least). We feel that those who work with us should assimilate to our norms. We believe that fair and equal treatment is to treat people the way we perceive fairness. We think that if there is a non-fit between our culture or organization and an outsider, it is the outsider who must change rather than our culture or organization.

In sum, there are tensions between assimilation and multiculturalism. What does the scientific literature say about the advantages and disadvantages of each?

A review of the topic of diversity [Triandis, Kurowski, and Gelfand, 1994] identified several advantages and some disadvantages. Among the advantages of heterogeneous groups is that they are more creative, they are more likely to solve difficult problems, and they are less likely to engage in "groupthink," a process in which the leader of the group convinces everyone else that his solution is the best. Groupthink often results in disastrous decisions as was discussed in previous chapters and again in this one.

However, not every kind of heterogeneity is desirable. Heterogeneity in ability within the work group is not desirable; it is only heterogeneity in attitudes, background, culture, and the like, that is desirable [Triandis, Hall, and Ewen, 1965]. Heterogeneity in ability can result in some members of the group looking down on other members. But heterogeneity in perspective means that different ideas can be placed in front of the group, and confronting different ideas can often generate new ones. In one study of the top management of 199 banks it was found that the more heterogeneous the management teams, the more innovative the bank. Diversity also increases the quality of ideas and decreases the chances that a major mistake will be made by the group. In a famous example, General Motors introduced a car called Nova to the Latin American market, with disappointing results. Had that name been determined by a team that included Spanish speakers, they might have detected a problem: No va (does not go) is not a good name for a car in that part of the world.

Research has found that the mental health of those who assimilate (give up their own culture to become indistinguishable from members of another culture) and those who separate (stay with fellow members of their culture in enclaves) is not as good as the mental health of those who are bicultural (retain their own culture, but also use most of the elements of the other culture). It is parallel to the comparison of a person who is monolingual with a person who is bilingual. If both languages are needed, the bilingual will behave more effectively.

On the other hand, diversity is often related to less interpersonal attraction, and that increases turnover. Group cohesion is lower in heterogeneous groups. The more diverse the team, the more communication problems are likely to emerge, and poor communication predicts low interpersonal attraction. Dissatisfaction with the job is likely, and that increases turnover. Heterogeneous groups often experience delays and distortions in communication. Language differences can result in misunderstandings. More than that, people engage in paralinguistic communication, such as touching or not touching, keeping a small or a large distance between their bodies when they talk to each other, using different body orientations, such as facing each other directly or at an angle, and looking or not looking in the eyes. Major differences in paralinguistic behaviors result in low group cohesion. Furthermore, differences in demographics and religion can result in increased stress.

When we look at the total picture we note that some teams manage diversity well and others do not cope with it. Those who manage it well have members with certain personalities, who have had a wide range of experiences, have traveled a lot, and have developed intercultural skills. People who are authoritarian, low in tolerance for ambiguity, and use narrow categories do not do well in intercultural situations. There are tests that can measure these personality attributes. For example, a test can be used to find out whether a person uses narrow or broad categories. The test consists of 10 pages. At the top of each page is a nonsense word (e.g., a *zupf*), and a nonsense shape. Under that line are 20 other nonsense shapes. The instructions tell the respondent that the nonsense shape is called a *zupf,* and ask him or her to circle all the shapes that appear to be *zupf*s. A broad categorizer would circle most of the shapes. A narrow categorizer would circle only the one or two shapes that are almost identical to the shape at the top of the page. Doing this ten times provides a score that is reliable. Broad categorizers do better in other cultures because when they see a behavior that is strange they fit it in their existing cognitive framework.

Special training programs that teach people how to deal with differences in culture are especially useful [Black and Mendenhall, 1990]. Consultants provide such training, but some of them are incompetent. We discuss, at the end of this chapter, some clues for identifying the most effective consultants.

7.2 UNDERSTANDING CULTURE

Culture is to society what memory is to individuals. It includes ideas, standard operating procedures, and unstated assumptions that have "worked" at some time in the history of a cultural group, thus becoming the standards for perceiving, thinking, and judging. It facilitates behavior, because people do not have to decide what to do; they do what is customary. It consists of shared beliefs, attitudes, self-perceptions, norms, role perceptions, and values that are transmitted from one generation to another among those who are able to communicate, because they share a language, historical period, and place.

Organizations also have a culture. If an idea is presented and all members of the laboratory agree without even thinking whether it is good or bad, the chances are very high that the idea is reflecting the culture of the laboratory or of the country.

Cultures have elements that are unique to them, and are usually expressed by words that are difficult to translate to other languages. But they also have elements that are universal. In order to understand the way people from other cultures look at the world it is important to learn something about the unique elements of their culture. For example, the Japanese have a concept called *amae,* reflecting a desire for close interpersonal relationships combined with a person's presumption that he/she can depend on the other. To understand relationships involving the Japanese it is very helpful to understand the meaning of this term. Nancy Sakamoto (1982), an American married to a Japanese, noted six underlying polite fictions that create problems in American–Japanese relationships.

1. Whereas Americans assume in social interactions that "you and I are equal," the polite Japanese assumes that "you are my superior."
2. Americans assume that "you and I are close friends." Japanese assume that "I am in awe of you."
3. Americans assume that "you and I are relaxed." Japanese assume that "I am busy on your behalf."
4. Americans assume that "you and I are independent." Japanese, reflecting *amae,* assume that "I depend on you."
5. Americans assume that "you and I are individuals" while Japanese assume that "you and I are members of groups."
6. Americans assume that "you and I are unique." Japanese assume that "you and I feel/think alike."

Language has some effects on the way people think. For example, in countries where the language forces people to distinguish male from female terms (where nouns have gender such as in Spanish or German), children learn to distinguish boys from girls earlier than in countries where this linguistic distinction is not made (e.g., Finnish).

In cultures where the language requires the use of different status terms (e.g., "tu" and "vous" in French or in Japanese several status-linked terms for "I," "me," and "you") people pay more attention to status differences than in languages where there are no such words. Some Japanese bilinguals prefer to use English, because that eliminates the complications of having to decide whose status is higher!

People can also differ in the way they develop an argument. For example, analyses of essays written by students from different countries have found that in English there is a preference for linear development, while in other cultures one can find other constructions (e.g., circular development).

Another difference is whether one states a conclusion first and then lists the facts that support it, or one lists a lot of facts first and then draws a conclusion.

Still another difference is whether one presents the best arguments first or last. For example, Western samples generally start with the most impressive arguments; Japanese samples keep their best ammunition for the last part of the argument.

The extent that people solve problems using visual versus language cues, analogies, and metaphors is also relevant. Male physicists have been found to use visual cues more often than female physicists. They also draw diagrams and use geometry in developing an argument more often than females. Female physicists have been found to use verbal cues and to use analytic geometry equations more than diagrams. Scientists who use analogies and metaphors have been found to be more successful than scientists who avoid these devices.

The extent that context is used in communication is also an important cultural difference. For example, Japanese speakers depend on paralinguistic cues and other context to interpret verbal exchanges much more than Western speakers, who depend on content and tend to ignore context.

All of these differences can create some interpersonal difficulties when people work together. Identifying these differences and training people to anticipate them and to adjust their behavior to fit the needs of the other can increase the effectiveness of heterogeneous work groups.

We will see below that Americans are high in individualism and Japanese are high in collectivism, and some of the differences mentioned above reflect these differences in cultural patterns.

7.3 CULTURAL DIFFERENCES

People from the West tend to be individualistic, while people from Asia, especially East Asia, tend to be collectivists. The differences between these two kinds of cultural patterns are reflected in several tendencies. Four will be mentioned here, and more can be found in Triandis (1995).

1. Collectivists define themselves as members of a collective (family, co-workers, neighbors, fellow countrymen, co-religionists, etc.). If you ask them who they are, for instance, they are likely to say "I am Japanese" or "I am an uncle." The individualists are more likely to say "I am responsible" or "I am kind" (i.e., use a trait).

2. Collectivist social behavior is predicted better from norms than from attitudes. Conversely, the social behavior of individualists can be predicted better from attitudes than from norms. In short, individualists more often do what they like than what they must do; collectivists more often do what is appropriate than what is enjoyable.
3. Collectivists have personal goals that are compatible with the goals of their collective, and when there is a discrepancy between the two kinds of goals they think that it is "obvious" that the goals of the collective should have precedence. Individualists often have goals that are not related to the goals of their collectives, and when there is an incompatibility between the two sets of goals they think that it is natural that their personal goals should have priority. For example, a research scientist working in a laboratory can spend time doing what is good for the laboratory or what is good for her. If she is a collectivist, she is likely to do more things that help the laboratory rather than herself; if he is an individualist, he is likely to do more things that help him rather than the laboratory. There is a slight tendency for women to be more collectivist than men.
4. Collectivists pay much attention to the needs of others, and they tend to stay in a relationship in which the other person needs them even if they do not feel that they are getting much out of the relationship. Individualists are more calculating. If they get enough out of a relationship, they stay; if the costs of the relationship exceed the benefits, they drop it.

Within any culture there are people whose personality is like the personality of collectivists (allocentrics) and people whose personality is like the personality of individualists (idiocentrics). Allocentrics in individualistic cultures are more likely to join groups, such as communes, gangs, unions, large organizations that provide job stability, the government, or the armed forces. Idiocentrics in collectivist cultures are most likely to try to leave their culture, so that they will not be pressured by their collectives to behave in ways they do not enjoy.

Differences between idiocentrics and allocentrics can be traced to upbringing. The parents of idiocentrics emphasize creativity, exploration, and self-reliance; the parents of allocentrics emphasize obedience, reliability, and duty.

Most cultures are collectivist, until their members become affluent and able to "do their own thing." The upper classes, the educated, and those who have traveled a lot tend to be individualistic. Exposure to the mass media increases individualism, because most of the programs emphasize pleasure rather than duty. Even in individualistic cultures there will be people who are allocentric among those who have been exposed to the traditions of particular cultures, have been raised in large families, or are financially dependent on others.

Research has shown that, on the whole, East Asian, Latin American, and African cultures are collectivist, while Western cultures are individualist. However, within the United States, Hispanics, Asians, and other minorities often have collectivist cultures. African-Americans have collectivist cultures when they have been raised by extended families, but have extremely individualistic cultures when they have been raised by single mothers.

All humans have both individualist and collectivist elements in their cognitive systems, and they sample the elements that fit the situations in which they find themselves. Thus, if the collective is under threat, members of all cultures are likely to sample the collectivist elements; when an individual is competing with members of his or her own collective, the sampling of individualist elements is extremely likely. When the situation is ambiguous, however, then allocentrics will sample the collectivist elements while idiocentrics will sample the individualist elements. Most situations are ambiguous. For example, in a negotiation, will the other person be perceived as "one of us" or as "one of them?" In collectivist cultures it is more likely that the "one of us" perception will occur, and there are customs that favor the development of that perception. For example, before the negotiation, people in collectivist cultures expect that they will "get to know each other" and exchange a lot of private, even intimate information (e.g., "how much do you make per month?"). In individualist cultures the exchange of such intimate information is taboo.

Cultural differences on individualism and collectivism often result in misunderstandings, and when a person from a collectivist culture works in an individualistic culture, or vice versa, there is often culture shock (e.g., people feel depressed, anxious, cannot sleep well, lose their appetite). Culture training can be used to expose people to different cultures, and research shows that this reduces culture shock.

7.4 WHAT HAPPENS WHEN PEOPLE FROM DIFFERENT CULTURES WORK TOGETHER?

When a person first sees another person, some of the attributes of the other person "stick out." According to recent research, people are likely to be categorized by whatever attribute makes them distinctive (Nelson and Miller, 1995). For example, a female engineer in a laboratory with many male engineers will be seen as especially female. Thus, she will be seen as a representative of the females of the world (a group), and the relationship with her is likely to be intergroup rather than interpersonal. Her unique attributes will not stand out. On the other hand, the same female engineer in a group of female engineers is likely to be seen interpersonally. When the other individual is seen as a person, unrelated to groups, the relationship is interpersonal. In this case it is the unique attributes of the person that stand out.

Collectivists are more likely to see relationships as intergroup, and individualists are more likely to see relationships as interpersonal. Intergroup relationships are also more probable when there is a history of conflict between the groups or when individuals are physically distinct or anonymous, have incompatible goals, and are strongly attached to their groups. Intergroup relationships occur because people have a tendency to categorize and stereotype others. Stereotyping decreases cognitive work. We do not have to discover who the other is, because we already know it from the category that we have assigned to that person. If we can recategorize the other person, from "one of them" to "one of us," we can begin to have an interpersonal relationship, in which the personal attributes of the other become important.

7.5 CULTURAL DISTANCE

The greater the cultural distance between the cultures of two people, the more likely it is that they will deal with each other in intergroup rather than interpersonal terms. When the elements of two cultures are very similar, there is a small cultural distance. Large cultural distances occur when people speak languages that belong to different language families (e.g., Chinese is a tonal language, whereas Indo-European languages are not), have different religions or different social structures (e.g., monogamous versus polygamous families), have different standards of living, have different political systems, and the like. The greater the cultural distance, the more difficult it is for people from the two cultures to communicate and form effective teams. Thus, there is a limitation to the degree of heterogeneity that is ideal. While a certain amount is desirable, extreme heterogeneity is not.

7.6 A MODEL FOR DIVERSITY IN GROUPS

Figure 7.1 shows a model that summarizes the factors that increase or decrease the effectiveness of relationships in work groups. The key idea is "perceived similarity." As we noted earlier, we are wired to prefer similarity. Similarity is rewarding and leads to attraction. Perceived similarity decreases where there is a large cultural distance and when there is a history of conflict between groups or individuals (the events in the former Yugoslavia show this clearly).

However, perceived similarity increases when people know a lot about each other, and that is what cross-cultural training courses try to accomplish. Also, when people are of approximately equal status or share a similar attribute, their perceived similarity increases.

Perceived similarity may not have much relevance if people do not have an opportunity to interact. If one person is on the 9th floor and the other on the 35th and they never meet, whether they are similar or different makes no difference. However, the more opportunities there are for people to have contact, the more perceived similarity will result in rewarding situations. The rewards will be even higher when people have superordinate goals—goals that they can reach only with the help of the other person. Furthermore, the rewards are greater when authorities support contact between diverse individuals. For example, where there is an official policy that supports multiculturalism, contact becomes even more satisfying. Rewarding contacts increase the chances that more interaction will take place between the individuals. The greater the interaction, the more likely it is that the individuals will develop an intimate relationship, revealing more about themselves and spend-

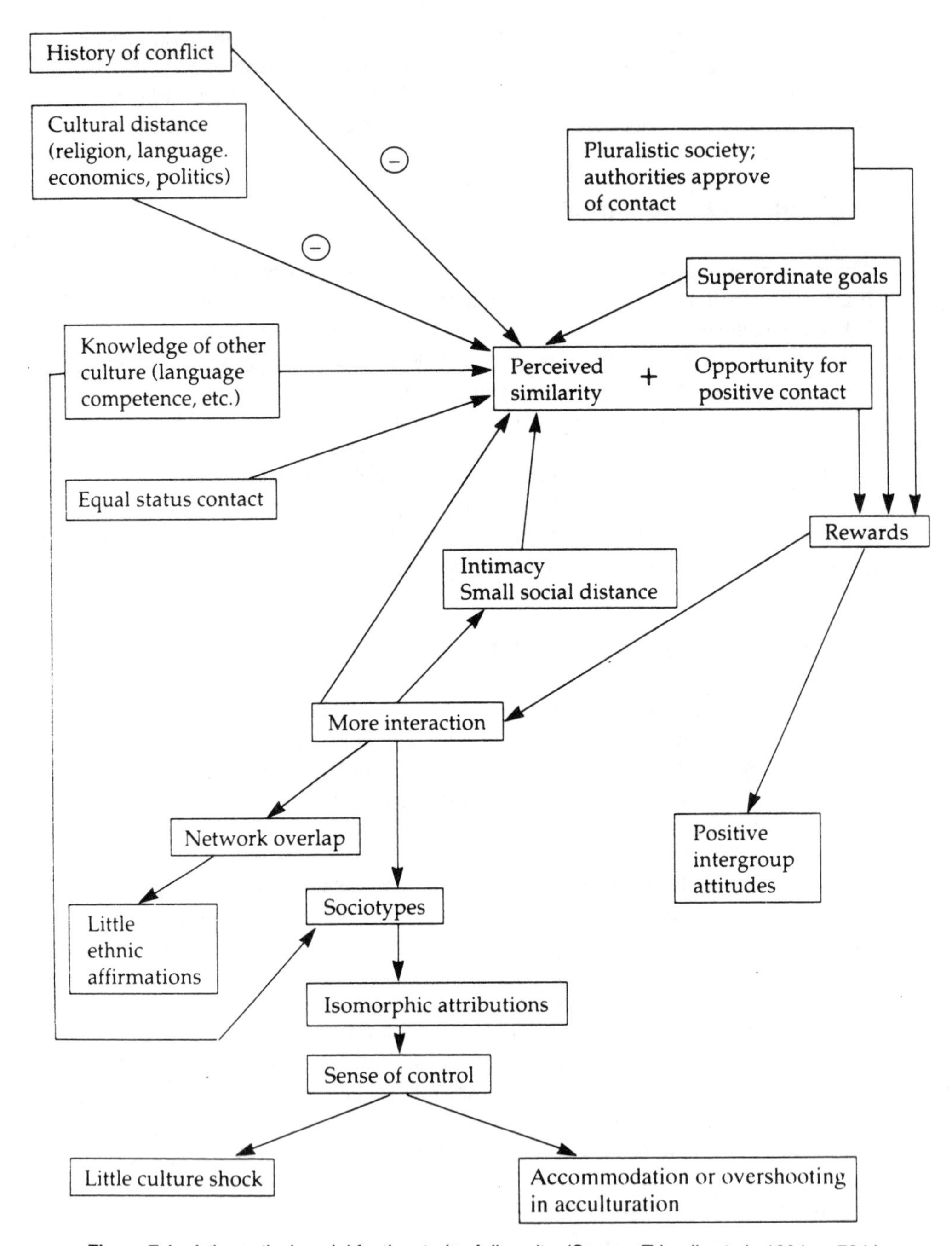

Figure 7.1. A theoretical model for the study of diversity. (Source: Triandis et al., 1994, p. 784.)

ing more time together. Such activities are likely to increase their perceived similarity. Also, the more interaction, the more people are likely to develop common friends and other indications of "network overlap." When there is much network overlap, people will avoid cutting themselves off from the other group and displaying signs of ethnic superiority.

More interaction also has the effect of developing accuracy in the perception of the other group. Stereotypes are corrected with increased interaction and become sociotypes (valid stereotypes). When there is accuracy in the way the other group is seen, people are likely to make correct attributions in explaining the behavior of members of the other group. For example, in some cultures people show respect by not looking another person in the eye. In others, not looking in the eye suggests that the person has something to hide. Thus the behavior of not looking in the eye can be interpreted very differently, depending on the attribution that is made: If I think that "you are not looking in my eyes because you are sly," I am making a very different attribution than if I think that "you are not looking in my eyes because you show respect." Making isomorphic attributions means that the observer makes attributions that are more or less like the attributions made by the actor of the behavior. Making isomorphic attributions means that each person understands correctly the behavior of the other. Then there is a match between the expectations of how the other will behave and the other's behavior. This match provides a sense of "control" over the social situation. "I know why he acted that way;" "I can make him act differently if I act this way." A sense of control results in far less culture shock when people from different cultures interact with each other.

Also, when a person from one culture feels "in control" of the social situation, the chances are that he will try to act in ways the member of the other culture will expect him to act. This will result in "accommodation" where behaviors that are acceptable to members of both cultures are especially likely to occur. In such a case, even overshooting might occur, where the member of one culture, X, behaves more like the member of the other culture, Y, than the norms of culture Y require. For example, a member from a culture that keeps considerable distance between the bodies when people are interacting (such as the Japanese) may interact with a member of a culture where people keep their bodies quite close and touch when they know each other (such as the Mexicans). Mexicans will normally give an abrazo (putting their hands around the body of the other person, and tapping him in the back) when they know the other person well. If a Mexican met a Japanese for the first time, he would not expect an abrazo. Thus, a Japanese who has studied about Mexican culture and who feels in control of the social situation and gives an abrazo would be "overshooting." He is not only friendly, but more friendly than required by the situation.

On the other hand, there is also "ethnic affirmation," when persons emphasize the merits of their own culture while putting down the other culture. This often happens when a group feels rejected: "If you do not accept me, you are the loser, because you do not realize how wonderful I am."

7.7 THE STATUS OF MINORITIES IN WORK GROUPS

There is strong evidence that minorities in a workplace do not get an equal share of the resources of the group. For example, in all countries, women get paid less than men, even when they do very similar work. One of the factors that seems responsible in part for such discrepancies is that minorities are less likely to have a mentor in the work group.

Mentoring is useful in the socialization of new employees, but it can also function as an elitist patron system that excludes the socially different. Mentoring offers opportunities for role modeling, acceptance by the work group, counseling, and friendship. Mentors provide invaluable inside information. Promotions often depend on knowing the right people and the subtle standard operating procedures for getting things done in the workplace that are not part of the instructions that one gets when joining the work group.

Many minorities find it difficult to find mentors, either because there are no members of their group in the workplace when they join the organization or because the potential mentors are already guiding many other members of the group and do not have the time to help them. Also, women often find it difficult to have male mentors because some males feel embarrassed about developing a close relationship with females that might be misinterpreted by spouses and others.

Similar to the problems of mentoring are problems of networking. Nontraditional employees often are excluded from networks and discussions that occur in clubs, golf courses, and other settings, where only white males get together. Some studies found that men use networks for both work-related matters and entertainment, while women use other women for entertainment and relate to men only on job-related matters. That means that women have less strong ties with male networks.

7.8 DEALING WITH PEOPLE FROM DIFFERENT DISCIPLINES, ORGANIZATIONAL LEVELS, AND FUNCTIONS

Disciplines develop their own terminology, which can sometimes make interdisciplinary communication very difficult. An engineer's definition of "attitude" is different from the definition given by a psychologist. Much interdisciplinary work involves defining terms and explaining the unstated assumptions of the discipline.

Triandis had, perhaps, the ultimate experience that illustrates the difficulties of such work. A committee was organized by UNESCO to develop a research project reflecting how values differ in different parts of the world. The group of experts met in Argentina at a villa that had been given to UNESCO. The members were a Filipino anthropologist, a Japanese statistician, a Polish sociologist, an Ivory Coast geographer, an Uruguayan jurist, a Brazilian philosopher, a Russian historian, and an American psychologist. A French intellectual presented a research plan. The proceedings were translated simultaneously into English, French, and Spanish. The whole week was spent discussing and trying to understand the plan, with each individual explaining what was meant by "values." However, one week proved insufficient to even reach agreement on the meaning of this term, let alone how to compare it in different cultures.

Similarly, members at different organizational levels such as top management and lower management often develop their own ways of thinking about problems. Those who have different functions, such as statisticians, computer experts, laboratory technicians, and so on, also have their own assumptions, their own levels of adaptation for making judgments, and their own ways of solving problems. In most cases the situation is not as extreme as the example from Argentina, but it often comes close.

7.9 INTERCULTURAL TRAINING

Intercultural training is a relatively new professional field that specializes in providing training to improve interactions across cultures, and can also be used to deal with all aspects of diversity. Unfortunately, the field is not regulated and is full of untrained trainers/consultants, many of whom do more harm than good. Before engaging such a consultant, it is good to check if he/she proposes to use a wide range of methods which will provide the training on a small scale first, and then evaluate it with multiple methods. It is not enough that the trainees like the course. They should also learn something. What they learn should change some behavior, and the change in behavior should be detectable in some favorable changes in organizational outcomes. After it has been shown that the training fits the particular organization, the consultant can then provide it to the entire group. The kind of evaluation mentioned is essential for the development of scientifically sound training that is tailored for the particular organization.

The mixture of training approaches provided by the consultant should include the following: some understanding of the trainee's own culture, some understanding of how members of different cultures make different attributions about the social behaviors that occur in that organization, some opportunities to practice insights gained about the way different cultures are similar and different, and some behavior modification techniques to change the behavior of the trainees.

Understanding the Trainee's Culture. One way to do this is to train an actor to behave in ways that contrast with the way people in the trainee's culture behave. When trainees interact with this actor, features of their culture become sharply clear. For example, Americans being individualists are likely to attribute their success to internal factors, such as their personal attitudes or abilities. Collectivists often attribute their success to the help they receive from others. When the actor attributes success to self instead of family, friends, co-workers, and so on, a discussion about this can help the trainees to better understand this aspect of American culture.

Understanding That Members of Different Cultures Make Different Attributions. A technique called the "culture assimilator" is used to achieve this goal. Trainees work through a book that consists of 50–150 "episodes" involving interaction between members of different cultures or different sexes.

Consider the following example of sexual harassment. A male colleague tells a woman that her body is exceptionally fit and attractive. This is followed by different "explanations" (attributions) of the man's behavior. One attribution may be that he is sexually harassing the woman. Another is that he is expressing an innocent compliment. Still other attributions are presented and the trainee is invited to choose the attribution most likely to be "correct."

During the construction of the training, a sample of men and women would have been presented with the specific episode and its attributions and would have been asked to select the correct one. If the women were more likely to select the sexual harassment attribution than the men, and the men were more likely to select the innocent compliment attribution than the women, there is an opportunity to teach men that such a statement may result in a "misunderstanding" that should be avoided. After trainees choose an attribution, they are directed to turn to a specific page of the training materials. In that page they receive feedback, such as "80% of the women who saw this episode chose this attribution." This way the trainees learn that members of other cultures or sexes see the world differently from the way they see it. By being exposed to many such episodes and their corresponding feedback, they get to understand the point of view of members of another cultural or sexual group.

Opportunities to Practice Insights Gained in Previous Training. Suppose the culture assimilator has taught that Arabs are very sensitive to being shown the soles of someone's feet or shoes—that is, that this is considered a great insult implying "I put you down." It is one thing to understand this intellectually and it is quite another thing to change one's habits of crossing one's legs, which is likely to result in exposure of the soles of the shoes. During training, when the trainees behave inappropriately they are reminded to change their behavior. Over time, they change their habits, eliminating offensive behaviors and boosting behaviors that are considered desirable in the other culture.

There are numerous additional techniques, such as simulations, working with video-computer systems, and so on, that can provide cross-cultural training. Different strategies, such as the following, can have beneficial effects on improving the way that people deal with diversity:

- Reducing the distinction between "one of us" and "one of them"
- Increasing the perceived heterogeneity of people from other cultures
- Providing positive experiences in the presence of members of other cultures
- Increasing the awareness of the similarities among people from different cultures
- Decreasing the emphasis on dissimilarities between people from different cultures

- Emphasizing the superordinate goals shared with members of the other culture
- Increasing the extent one values diversity (e.g., explaining that when the group is heterogeneous it is less likely that a major mistake will be made, because someone is likely to identify the mistake and thus avoid "groupthink" [Janis, 1972])
- Learning to look for win–win solutions in situations where people disagree

More details on such techniques can be found in Triandis [1994, Chapter 10], and additional details can be found in Landis and Bhagat's [1996] *Handbook of Intercultural Training*.

7.10 SUMMARY

The optimal utilization of personnel in R&D organizations requires (a) the identification of the way different groups are similar and different and (b) the development of training programs that will make both the similarities and differences salient. The crew then needs to learn how to manage diversity. This requires the development of insights about the way people from diverse groups are similar and different and how to deal with each kind of difference.

7.11 QUESTIONS FOR CLASS DISCUSSION

1 Examine Figure 7.1 from the point of view of what a manager can do to improve the effectiveness of intercultural relations among subordinates.

2 Collectivists differ from individualists in the standard operating procedures that make sense to them. Look at pages 120 and 121 and identify the implications of these differences for (a) employee selection, (b) motivation, and (c) appraisal. What do such cultural differences imply for the leadership behaviors of an executive from an individualistic culture working in a collectivist culture?

7.12 FURTHER READINGS

Chemers, M., S. Oskamp, and M. A. Costanzo (Eds.) (1995). *Diversity in Organizations.* San Diego: Academic Press.

Jackson, S. E. and M. N. Ruderman (1995). *Diversity in Work Teams: Research Paradigms for a Changing Workplace.* Washington, D.C.: American Psychological Association.

Landis, D. and R. S. Bhagat (1996). *Handbook of Intercultural Training,* 2nd ed. Thousand Oaks, CA: Sage.

Triandis, H. C., L. Kurowski, and M. Gelfand (1994). Workplace diversity. In H. C. Triandis, M. Dunnette, and L. Hough (Eds.) *Handbook of Industrial and Organizational Psychology,* Vol. 4, 2nd ed. Palo Alto, CA: Consulting Psychologists Press, pp. 769–827.

8

LEADERSHIP IN R&D ORGANIZATIONS

During the past 30 years the study of leadership has utilized a variety of approaches. Some researchers have spent a good deal of time observing the behavior of groups and the emergence of leaders. As a result, they have seen that the activities of leaders fall into two general categories. The first involves maintaining (M) the group by paying attention to the needs of the members and making sure that conflicts do not become serious. The second involves that actual task that the group must perform (P), the definition of the task, how and when it is to be done, and so on. We can label these two types of activities *consideration* and *structure*.

"Consideration" involves paying attention to people, being considerate of their needs and goals, being employee-oriented, and paying attention to the human factor. "Structure" refers to what is to be done and to where the group is going. What is to be accomplished? How is it to be accomplished? How can the activities of the members be controlled?

Parallel to these is a new set of concepts that are currently more popular. These were developed by the Japanese psychologist Misumi [1985]. He has shown that a supervisor might emit a lot of behaviors that are M (such as paying attention to people, being considerate of their needs and goals, making people feel significant, helping people value learning and competence, helping people feel part of the laboratory community, and inspiring them to find the work stimulating) or a lot of behaviors that are P (scheduling work, defining goals, telling people how to attain the goals, and making sure that people do what is expected of them). However, Misumi found that great leaders emit both a lot of M as well as a lot of P behaviors.

For each job setting, Misumi identifies those behaviors that are P or M for that particular job. What fits one laboratory does not necessarily fit another. Misumi talks to people in the job setting and asks each subordinate to describe the behavior of the leader and rate him/her on their M or P behaviors. Misumi uses the symbols M and P for those who use many behaviors, and he uses m and p to indicate that the leader does few maintenance or production behaviors. This way, in each setting, Misumi identified four kinds of leaders:

mp = little maintenance, little production
mP = little maintenance, a lot of production
Mp = a lot of maintenance, little production
MP = a combination of high maintenance and high production

An interesting finding is that a leader who is high in production behaviors and also does many maintenance behaviors is seen as providing "planning" or "expertise;" but the leader who does a lot of production behaviors and few maintenance behaviors is perceived as "pressuring for production." Pressure for production is resisted. In short, the same behavior (production) is perceived differently depending on the context within which it appears.

In different cultures the behaviors that express M can be quite different. Research has shown, for instance, that "to criticize a subordinate directly, privately in your office" is seen as high M in the West and low M in Japan. In Japan one is supposed to criticize indirectly—for instance, by asking a colleague of the subordinate to convey the manager's criticism to him/her—so that the subordinate will not lose face.

While both behaviors are important, depending on the situation and one's natural inclinations, individuals have personalities that incline them either toward M or toward P behaviors. A person for whom M behaviors are "totally unnatural" would seem to be "putting on a show" if he did a lot of M behaviors. For such a person, a change of the work environment is indicated to make it match the leadership style. Fiedler's leadership theory explains how to do this. Fiedler calls people who do a lot of M behaviors high LPCs (you will soon see why) and people who do a lot of P behaviors low LPCs. He has a way of finding out what your natural inclination is and then recommending specific changes to your work environment to make it compatible with your leadership style.

People who observe groups note that leaders may specialize in one of these activities or may sometimes engage in both; or in the case of "great leaders," they will perform both activities with great frequency.

First providing general theory and then focusing on R&D organizations, this chapter covers:

- Theories of leadership and leadership styles
- Leadership in R&D organizations
- R&D leadership—a process of mutual influence
- A leadership style case (where the problem of abdication style of leadership is presented)
- Leadership in a creative research environment

8.1 IDENTIFYING YOUR LEADERSHIP STYLE

In characterizing the behavior of their supervisors, subordinates used similar ideas—for example, bossy or structured versus people-oriented or considerate. Similarly, when leaders are questioned, some claim that they pay attention to people and others say they focus on the task.

As it turns out, however, the distinctions are not so clearly drawn. Extensive research by Fiedler [1967, 1986a] found that some people are task-motivated when they are relaxed but person-motivated when they are under stress, while others show the opposite pattern—that is, they are person-motivated when relaxed and task-motivated when under stress. It might be useful to find out for yourself what kind of leader you are. To do that, look at page 133. Follow Fiedler's instructions in *"Identifying Your Leadership Style"* [Fiedler et al., 1977].

You can score your own Least Preferred Co-Worker test. If your score was 64 or more on that test, Fiedler's evidence is that you are person-oriented under stress and task-oriented when relaxed. A score of 53 or less is evidence that you are task-oriented under stress and person-oriented when relaxed. If you got more than 64, you are a high LPC (least preferred co-worker); and if you got less than 53, you are a low LPC. If you scored between those two numbers, Fiedler's data do not have anything to tell you about your leadership style.

Fiedler argues that people are difficult to change, and that it is easier to change the situation in which people find themselves than to change the people. At least for routine, everyday behaviors that are normally under habit control, people act in ways over which they do not have much control. So, rather than change themselves they should try to change their leadership situation. Fiedler has provided us with ways to measure the situation. On pp. 136–139 you will find the Leader–Member Relations Scale, the Task Structure Rating Scale, and the Position Power Rating Scale. You can answer these scales and score them by following the instructions on the forms. Next comes the Situation Control Scale (p. 140). Follow the instructions and get your score. If your score is 51–70 you have high control; if it is 10–30 you have low control. Fiedler and others have done literally hundreds of studies linking LPC and the Situational Control Scale on the one hand, and group effectiveness (profits, high productivity, speed in getting the job done, accuracy) on the other hand. The findings from these studies fall into a pattern. It turns out that low LPCs do well in situations in which they have either high or low control, but do not do well in situations where they have intermediate control. On the other hand, high LPCs do well in situations where they have intermediate control. So, first find out how much control you have in your particular job situation.

Now you know your LPC score and your situational control score (based on your particular leadership situation, your team, group, department, or division). Do they match? That is, if you are high LPC, are your situational control scores in the 31–50 range; or if you are a low LPC, are they in either the 51–70 or the 10–30 range? If they match, you need not do anything. But if they do not match, Fiedler suggests making some changes. For example, if you want to increase your Leader–Member Relations you may make a special effort to communicate with your subordinates to

(text continues on p. 141)

Identifying Your Leadership Style*

Your performance as a leader depends primarily on the proper match between your leadership style and the control you have over your work situation. This section will help you identify your leadership style and the conditions in which you will be most effective. *Carefully read the following instructions and complete the Least Preferred Co-worker (LPC) Scale (page 135).*

INSTRUCTIONS

Throughout your life you have worked in many groups with a wide variety of different people—on your job, in social groups, in church organizations, in volunteer groups, on athletic teams, and in many other situations. Some of your co-workers may have been very easy to work with. Working with others may have been all but impossible.

Of all the people with whom you have ever worked, think of the one person now or at any time in the past with whom you could work *least well.* This individual is not necessarily the person you *liked* least well. Rather, think of the one person with whom you had the most difficulty getting a job done, the *one* individual with whom you could work *least well.* This person is called your *Least Preferred Co-worker* (LPC).

On the scale below, describe this person by placing an "X" in the appropriate space. The scale consists of pairs of words which are opposite in meaning, such as *Very Neat* and *Very Untidy.* Between each pair of words are eight spaces which form the following scale:

Very Neat	___	___	___	___	___	___	___	___	Very Untidy
	8	7	6	5	4	3	2	1	

Think of those eight spaces as steps which range from one extreme to the other. Thus, if you ordinarily think this least preferred co-worker is *quite neat,* write an "X" in the space marked 7, like this:

Very Neat	___	X	___	___	___	___	___	___	Very Untidy
	8	7	6	5	4	3	2	1	
	Very Neat	Quite Neat	Somewhat Neat	Slightly Neat	Slightly Untidy	Somewhat Untidy	Quite Untidy	Very Untidy	

* Material on pp. 133–140 is from Fiedler, F. E., M. Chemers, and L. Mahar, *Improving Leadership Effectiveness: The Leader-Match Concept.* Copyright 1977 John Wiley & Sons. Reprinted by permission of the author and publisher.

However, if you ordinarily think of this person as being only *slightly neat*, you would put your "X" in space 5. If you think of this person as being *very untidy* (not neat), you would put your "X" in space 1.

Sometimes the scale will run in the other direction, as shown below:

Frustrating	___	___	___	___	___	___	___	___	Helpful
	1	2	3	4	5	6	7	8	

Before you mark your "X," look at the words at both ends of the line. *There are no right or wrong answers.* Work rapidly; your first answer is likely to be the best. Do not omit any items, and mark each item only once. Ignore the scoring column for now.

Now go to the next page and describe the person with whom you can work least well. Then go on to page 136.

LEAST PREFERRED CO-WORKER (LPC) SCALE

										Scoring
Pleasant	8	7	6	5	4	3	2	1	Unpleasant	___
Friendly	8	7	6	5	4	3	2	1	Unfriendly	___
Rejecting	1	2	3	4	5	6	7	8	Accepting	___
Tense	1	2	3	4	5	6	7	8	Relaxed	___
Distant	1	2	3	4	5	6	7	8	Close	___
Cold	1	2	3	4	5	6	7	8	Warm	___
Supportive	8	7	6	5	4	3	2	1	Hostile	___
Boring	1	2	3	4	5	6	7	8	Interesting	___
Quarrelsome	1	2	3	4	5	6	7	8	Harmonious	___
Gloomy	1	2	3	4	5	6	7	8	Cheerful	___
Open	8	7	6	5	4	3	2	1	Guarded	___
Backbiting	1	2	3	4	5	6	7	8	Loyal	___
Untrustworthy	1	2	3	4	5	6	7	8	Trustworthy	___
Considerate	8	7	6	5	4	3	2	1	Inconsiderate	___
Nasty	1	2	3	4	5	6	7	8	Nice	___
Agreeable	8	7	6	5	4	3	2	1	Disagreeable	___
Insincere	1	2	3	4	5	6	7	8	Sincere	___
Kind	8	7	6	5	4	3	2	1	Unkind	___
									Total	________

LEADER–MEMBER RELATIONS SCALE

Circle the number which best represents your response to each item.

	strongly agree	agree	neither agree nor disagree	disagree	strongly disagree
1. The people I supervise have trouble getting along with each other.	1	2	3	4	5
2. My subordinates are reliable and trustworthy.	5	4	3	2	1
3. There seems to be a friendly atmosphere among the people I supervise.	5	4	3	2	1
4. My subordinates always cooperate with me in getting the job done.	5	4	3	2	1
5. There is friction between my subordinates and myself.	1	2	3	4	5
6. My subordinates give me a good deal of help and support in getting the job done.	5	4	3	2	1
7. The people I supervise work well together in getting the job done.	5	4	3	2	1
8. I have good relations with the people I supervise.	5	4	3	2	1

Total Score ☐

TASK STRUCTURE RATING SCALE—PART I

Circle the number in the appropriate column.	Usually True	Sometimes True	Seldom True
Is the Goal Clearly Stated or Known?			
1. Is there a blueprint, picture, model or detailed description available of the finished product or service?	2	1	0
2. Is there a person available to advise and give a description of the finished product or service, or how the job should be done?	2	1	0
Is There Only One Way to Accomplish the Task?			
3. Is there a step-by-step procedure, or a standard operating procedure which indicates in detail the process which is to be followed?	2	1	0
4. Is there a specific way to subdivide the task into separate parts or steps?	2	1	0
5. Are there some ways which are clearly recognized as better than others for performing this task?	2	1	0
Is There Only One Correct Answer or Solution?			
6. Is it obvious when the task is finished and the correct solution has been found?	2	1	0
7. Is there a book, manual, or job description which indicates the best solution or the best outcome for the task?	2	1	0
Is It Easy to Check Whether the Job Was Done Right?			
8. Is there a generally agreed understanding about the standards the particular product or service has to meet to be considered acceptable?	2	1	0
9. Is the evaluation of this task generally made on some quantitative basis?	2	1	0
10. Can the leader and the group find out how well the task has been accomplished in enough time to improve future performance?	2	1	0
Subtotal			

TASK STRUCTURE RATING SCALE—PART 2

Training and Experience Adjustment

NOTE: Do not adjust jobs with task structure scores of 6 or below.

(a) Compared to others in this or similar positions, how much *training* has the leader had?

3	2	1	0
No training at all	Very little training	A moderate amount of training	A great deal of training

(b) Compared to others in this or similar positions, how much *experience* has the leader had?

6	4	2	0
No experience at all	Very little experience	A moderate amount of experience	A great deal of experience

Add lines (a) and (b) of the training and experience adjustment, then *subtract* this from the subtotal given in Part 1.

Subtotal from Part 1.	
Subtract training and experience adjustment	–
Total Task Structure Score	

POSITION POWER RATING SCALE

Circle the number which best represents your answer.

1. Can the leader directly or by recommendation administer rewards and punishments to his subordinates?

2	1	0
Can act directly or can recommend with high effectiveness	Can recommend but with mixed results	No

2. Can the leader directly or by recommendation affect the promotion, demotion, hiring or firing of his subordinates?

2	1	0
Can act directly or can recommend with high effectiveness	Can recommend but with mixed results	No

3. Does the leader have the knowledge necessary to assign tasks to subordinates and instruct them in task completion?

2	1	0
Yes	Sometimes or in some aspects	No

4. Is it the leader's job to evaluate the performance of his subordinates?

2	1	0
Yes	Sometimes or in some aspects	No

5. Has the leader been given some official title of authority by the organization (e.g., foreman, department head, platoon leader)?

2	0
Yes	No

Total ______

SITUATIONAL CONTROL SCALE

Enter the total scores for the Leader–Member Relations dimension, the Task Structure scale, and the Position Power scale in the spaces below. Add the three scores together and compare your total with the ranges given in the table below to determine your overall situational control.

1. *Leader–Member Relations Total* ☐

2. *Task Structure Total* ☐

3. *Position Power Total* ☐

Grand Total ☐

Total Score	51–70	31–50	10–30
Amount of Situational Control	High Control	Moderate Control	Low Control

decrease the level of conflict among them, and to be accessible to them. If you want to increase your task structure, you may make a special effort to develop procedures for doing the job. If you want to increase the power, you might ask for more power from your supervisor. Similarly there are things to do to decrease structure (design the task so that subordinates can decide how to do the job) and power (let subordinates make more of the important decisions). You can train subordinates, rotate them into other jobs, and so on. The point is to do things to change your environment so it will match your leadership style.

The emphasis on Fiedler's leadership theory is based on the fact that Fiedler, more than any other researcher, has tested his theory with a variety of methods, in a variety of realistic settings. For example, Fiedler et al. [1984, 1987] reported on a study of several mines, where a particular management training program based on a theory [Fiedler et al., 1977] was compared with a widely used program to develop supervisory skills that used organizational development approaches requiring consultant expenses from $80,000 to $150,000. The management training program, which was estimated to cost between $4,000 and $10,000 at most sites, was more effective than the other methods in improving both productivity and the mine's safety record.

While Fiedler's is by far the best-researched theory of leadership, there are a number of other theories that should be noted.

8.2 THEORIES OF LEADERSHIP AND LEADERSHIP STYLES

No leader can afford to ignore M and P behaviors. Ideally, leaders should do a lot of both. However, there are other leadership theories that suggest that in some situations the leader should emphasize one or another even more than is usual.

Another way of looking at leadership is to say that the leader is supposed to supply what is necessary for the followers to reach their goals. This is called the *path–goal theory of leadership*. Basically, this theory argues that the way a leader acts should be determined by what the followers need. For example, if the followers do not know how to do the job, then it is necessary for the leader to be very structuring. If the followers have several needs that are not being met, then it is important for the leader to be especially considerate.

Consider another example. If the job is very monotonous, then the leader must provide some excitement, some change. Obviously, this is not necessary if the job already has variety. In one study, it was found that when the job had a lot of structure and people knew what they were supposed to do, considerate leaders were particularly effective.

There are a number of other factors that interact with the ones just mentioned. For example, if the task is very complex or requires creativity, then it is better to let the employees decide for themselves what to do, and, therefore, consideration is more important. If the abilities of the subordinates are very highly developed, then it is desirable to leave them alone. On the other hand, if they do not have much ability, then a certain amount of structure is appropriate. Several other factors, such as the needs for independence of the subordinates, their readiness to assume responsibility for decisions, their tolerance for ambiguity, their interest in the problem, and their feeling that the problem is important, make the less bossy supervisor more effective. When there is identity between the goals of the subordinates and the organization, when the subordinates have the skills and knowledge, and when their expectations are that they should participate in decisions, it is again important for the leader to emphasize consideration rather than structure.

The age of the relationship between a leader and subordinates will influence whether one or another leadership pattern may prove more effective. Hersey and Blanchard [1982] have argued that in the beginning of the relationship the leader is supposed to tell, later to sell, still later to use participation, and finally to use delegation.

If the leader must make a very important decision, his behavior will be under intentional control. He will have the time to think about what to do. This contrasts with the situation when the leader behaves under habit control. In the case where time is available, the leader can change his behavior according to a scheme developed by Vroom and Yetton [1973].

Consider the following different kinds of leadership styles:

1. *The directive style,* in which the leader simply makes the decision and tells the subordinates what to do.
2. *The negotiator style,* in which the subordinates give the information that the leader needs in order to make the decision, but then the leader makes the decision.
3. *The consultation style,* in which the leader asks for information and suggestions on what to do and makes the decision on the basis of these suggestions.
4. *The participative style,* in which the subordinates provide information and suggest solutions, the leader negotiates with them, and together they reach a mutually satisfying agreement and the best decision.
5. *The delegation style,* in which the leader provides information to the subordinates about the problem and suggests possible solutions. The responsibility for the decision is ultimately given to the subordinates. In this case, the leader does not even ask the subordinates to report what solutions were adopted.

Vroom and Yetton [1973, pp. 13 and 194] provide a decision tree that indicates when each of these five leadership styles is appropriate. It consists of a number of questions, and, depending on the answers to these questions, it recommends a particular leadership style.

There are seven questions that are arranged in a particular order. The first question is, "Are there quality requirements that one solution is likely to be more rational than another?" Depending on the answer (yes or no), one goes on to the second question: "Do I have sufficient information to make a quality decision?" Depending on the answer, one asks a third question: "Is the problem structured?" This process continues until the leader has been directed to the best leadership style.

It is useful to say a word about the difference between the approach of Fiedler and the approach of Vroom and Yetton. The Fiedler approach assumes that leaders have fixed personalities. If they discover that the conditions within which they operate are not consistent with their style, they "engineer the environment" to make it consistent with their style. By contrast, in the Vroom and Yetton approach, the individual uses different leadership style, depending on the situation. He or she may decide to delegate in one case or to be directive in another case. The leader's style is decided through an analysis of the situation and on the basis of the answers to specific questions.

Both approaches assume that there is no "best" leadership style. Leadership effectiveness depends on the situation. The Fiedler approach, in fact, is called the "contingency model," since it states that effective leadership behavior is contingent on the situation. Vroom's approach also is a contingency theory, but while Fiedler's is based on personality, Vroom's relies on logical analysis of the situation.

In some ways both the Fiedler and Vroom–Yetton viewpoints are correct. In Chapter 6 on motivation we discussed the importance of habits and behavioral intentions as determinants of behavior. When the job is well-learned, if there is an emergency or time pressure is high, habits are likely to be the major determinants of behavior. In situations when habits are all-important, Fiedler is likely to have the correct theory because he assumes that the behavior of the leader is fixed—that is, under the control of habits or a deeply ingrained personality. When behavioral intentions are the important determinants of behavior, then Vroom and Yetton are likely to give the best guidance. That is the time to use the decision tree and to teach oneself to use the correct decision-making style.

It seems likely that for everyday decisions and the sort of routine day-in–day-out behavior that is typical of leaders, Fiedler's point of view is more likely to be descriptive of the realities of leadership behavior. On the other hand, when the leader is just starting on a job or when the decision is very important and there is the time to think carefully about it, the Vroom and Yetton analysis can be helpful.

As we stated earlier, if habits are important, then the leader's experience is all-important too; when behavioral intentions are important, the leader's intelligence is all-important. Research shows that the leader's intelligence is not correlated with group effectiveness, or, to put it more accurately, the correlation is so low that it is not of practical significance. However, there is one condition when the correlation of leader intelligence and group effectiveness is high: when the leader is very dominant *and* the subordinates respect and admire him. Fiedler [1986a] reported correlations around 0.70 in that condition and around 0.10 in all other conditions. This finding is particularly relevant in R&D organizations, since the subordinates are likely to be very intelligent. If they have IQs around 130, the leader would have an IQ around 140 (something quite rare, since it occurs only among three people in a thousand) *and* have a record and personality that inspires respect and admiration to get away with being dominant. On the other hand, if the leader is participative, he/she can use the intelligence of the followers to increase the quality of the group's output.

In summary, while the dominant, structured behavior of the leader can be effective, this is only so under relatively rare conditions. On the other hand, consideration behavior is effective under a relatively wide range of conditions. This is even more likely to be true in R&D labs than in industrial settings, because the subordinates are highly intelligent, want to be autonomous, and often can do very good work when left alone.

Yet there are important additional jobs for the leader. As Bennis [1984] has suggested, the manager's four competencies are *attention* (making people attend to goals that serve the organization), *providing meaning* (using metaphors to communicate these goals), *creating trust* (being predictable, reliable, consistent), and *managing the self concept of subordinates* (making them feel significant, enjoy work, feel like a community or team). Even better is the leader who can inspire (is a model for subordinates) and who provides individualized consideration (gives personal attention to members who seem neglected), rewards frequently, and provides intellectual stimulation (enables subordinates to think of old problems in new ways) [Bass, 1985].

Some special problems that have occupied researchers in R&D labs will now be examined.

8.3 LEADERSHIP IN R&D ORGANIZATIONS

While P behaviors of the leader are needed, most leaders do P, but many do not do enough M. M behaviors are especially important in R&D labs. However, subordinates still require a certain amount of guidance from the manager; otherwise their activities will become unrelated to the needs of the organization. Pelz and Andrews [1966a,b] have shown that when there is either excessive or insufficient autonomy, the contributions of the professional to the research organization are minimal. An intermediate amount of autonomy provides optimal conditions for the professional. Only then can the contributions of the scientist to the organization be maximized.

Some R&D managers feel that administration is just paper-pushing and that the "real" work is technical. Thus, they miss the point that consideration is needed to develop the right kind of environment for subordinates. Also, some of these managers feel that "holding hands" (an aspect of *consideration*) is not consistent with their self-image; it is too soft or feminine an activity. Perhaps it would help such managers to know that research on psychological adjustment suggests that better-adjusted people have traits that traditionally were considered both masculine *and* feminine ones. That is, they are independent and self-reliant, but also warm, supportive, and nurturing. A person who has trouble relating to others will be better off restructuring the environment to make it compatible with his leadership style, as suggested by Fiedler, or he may find it is better to limit himself to technical work.

One way to paraphrase this is to say that good managerial policy requires "controlled freedom." This view is also consistent with the writings of Andrews and Farris [1967], Fisher [1980], and Smith [1970]. Some examples relevant to R&D organizations follow.

Research by Pelz and Andrews [1966b] also suggests that the most effective scientists in R&D laboratories are those who are allowed to do some basic research in addition to their applied research. It is frustrating for a scientist to come up with an idea that requires basic research and then not be allowed to pursue it because it is not obviously linked to the needs of the organization. A manager who protects his subordinates from this kind of frustration is a good manager.

A good manager also makes sure that his subordinates do not become overspecialized. One of the problems in many laboratories is that some people become so specialized that when their specialization becomes obsolete, so do they. Further research by Pelz and Andrews [1966b] suggests that the effectiveness of a scientist increases with the number of demonstrated areas of specialization.

According to Pelz and Andrews, the scientist who spends about 50% of his or her time in research and 50% doing other things often is more effective than the one who spends 100% of the time on research. The manager who is sensitive to these issues and makes his assignments so that they take into account this fact is likely to be more effective.

A large dose of delegation is essential in the case of research scientists. Praise, recognition, and feedback are also extremely important. The manager who rewards, praises, and recognizes good work is more effective than the supervisor who simply grins when he sees good work, but says very little. On the other hand, the good manager should be able to identify incompetent work and to make sure that it is not rewarded. A good manager encourages subordinates to take sabbaticals, to develop and apply new skills, and to set difficult but achievable goals. Goal achievements are reexamined every 6 months or so, and rewards are given.

Another problem that is unique to many research and development organizations is that people often have two bosses. Usually, there is both a functional supervisor (who is a specialist in the particular field that the scientist has been trained in) and a project supervisor (who focuses on a particular problem that has to be solved). Classic organization theory warns against arrangements in which there are two supervisors, but this arrangement can be made to work.

In such cases, the effectiveness of the scientists often depends on the balance between the influence of the two supervisors [Katz and Allen, 1985]. The best performance in the studies by Katz and Allen occurred when the project manager was mostly concerned with relating the project to the outside world (i.e., the suppliers, the customers, the organization), while the functional manager did most of the inside work. These authors say "project performance appears to be higher when project managers are seen as having greater organizational influence." This is an outward orientation, and, as a result, they should be concerned with gaining resources and recognition for the project, linking it to other parts of the business, and ensuring that the project's direction fits the overall business plan of the organization. According to Katz and Allen, functional managers, on the other hand, should be concerned with technical excellence and integrity—that is, seeing that the project is scientifically sound and includes state-of-the-art technology. Their orientation is inward and focuses on the technical content of the project. The technical decisions should be made by those who are closest to the science and technology.

The location of technical decision-making in functional departments, however, implies important integrating roles for project managers, who are responsible for ensuring that the technical directions overseen by several different functional managers all fit together to yield the best possible end result. Clearly, the greater the influence of project managers on the organization, the easier it is to integrate and negotiate with the various functional managers, whose technical goals are often in conflict.

From this study one might conclude that project managers should have more organizational experience and more status than functional managers. A good tactic might be to deliberately place highly competent young professionals in the role of functional manager to supervise the technical aspects of the work, while having the older ones act as project managers.

A study of 66 industrial R&D project groups found transformational leadership to account for higher project quality in research projects [Keller, 1995]. What is a transformational leader? According to Keller, a transformational leader strives to achieve results beyond what is normally expected by (a) inspiring a sense of importance about the project group's mission by stimulating professional employees to think about the problem or task in new ways and (b) emphasizing group goals. In development projects, however, a more directive style of leadership was responsible for higher project quality.

8.4 R&D LEADERSHIP: A PROCESS OF MUTUAL INFLUENCE

Based upon a study of R&D organizations, Farris [1982, p. 344] states:

> In the high innovation groups the supervisors were more active participants in the informal organization. They were especially helpful to group members for critical evaluation, administrative aid, and help in thinking about technical problems. In addition, group members were more helpful to these supervisors for providing technical information, aid in thinking about technical problems, critical evaluation, and original ideas.

Leadership in an R&D organization is essentially a process of mutual influence between the supervisor and the employees. Based on this approach, Farris suggests four styles of leadership or supervision [1982, p. 344]:

Collaboration: Both the supervisor and the employees have a great deal of influence in making decisions.

Delegation: The employees are given considerable responsibility for the decisions, and the supervisor has little influence.

Domination: The supervisor has a great deal of influence, and the employees have very little input.

Abdication: The supervisor neglects to assign a particular task to the employees and neglects to work on it himself. In this case, neither the supervisor nor the employees have much influence on a particular decision.

Studies clearly indicate that the collaborative style, both in terms of setting schedules and in terms of informal organization, is most conducive to higher performance in innovation. There are, however, situations in which other styles of leadership and supervision have to be used. For example, when there are time constraints in making a decision, there may not be sufficient time to seek extensive input from the employees. Consequently, the collaborative style is not possible. Depending on the situation, domination or delegation would be preferred alternatives. There are also situations in which an R&D manager lacks the competence or ability to provide the necessary leadership. In these cases, it is very common to use extensive delegation or abdication.

Abdication takes place more often than most organizations would like to admit. It is likely to happen when supervisors who are either technically incompetent or lacking in intellectual abilities find themselves "surrounded" by more competent people, above (higher-level managers) and below (lower-level researchers) them. In hierarchical bureaucratic settings, situations can easily arise in which individuals who are technically incompetent and unable to provide effective leadership can still occupy important managerial positions in R&D organizations. This may also occur as a result of organizational growth or change.

In summary, studies by Farris [1982, pp. 345–346] show that collaboration, not delegation, is likely to produce the most successful innovations. The supervisors of low-performing innovation teams frequently think they are delegating when in fact they are abdicating.

8.5 A LEADERSHIP-STYLE CASE

In governmental and large nongovernmental (industrial or academic) research organizations that are hierarchical in nature, lack of effective leadership characterized by abdication exists quite frequently. How does the organization cope with this?

Let us take the actual case of a research laboratory,* where a research division director, Mr. Lewis, manages seven departments with a total research staff of 150. Mr. Lewis has no research training: His educational background is limited to an undergraduate engineering degree, he has made no attempt to continue his technical training, and he has not been active in scientific and professional organizations. The personnel records show that he is considered one of the high performers; and any time there is organization growth or change, he acquires some new functions. And when there is organization contraction, he maintains his position while technically more competent division directors are sent back to the bench. Analyzing this case, let us review these questions:

- What is the leadership style and behavior pattern of the division director?
- How is the organization performing?

*The situation is real, but all the names and facts in this case are fictional.

- What clues does one have that there is a substantial leadership problem?
- How does the organization cope with it?

Leadership Style

The leadership style is a combination of abdication and delegation. All technical responsibility is delegated to the department heads, integration at the division level is minimal, and responsibility for decisions normally made at the division level is abdicated and passed on to a higher level—the laboratory director, Ms. Himler. Her views are sought and essentially all decisions normally made by the division director are in fact made by the laboratory director. The behavior pattern of the division director is characterized by his readily admitting lack of technical competence, building strong alliances with selected research divisions, degrading division directors who may have distinguished scientific records, doing what the laboratory director, Ms. Himler, wants him to do, getting marching orders on all division operations from Ms. Himler, and taking zero risk.

Organization Performance

The organization performance is quite adequate. Increasingly, emphasis is on technical assistance instead of research. Innovative research programs never reach fruition; they are downgraded to technical assistance activities.

Leadership Problems

Clues to the existence of the problem manifest themselves via the leadership style and the behavior pattern of Mr. Lewis (the division director) and via lack of substantial innovation by the division over the years.

Organization Response

An organization can cope with such problems in a number of ways. Before discussing that, it might be useful to see how such a situation arose.

Mr. Lewis started managing a group of three to five technicians providing laboratory analysis of building materials. As the parent organization grew, the research needs of the organization grew substantially. Initially, a small research and technical assistance group was formed under Mr. Lewis. Later, a separate research laboratory with three divisions was created. A new geographic location was selected for the laboratory, and a proper research facility was built for the purpose. Because of Mr. Lewis' pleasant and nonthreatening personality, he continually assumed increasingly more important managerial responsibility. During organization contraction, Mr. Lewis maintained his higher managerial position because of his inability to make a meaningful contribution at the lower or research performance levels.

Now, how best to cope with the situation?

Discharging or demoting Mr. Lewis may seem like one of the choices. Often, in hierarchical bureaucratic organizations, this action would cause more trouble than it is worth. The personnel process is often cumbersome and time-consuming; and even if management is successful, Mr. Lewis could end up at a lower, research-execution level, where he is clearly unable to perform. Mr. Lewis never claimed to be technically competent and never falsified or exaggerated his record of accomplishments. He is a dedicated, hard-working employee whose performance, considering his lack of technical and intellectual background, has been adequate. It would seem the problem lies less with Mr. Lewis than with "the system." Consequently, any adverse action against him would seem unfair.

Other choices might be assigning a high-level, technically competent assistant to Mr. Lewis to carry out the necessary integration of effort; assigning technical assistance activities to Mr. Lewis' division instead of innovative research programs; or assigning Mr. Lewis to another part of the organization where his pay level is preserved and where his long association with research could benefit the organization in terms of transfer and research liaison activities.

8.6 LEADERSHIP IN A CREATIVE RESEARCH ENVIRONMENT

In an R&D organization, a person holding an important leadership position would normally have a significant research program. In many U.S. government departments and in industry, some individuals have oversight responsibility for research organizations, although they are not involved in research program execution. It is therefore useful to focus on those leadership and managerial aspects that are directly involved in managing and executing an important research program involving a significant number (say 50 or more) of scientists and engineers.

Mintzberg [1975, p. 61] suggests a number of leadership or managerial skills that are important. These are: developing peer relationships, conducting negotiations, motivating employees, resolving conflicts, obtaining and disseminating information, making decisions in conditions of extreme ambiguity, and allocating resources. In an R&D organization, some additional activities are important, such as establishing information networks in order to relate to the wider scientific community and to attract and recruit highly qualified personnel.

Persons who reach the position of managing a significant research program are precluded from thinking deeply or broadly about anything because of time constraints. In most R&D managerial positions, managers are not able to do any serious, original, or conceptual work. Most people in these positions work long hours and are very busy. They must deal with a number of constituencies, which include such typical individuals or groups as the director of the research organization, research sponsors (who normally provide funding for the research effort), the user community (who might use the research effort and have some effect on research funding), researchers within the group, peers in the scientific community, and, last but not least, the nonscientific bureaucracy (this may include the comptroller, personnel office, and contracting office). Any experienced R&D manager knows that difficulties are likely to arise if any one of these constituencies is ignored. Effectively executing the research program and meeting demands placed by individuals in these constituencies can keep an R&D manager quite busy indeed. A quotation attributed to Thoreau states: "It is not enough to be busy . . . the question is, what are we busy about?"

Thus the question is, What is a manager to do? It is important to realize that a manager in an important leadership position must continue to meet the day-to-day responsibilities placed on him or her by the job. In addition, he or she must focus on the long-term requirements of the organization and must consider strategic and policy implications for the research group. In terms of individual performance and organization effectiveness discussed earlier, the manager has to look at issues not just in terms of process and result indicators, but also in terms of strategic indicators. Focusing on strategic issues requires creative thinking and serious reflection because it has the added problem of inherent uncertainties, risks, and delayed gratification (if any). This focus, perhaps more than anything else, provides for the long-term productivity of an R&D organization and for the excellence for which a leader may want to strive. Being busy doing day-to-day things simply is not enough. An effective R&D manager has to integrate the efforts of others, provide foresight for strategic issues, and, at the same time, make technical contributions in the area of his/her specialty.

8.7 SUMMARY

For everyday decisions and actions, your personality is likely to determine how you act. To make sure that your actions will be effective, you may want to change your work environment so that it matches your personality [Fiedler et al., 1977].

If you are a high LPC (see Section 8.1), avoid environments that are too easy (you have much control) or too difficult (you have little control) for you as a leader. In other words, foster an environment in which you have moderate control and influence, which is common in R&D organizations. Then, your leadership style should be participative or relationship-oriented.

If you are a low LPC, your ideal, then, is to be liked, to have no conflict among your subordinates, to have a clear task, and to have a lot of power. If, on the other hand, you have no power and your subordinates hate you, and you do not know what they are supposed to be doing, do not panic. To get high performance, you will have to rely on task orientation and a directive form of leadership.

Since you are dealing with rather bright and autonomous people, task structure is going to be low; since you want to minimize stress, you want the group to like you; since your power and influence in an effective R&D organization are going to be moderate, you really need to work toward a relationship-oriented management style.

For important decisions, when you have a lot of time, ask the questions Vroom and Yetton have recommended, go through the decision tree they have developed, and use the leadership style that is recommended.

If other factors are critical, the above recommendations need to be taken with a grain of salt. You should emphasize participation if the following hold true:

- You have egalitarian values.
- You respect your subordinates' skills.
- You know too little about their jobs.
- You can live with uncertainty.
- You feel it is very important that your subordinates like you and that they get along with each other and know much about the job.
- The job is interesting and requires many solutions and high quality.
- There is no crisis, the job does not involve conflicts of interest and permits interpersonal interactions, and you work in an environment where things are changing fast.

If these factors do not hold, you might shift toward more directive management styles.

When you have a new employee, you can get away with a style that involves telling people what to do; with a more mature subordinate, selling is better; with a still more mature subordinate, participation is highly desirable; and with a subordinate who knows a lot and has been around a long time, delegation may be ideal [Hersey and Blanchard, 1982].

Look at your subordinates' job situation. What is missing from the ideal work environment? Suppose they do not know what they are to do; then tell them. Suppose they are bored; then entertain them. In other words, your job is to help them reach their goals, to supply the missing resources (path–goal theory).

There are other factors that moderate what was just said: You can be more bossy if you have deadlines or limited resources, or if you are expected to be bossy (expectation can come from the culture, the organization, your supervisor, your subordinates, your peers), or if your followers are incompetent or inexperienced.

Finally, there are some unique situations related to R&D organizations. Thus, leadership concepts and styles need to be explored in the context of R&D organizations.

8.8 QUESTIONS FOR CLASS DISCUSSION

1 Under what conditions is Fiedler's theory of leadership likely to predict high effectiveness in an R&D lab?

2 Under what conditions should Vroom's theory be used in an R&D lab?

8.9 FURTHER READINGS

Campbell, D. (1993). Good leaders are credible leaders. *Research-Technology Management*, **36**(5), 29–30 (September–October).

Clement, R. W. (1993). Teaching leadership theory: reconsidering the situational view. *Management Research News,* **16**(2 and 3), 12–18.

Fiedler, F. E. (1967). *A Theory of Leadership Effectiveness.* New York: McGraw-Hill.

Fiedler, F. E. (1986a). The contributions of cognitive resources and leader behavior to organizational performance. *Journal of Applied Social Psychology,* **16,** 532–548.

Green, S. G. (1995). Top management support of R&D projects: a strategic leadership perspective. *IEEE Transaction on Engineering Management,* **42**(3), 223–232.

Vroom, V. and P. W. Yetton (1973). *Leadership and Decision Making.* Pittsburgh: University of Pittsburgh Press.

9 MANAGING CONFLICT IN R&D ORGANIZATIONS

There are three kinds of conflict that we need to discuss in this chapter: intrapersonal conflict, interpersonal conflict, and intergroup conflict. The first occurs within the individual, the second between individuals, and the third between groups.

9.1 CONFLICT WITHIN INDIVIDUALS

There are many kinds of conflict that occur within individuals. The first one that will be discussed is *role conflict*. Roles are ideas about correct behavior for a person holding a position in a social system. For example, the position of chief engineer specifies particular activities that are appropriate for the position. When analyzing roles, it is important to talk about *prescribed, subjective,* and *enacted* roles. A prescribed role is a role that is prescribed by other people. In other words, the chief engineer usually receives definitions of what he is supposed to do from his boss, from his subordinates, and from his peers, and each has specific ideas about the engineer's role, which are integrated into a concept of what he is supposed to be doing. In other words, when the chief engineer says, "I am doing this because I am the chief engineer," that is an element of the *subjective role*. Finally, we have the *enacted role,* which is the actual role behavior of the chief engineer.

It is useful to look at the enacted role and see if it corresponds to the subjective role or to the prescribed role. According to research, the three kinds of roles—the prescribed, the subjective, and the enacted—frequently do not match very well. In other words, what is prescribed may actually not be similar either to the subjective role or to the enacted role.

In one of the situations of role conflict, the various prescribed roles are usually quite different. That is, the person receives prescribed roles from a variety of individuals, and these role senders disagree among themselves about what that particular individual's role should be. As a result, the *subjective role* the person develops is confused or contains elements that are in conflict because the boss says one thing, a subordinate says a different thing, and a colleague says yet another. Other situations of role conflict occur when a person does not develop a subjective role that resembles the prescribed role or when the person developing an enacted role fails to match the prescribed one.

Research done by Kahn, and summarized in Katz and Kahn [1980], indicates that about 50% of the people studied in various organizations have experienced a great deal of role conflict, much of it due to conflict with the hierarchy. That is, the person's definition of what he is supposed to be doing is different from the definition that his boss is sending or that the top manager is sending to the boss.

A second kind of role conflict is related to *workload*—in other words, how much is one supposed to do. Given a particular role, there are different definitions of how much one should do.

A third conflict has to do with *creativity*. Who is supposed to initiate what or who is supposed to do new things does vary according to role senders.

Finally, there are conflicts that have to do with *organizational boundaries* and who has responsibility for what activity—for example, who must decide whether a laboratory member is to go to a conference.

The research by Kahn shows that the greater the role conflict: (1) the greater the dissatisfaction of the individual, (2) the more frequent the physical symptoms of the individual, (3) the greater the number of hospital visits the individual undertakes, and (4) the less confidence the individual has in the organization.

A good example of role conflict in research and development organizations is the conflict that occurs when the person is part of a team that is developing a new product that involves both research people and marketing people. Depending on the structure of such teams, a person may experience varying degrees of conflict. In the chapter on the design of jobs, we discussed a very interesting project by Souder and Chakrabarti [1980] in which they identified three kinds of relationships between a research and development team and a marketing team. They were the *stage-dominant,* the *process-dominant,* and the *task-dominant* structures within the organization. We identified the conditions under which the various forms of organization might be more or less effective. The greatest role conflict is apt to occur in the task-dominant form of organization, since it is in that particular structure that the individual is *both* a marketing person and a scientist. In the stage-dominant structure, there will be a minimum amount of conflict, since in that form of organization there is a very clear separation between the scientist and the marketing team. In the process-dominant form there will be an intermediate amount of conflict.

Technicians versus Researchers

Other kinds of intrapersonal conflict occur when certain technical employees have problems with the way they are perceived by members of the organization. A good example is provided by Fineman [1980], who discusses the problem of technicians in large R&D organizations. They are often in a supportive role; in other words, they are supposed to be helping the researcher do the work. This frequently makes them feel like second-class citizens who are being "used" by the researchers as servants rather than as co-workers. Furthermore, their job appears to lack creativity, since it is the researcher who does all the original work and they are only providing the technical support. Naturally, such people often feel that their technical skills and qualifications are underutilized and that their superiors do not take their personal needs into account. In R&D organizations, quite often, support personnel experience helplessness and lack of power and influence. Managers must find ways to integrate support staff by providing common goals for them and for the researchers.

Furthermore, in research organizations people are likely to get promoted primarily on the basis of technical excellence rather than managerial skills. As a result, the managers tend to be technically competent but rather poor administrators. Managers who focus on technical problems often do not take the personal needs of their subordinates very seriously, and, as a result, their subordinates are upset and unhappy and experience a good deal of conflict about staying in the organization.

Fineman [1980] provides a number of suggestions that may reduce some of these problems. For example, organizations might give higher-sounding titles to the support personnel and develop promotion policies that allow them to feel better integrated into the organization. In addition, they can enrich their jobs by doing a greater variety of activities. Organizations may also find it useful to pay more attention to the training of their R&D management personnel. Finally, selecting technicians and support personnel who are "thick-skinned" enough to put up with poor managers might be a good strategy for this kind of situation.

Supervisor–Subordinate Expectations

A frequent problem in most organizations is that the expectations of one's supervisor and of one's subordinates may be quite different. This problem becomes especially difficult to solve when the training backgrounds of the supervisor and the subordinates are very different. For example, in some commercial organizations the top management has M.B.A. training or degrees in law or accounting. Managers of the R&D functions may report to an M.B.A. while their subordinates might be physicists or engineers. The expectations of people with such varied kinds of training can be very different. As a result, the managers find that their supervisor expects a particular set of behaviors while their subordinates expect a different set with minimal overlap between them. Such "role conflicts" have been found to result in health problems (e.g., ulcers), job dissatisfaction, and even depression.

It is important for people who find themselves in role conflict situations to first identify that they are facing a role conflict, and second to bring the relevant parties to a conference to "negotiate" the kind of role that they should have. Generally, when such a problem is identified and discussed, solutions can be found. One technique that is especially helpful is to discuss with co-workers what they expect the manager to do "more of" and "less of." For example, subordinates may indicate that they want to be evaluated more frequently, but be given explicit directions less often about how to do their jobs. It is through such discussions that roles can be clarified, negotiated, and agreed on. Role conflict can, in principle, be eliminated if reasonable people are allowed to discuss the problem and to seek constructive solutions.

Engineers' Status and Organizational Conflicts

An analysis of the kinds of stresses that professional engineers face is provided by Keenan [1980]. Keenan also identifies, as a problem, the fact that professional engineers have a relatively low status in society (mostly the case in the United States, not so in Japan and Germany) despite their academic level of qualifications and their level of contributions to society. He has reviewed studies that found that a substantial percentage of engineers feel they are not sufficiently high in status and are dissatisfied and frustrated by this. A number of scholars have pointed out that scientists and engineers who work in industrial organizations are likely to experience strains due to the conflict between their professional values and the goals of the organization for which they work. Conflicts between the technologist and the organization over issues such as which project to focus on and how and in what way to do them can drain the engineer's energies.

An important basis for these conflicts is the fact that technologists generally desire to be involved in projects based on their technical and scientific merit, whereas the primary consideration of the organization is product marketability. Keenan further summarizes a number of studies that show that there is a good deal of stress among technologists and scientists in various organizations. Among the complaints of engineers and scientists is that they do not have enough job autonomy and that they lack the opportunity to use their research skills. This dissatisfaction is particularly high among the younger engineers and scientists. There is also some evidence that those organizations that provide freedom to do research, promote personnel on the basis of technical competence, and allow individuals to attend scientific meetings to improve their professional knowledge and skills generally have scientists and engineers who are less dissatisfied than those in organizations that do not have such policies.

Role Ambiguity

In addition to role conflicts, some of the scientists and engineers experience *role ambiguity*—that is, uncertainty about the meaning of communications received from a variety of important "others" in the organization. About 60% of the engineers and scientists in one study experienced role ambiguity. In other words, they did not really know what the boss wanted.

Role Overload and Underload

In some cases, there is *role overload;* that is, the work that needs to be done is too difficult and exceeds the individual's abilities, skills, or experience. In a study summarized in Keenan's [1980] paper, French and Caplan [1973] found that engineers and scientists more frequently experienced situations in which the job was too difficult than did administrators. Another problem is role *underload;* the demands made by the job are insufficient to make full use of the skills and abilities of the scientist. The Keenan paper suggests that this is a frequent problem among engineers. Engineers receive sophisticated training (e.g., in mathematics) that results in skills often not required by their job. In one study, more than half of the engineers complained that many aspects of their jobs could be handled by someone with less training.

Boundary Role

Another source of stress or interpersonal conflict comes from occupying a *boundary role,* one that connects the organization with the external environment. There is some evidence that engineers who are in such roles experience more stress and strain than other engineers. Individuals in boundary roles frequently complain that they experience greater deadline pressure, fewer opportunities to do the work they prefer, and less opportunity for advancement. They also claim that they are not attaining the maximum utilization of their professional skills.

Coping with Conflict and Stress

The ways engineers cope with work-related stress is discussed by Newton and Keenan [1985], who point out that there are different ways in which one can cope. For example, one can talk with others, take direct action, withdraw from the situation, or simply resent it. Exactly what is done depends on (1) individual differences (for example, people who are characterized as having a Type A personality are most likely to be resentful) and (2) situational variables (for example, withdrawal or doing as little as possible occurs more frequently among those who work in organizations that lack a supportive climate). Withdrawal appears to be more common in some fields of engineering than in others. Also, the way the person looks at the stressful situation determines whether the person will talk to others or take action, such as quitting. One cannot generalize and say that there is an effective coping technique that should be taught to everyone, because coping differs from person to person and from organization to organization. It also depends on the way the person perceives the conflict situation. Nevertheless, in training engineers and scientists, we can sensitize them to intraperson conflict and teach them stress-reduction techniques (such as biofeedback). Often being able to understand that role conflict and role ambiguity are "normal" in organizations makes dealing with such conflict more manageable. Facing the conflict squarely by "negotiating" one's role is most helpful.

It is important to realize that one of the best ways to reduce role ambiguity and role conflict is to use participative management. In participative management situations, employees determine what they are to do, when they are going to do it, and how they are going to do it. When such factors are decided either by the supervisor or by the job itself (i.e., by "external" determinants), there is more role conflict and role ambiguity [Jackson and Schuler, 1985]. In this study, which analyzed 29 correlates of role ambiguity and conflict, it was found that across a large number of empirical investigations the best correlates of low conflict were participation and feedback. In other words, when the employee sets the task cooperatively with the supervisor and the supervisor (or the task) provides feedback to the employee, there is minimal role conflict. Incidentally, the same study showed that when there is conflict there is tension, dissatisfaction, and low self-ratings of performance.

9.2 CONFLICT BETWEEN INDIVIDUALS

Chan [1981] has studied conflict between R&D managers and nonmanagers in four organizations, and he found that they perceived conflict as generally having negative consequences. Most conflict occurs in the areas of reward structure (most important), control of goals, authority, and insufficient assistance. Most respondents saw a negative link between conflict and performance and job satisfaction, but a few respondents saw conflict as having positive consequences such as increased performance. Reactions to conflict were perceived as quite different. Competition and avoidance reactions were seen as most detrimental to the effectiveness of the work group; cooperation was seen as the most desirable reaction to conflict.

In general the ideal way to deal with conflict is to be creative and try to reach win–win solutions. For example, if authorship of a paper is a disputed issue, arranging for one of the persons to do extra work on it, in order to justify joint authorship, can result in a win–win situation.

Evan [1965a,b] has developed a typology of interpersonal conflict in organizations. Three types of conflict in two distinct areas can be defined. There can be conflict with peers, with supervisors, or with subordinates, and this conflict can occur in the technical area and in the interpersonal area. Conflict with peers, supervisors, or subordinates in the technical area involves technical goals, milestones, the means of reaching a particular goal, and interpretation of data. Conflict with peers or subordinates at the interpersonal level involves personal likes and dislikes, trust, and fear that the other person misperceives what one is doing. The conflict with the supervisor usually deals with project administration or with power relationships. This includes conflict over who is supposed to decide what to do or what rules or procedures are in effect.

Research shows that conflict is more likely to occur in those situations in which two individuals have different attitudes and values. In one such situation, for instance, one person believes it is very important to keep a certain distance between the supervisor and the subordinate, while the other person may not think that a large distance is appropriate. Similarly, one person may require an exact clarification of rules or specifications of what is to be done, while the other person does not feel that such clear statements are necessary. Conflict can also arise if one person believes that people should be independent, while the other person thinks that there should be greater interdependence and coordination in activities, and that the most important thing is to have good interpersonal relationships.

Such conflict is more acute when one or both of the individuals are cognitively simple and tend to see things in black and white, in stereotypes, or in a very simple manner. Generally, the conflict is less important when the people are cognitively complex. Some conflict can be traced to incompatible personalities. It is beyond the scope of this book to discuss this type of conflict, but when it occurs, counseling or the use of a clinical psychologist ought to be investigated by management.

Conflict in general is more difficult to reduce when there are major discrepancies of power. If one person can totally dominate the other, it is possible for the lower-status person to take the view that he has nothing to lose if he makes a tremendous mess of the relationship. When the relationship has a reasonable balance of power (in other words, the subordinate has some power) the relationship is likely to allow reductions of conflict. It is also obvious that when two people have a history of bad relationships with others and with each other, it is much more difficult to improve these relationships in the future. One of the ways to get around poor relationships is to put the two people into a situation in which they have what is known as a *superordinate goal*—that is, a goal that both of them want to attain and that neither can reach without the help of the other. Creating opportunities for communication, getting help from professional counselors, and organizational restructuring should all be considered as alternative options to reduce this type of conflict.

9.3 CONFLICT BETWEEN GROUPS

Conflict between groups is very common in organizations. In what follows, we will summarize some of the major findings in social psychology concerning the study of intergroup relationships [Worchel and Austin, 1985].

In-Groups, Out-Groups

The first point is that it is very easy to create confrontations between *in-groups* and *out-groups*. An in-group is one with which the individual is ready to cooperate and whose members consist of individuals who trust each other. An out-group consists of people one distrusts.

It is very easy to create in-group/out-group distinctions. For example, in a laboratory experiment, one can take teenagers and say to them "You belong to the yellow group" and the others constitute "the red group." With no other visible distinction, one says: "All right, you yellows, here is a pile of money. Divide the money between your group and the other group." This simple manipulation is sufficient to make the individuals who are doing the dividing favor their in-group. For instance, they may give 60% of the money to the in-group and 40% to the out-group. It is as if there were a natural way of thinking that "since I belong to this group and the other group is my 'enemy,' it is natural for me to give more to my group and to be a little distrustful of the other group." The research also shows that out-groups are perceived as more homogeneous than in-groups. In other words, the "other" people are "all the same." By contrast, in-groups are perceived as relatively heterogeneous. The members of one's in-group are perceived as "all different" from one another. These tendencies imply that we stereotype members of out-groups and may perceive them more inaccurately than we perceive the in-group.

It is useful to distinguish relationships that are *intergroup* from those that are *interpersonal.* In an interpersonal relationship, the individual is very much aware of who the other is. In the intergroup relationship, the individual is not aware of the other's personal characteristics. For example, when soldiers shoot at the enemy they do not care who that particular individual is. It is just a global reaction or judgment about the other person as a representative of a group. Intergroup relationships are more likely to develop than interpersonal relationships under the following five conditions: (1) when there is intense conflict, (2) when there is a history of conflicts, (3) when there is a strong attachment to the in-group, (4) when there is anonymity of membership in the out-group, and (5) when there is no possibility of moving from the in-group to the out-group.

Let us examine these conditions with an example from the relationship between researchers and marketing specialists. If the marketing specialists look at the world in a different way from the way the researchers do, then it is far more likely that the researchers will say that the marketing people are "all the same." On the other hand, if there is less conflict they may see differences between various members of the marketing group. Second, if there is a history of conflict between the two groups, then they are much more likely to look at each other in terms of their group rather than as individuals. Also, if the researchers feel very strongly about being researchers or the marketing types feel very strongly about being marketing types, they will perceive the people in the other group as undifferentiated. Anonymity means that you do not really know who the other people are. A marketing committee will make a decision and say "No!" or will send a letter to the research people that says "we have decided not to support your proposal." There is no indication of who the people are that made the decision, and this increases the tendency to perceive them as all alike, as a group, and not as individuals. Finally, perception also tends to be intergroup when there is no possibility of moving from one group to another, as happens when the organizational structure is such that engineers and scientists never work in marketing or market department members do not work in research.

The interesting thing is that in-group favoritism occurs even when these five conditions do *not* operate! In other words, intergroup favoritism is such a common and fundamental idea (given that you are in my group I *must* favor you) that people are not critical of their own actions. In order to show favoritism, people do not need tension or conflict, a history of conflict, strong attachment to their group, anonymity of the out-group, or the inability to move to the other group.

The fact that the out-group is seen as a homogeneous entity means that stereotypes increase. Stereotypes are overdetermined because they occur for at least two reasons: (1) It is easier (requires less cognitive work) to *see* others as being more or less alike, and (2) it is so much simpler to *deal* with others as if they were alike. Studies have shown that people are more likely to generalize in the negative direction (toward criticism) from the behavior of one individual who is a member of an out-group to the whole out-group than to do so in the case of the in-group. In our example, our researchers are much more likely to stereotype and evaluate unfavorably the whole marketing department on the basis of the behavior of one of its members than they are likely to change their view of the research department on the basis of the behavior of one researcher.

Biased Information-Processing

An interesting example of biased information processing is that people attribute positive actions by the in-group to internal aspects (e.g., they are honest), but positive actions by out-group members are considered due to external aspects (e.g., they were forced to act that way). In other words, suppose the marketing people unexpectedly did something very nice for the researchers. The researchers would claim that the marketing department members were forced to act in this way by outside circumstances. On the other hand, if they did something nasty, it would be explained as being "their nature"; that is, people make dispositional attributions (they were nasty people) when negative behavior of out-groups is perceived. Conversely, if the researchers (the in-group) did something nice, it would be explained as "their nature," but if they did something nasty, they would be perceived to have acted that way as a result of external circumstances.

Coping with Conflict between Groups

What we said in the case of interpersonal conflict also applies to intergroup conflict: If superordinate goals (goals of *both* groups that neither group can reach without the help of the other) can be found, the relationship can be improved.

There are two orientations that one can adopt in an intergroup situation: One is called a *win–lose* orientation and the other is called a *win–win* orientation. In the win–lose orientation, one tries to win for one's in-group something that the out-group loses, while in the win–win orientation, one tries to win something for both groups. Another way to look at conflict is to examine the Conflict Resolution Grid of Blake and Mouton [1986, p. 76]. The win–win orientation corresponds to position 9.9. The win–lose orientations are 1.9 and 9.1. Two other orientations—compromise and all lose, both less satisfactory than the win–win—are also shown in Figure 9.1.

For example, suppose the researchers have a design that satisfies many technical and production criteria, but which the marketing people find almost impossible to market. In a win–lose orientation either the technical people manage to impose the design or the marketing people manage to eliminate it from further consideration. In a win–win orientation a new design is developed that has the advantages visualized by the research group, but also incorporates the advantages of the ideal marketing design. Other positions shown in the diagram would be (a) the lose–lose posi-

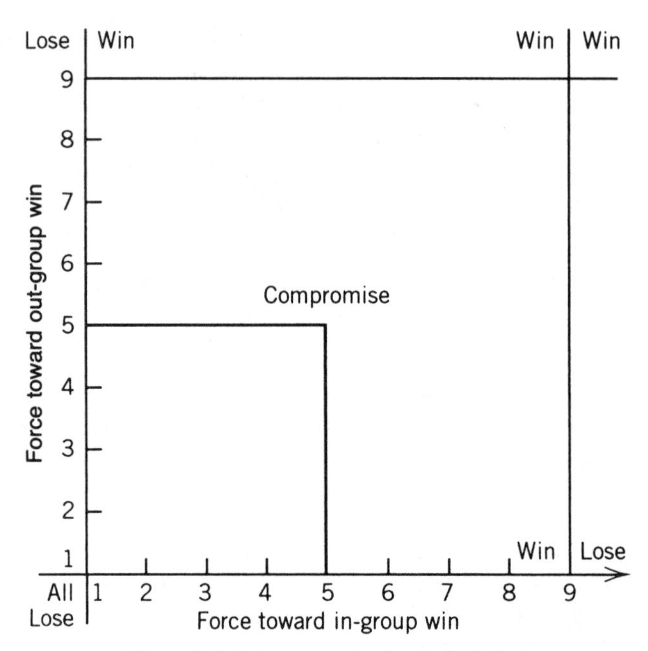

Figure 9.1. Intergroup win–lose orientation.

tion in which no design is adopted and (b) the compromise position in which a design is adopted that has some, but not all, of the desirable elements of the designs of each of the groups in conflict. It is obvious that the win–win orientation is the most desirable. However, it should also be obvious that in order to reach that design one has to be very creative and come up with a very new concept. Such a concept may have either (a) none of the elements of the original design of the technical people or (b) none of the elements of the original ideas of the marketing people. It is a fact that the win–win solution requires "insights" not available before the confrontation took place that leads to successful conflict resolution. So, in this case we can talk about "productive" or "constructive" conflict, as opposed to talking about "destructive" conflict.

Research shows that the win–lose orientation is associated with cognitive distortions, that make the outcome of the conflict undesirable for both sides. The "product" that is an outcome of this conflict (e.g., a negotiated agreement) is likely to be poor. In such cases the position of the in-group is perceived as very much more desirable than the position of the out-group. While the in-group knows its position well, it does not know or fully understand the position of the out-group. The position of the in-group appears to be much more desirable than it is. The "common ground" between the two positions is seen as belonging to the in-group's solution, and the in-group perceives only its own position as acceptable, using a "narrow cognitive field" to understand the positions of the various parties. In other words, in the win–lose orientation, the groups look at the conflict in a distorted, overly simple way.

In the case of a win–win orientation the product of the conflict (e.g., the negotiated agreement) often shows much creativity. It is more insightfully conceived and therefore more valuable. The in-group does not distort its view of the desirability of its own solution and understands much better the position of the out-group than is the case in the win–lose orientation. The full complexity of the issues is perceived. In other words, in the win–win orientation there is little distortion and the perception of the positions and values of the proposals of each party in the conflict is realistic.

When a win–win orientation is used, there usually is acceptance of the other side. There is trust, confidence, and communication between the two groups. Usually there are no threats, there is less cognitive distortion, and people do understand that there is common ground between their position and that of the other side. Also, they look for and are likely to adopt a creative orientation. In other words, they say, "Let's solve this problem together and let's solve it creatively. Let us find a solution that we have not thought about before, that will be satisfactory to both sides."

The distortions of information processing are particularly strong when there is no objective way to evaluate the information each side presents to the other. Distortions are minimized when the behavior of the other group is predictable. Predictability also has impact on the trust that each group feels toward the other. We trust people whose behavior we can anticipate.

One of the issues that comes up quite often when there is intergroup conflict concerns the question, Who should go to a negotiation session? There is research showing that a satisfactory agreement is less likely to be reached when the negotiation session is attended by representatives of the group and not by the whole group. In other words, the representatives feel constrained by the fact that they are representing a group and thus have little flexibility to move. They freeze in a particular position and the two sides become deadlocked. On the other hand, when the whole group participates in the discussions there is usually some movement. If there is a choice, it is, therefore, much better to have a session in which both groups are present in the negotiations.

When a representative is given full power to represent the group and reach an agreement as he/she sees fit (in other words, when he does not need to go back to the group to convince its members that a particular decision is the best one obtainable), deadlock can be avoided. Such representatives speak for their group and also for themselves. When they see that there is a possibility of an agreement, they agree. It is a desirable (low conflict) condition when the meeting becomes an interpersonal one involving two people, each representing a group honestly seeking to reach the best agreement that they, as individuals, can reach. The chances of creative solutions increase in such situations. The disadvantage of this solution, however, is that the group often feels the representative did not get a "good enough" agreement.

There are certain conditions that increase the probability that the intergroup conflict will become productive rather than destructive: First, when there is a perception that cooperation is highly desirable. Second, when the out-group is seen as being heterogeneous rather than homogeneous (in other words, when we teach members of the in-group that there are people with different views in the out-group). Third, when the in-group and the out-group have common goals. Fourth, when there is mobility and some of the members of the in-group were members of the out-group in the past.

This is one of the rationales for job rotation in an R&D organization. By having people in the organization change their jobs and moving them from one department to another, they become able to deal with members of these other departments in conflict situations. Finally, it is helpful to have a condition in which the in-group is not using the conflict with the out-group as a means of consolidating the leadership position of the in-group's leader. In some situations the in-group is not sufficiently cohesive and its leader tries to create unity by leading the in-group into battle. This, however, is highly undesirable since it makes conflict difficult to resolve.

Conflict can become destructive when one or both sides see only a single issue as important. Usually when several issues are involved, it is possible for each side to concede on some issues; but when only one issue is critical, it is difficult for one or the other of the two sides to yield.

Blake and Mouton [1986] have described in detail their approach to conflict resolution. They utilized groups of 20–30 executives from industry who came together for 2 weeks to discuss interpersonal and intergroup relations. They were first exposed to controlled laboratory experiments, and then to the results of the experiments.

In Phase I, in-groups were formed and in-group cohesiveness was developed. Usually each in-group was concerned that another group might "do better," and these feelings were expressed kiddingly during coffee breaks. At that point, each group was provided with an identical human relations problem. The solutions were then compared by the researchers, who explicitly stated which solution was better and why, thus creating competition and a win–lose orientation. In some cases a group of outside judges was used or each group was invited to do the judging. When a group evaluated its own solution in relation to the others, it typically focused on the differences between the solutions and the good points of its own solution, ignoring the strengths of the other group's solution.

Blake and Mouton discuss the barriers to cooperation that arise from the win–lose orientation. Of special interest to R&D managers who may be in conflict with accountants, finance experts, or marketing specialists is that these and other [e.g., Davis and Triandis, 1971] studies show that one can resolve conflict better when the entire groups meet together rather than when each group elects a representative to negotiate with the other side. In short, have all the people in your department who are affected meet with all the accountants or the marketing people who are relevant.

Conflict is usually resolved better when the solution is arrived at by the two parties in conflict than when the solution is imposed by outsiders. It is very helpful to identify distortions in perception, which occur when one's perspectives become narrow, as usually happens when one is tense, or when one thinks the in-group products are much more desirable and of higher quality than they really are, or if the common ground between the solutions of the two sides seems to be part of the in-group's solution exclusively. This job can often be helped by having a third party listen to the positions of the two sides.

Superordinate goals have been found to be helpful in reducing conflict. As already mentioned, such goals are common goals that one side cannot reach without the help of the other. Experiments, field studies, and consultant experiences suggest that such goals are highly desirable. Managers will do well to search for and foster such goals for groups that might be in conflict.

9.4 INTERCULTURAL CONFLICT

Intercultural conflict is a special case of intergroup conflict. "Culture" here is defined as unstated assumptions, beliefs, norms, roles, and values found in a group that speaks a particular language and lives in a specific time period and place. Potentially, there can be cultural conflict whenever people speak a different language including dialects (e.g., black English), live in a different place (e.g., Australia versus Canada), or have been socialized in different time periods (e.g., old versus young). Other contrasts, such as differences in religion, social class, and race, can also create intercultural conflict.

Socialization in a particular culture results in a specific "world view." Unstated assumptions (e.g., one must not start a new venture without consulting an astrologer), customs, and ways of thinking (e.g., starting with facts and abstracting a generalization versus starting with a generalization or an ideological position and finding facts that fit it) can create more trouble in interpersonal or intergroup relationships than even having something valuable to divide. This is because unstated assumptions appear so natural to the thinker. Intercultural disagreement can be more damaging to interpersonal relationships than disagreement within culture situations because rational arguments are not particularly helpful.

One often finds a cultural mixture of people working together in R&D laboratories since selection is usually on the basis of competence. For example, in the U.S. observatory in Chile, the staff is largely Chilean, but many of the scientists are from North America. Thus, it is important to consider what to do to improve intergroup relationships in situations of intercultural conflict.

There are numerous techniques, most of which involve training, that are designed to improve intercultural relationships. Triandis [1977] describes them in some detail, and Landis and Brislin [1983] and Landis and Bhagat [1996] do so in even greater detail. These publications are designed for social scientists who will train others. Managers of R&D laboratories will want to know only that these techniques exist and what they do.

There are basically four approaches to intercultural training: the cognitive approach, the affective approach, the behavioral approach, and self-insight.

The Cognitive Approach. The cognitive approach teaches people the world view of the other culture. This is done with a series of "critical incidents" in which interpersonal behaviors between members of cultures A and B are described. After each incident, there are four explanations of the behavior of the people in the incident. If one is training a person from culture A to understand the point of view of persons from culture B, three of the four explanations are commonly given by people in culture A and one by people from culture B. The incident and the explanations, presented in multiple choice format, are shown to the trainee, who chooses one explanation, turns to the appropriate page, and receives feedback on the choice. If the trainee has chosen the correct explanation, about a page of feedback is given that explains the particular cultural point of view and the probable reasons for its emergence in the culture (i.e., it is usually functional to have that point of view in that culture). When the trainee chooses an "incorrect" explanation, the feedback simply instructs him or her to read the episode one more time and choose another explanation. Thus, the trainee is gradually exposed to the point of view of the other culture. The construction of these training materials has been streamlined, and the determination of what answers are to be included is done empirically. Albert [1983] provides a detailed explanation of how to construct such training materials. Evaluation studies have shown that people trained with such materials feel better about their relationships with members of other cultures.

The Affective Approach. The affective approach involves exposing trainees to situations in which their emotions are aroused when in interaction with members of the other culture. This can be done by having them interact with members of the other culture in specific situations. When negative emotions develop, they are exposed to a positive experience that competes with the negative emotions. In some cases, simply breathing deeply or doing some exercise that reduces stress in the presence of the negative emotion is helpful. In other cases, arranging for pleasant experiences, such as the sharing of tasty food, listening to enjoyable music, or being exposed to agreeable perfumes, can create the right mood.

The Behavioral Approach. The behavioral approach involves shaping the behavior of the trainee to make sure that behaviors that are objectionable in the other culture do not occur. For example, crossing your legs and showing the bottoms of your shoes is absolutely insulting in some cultures, but many Americans do this and are not even aware of it. Simply telling them that they must not do it (the cognitive approach) is not effective. They have to experience rewards and punishments that will change their habits. The best way to accomplish this is to reward a competing behavior, such as keeping one's shoes on the ground.

Self-Insight. Self-insight is an approach designed to make the trainee understand how much culture influences behavior. The aim in this case is to give the trainee a chance to analyze his or her own culture. Understanding how much of one's own behavior is under the influence of norms, customs, and values unique to one's culture can be very instructive. The technique used in this kind of training is to have the trainee interact with a person who is a trained actor and who acts in the *opposite* way from the way people in the trainee's culture usually act. The experience of interaction with such a person and discussion of the experience with the trainer makes very clear that one's behavior and feelings are shaped by culture. When people know how culture influences their behavior, they are able to be more sensitive to culture as a variable affecting social behavior and interaction.

These four ways of training are not incompatible. On the contrary, they are complementary. A good program of training will use all of them in some mixture. It is beyond the scope of this book to discuss how to develop such training. The important point is for the manager to know that such training does exist, that it can be provided by social scientists specializing in interpersonal and intercultural behavior, and that it does help develop better relations and reduce intergroup conflict.

9.5 PERSONAL STYLES OF CONFLICT RESOLUTION

Each person has a particular style for solving conflict. Do you want to know what your style is? If you do, respond to the statements in Table 9.1 by strongly agreeing (score it a 5), agreeing (score it a 4), disagreeing (score it a 2), or strongly disagreeing (score it a 1) with each item. If you are not sure, use a score of 3. In the blanks imagine the words *boss, subordinates,* or *peers.* You can do this three times, to get an idea of your style when you are dealing with these three types of people. It is likely that your style will be a bit different when you deal with your subordinates or peers. *Please do this before you read the next paragraph.*

Now take another look at Figure 9.1. The win–win style of this figure corresponds to the *integrative style* of conflict resolution—that is, when you try to get all the information and jointly come up with a creative solution that satisfies all parties in a dispute. Sum your scores of items 1, 4, 6, 15, 28, 29, and 35 and divide by 7 to get your score on the integrative style.

The lose–win corner of Figure 9.1 corresponds to the *obliging style,* when you try to give in to the other. To measure that style, sum your scores of items 2, 12, 13, 17, 25, and 30 and divide by 6.

The win–lose corner of Figure 9.1 corresponds to the *dominating style,* when you try to impose your views on others. To measure this style, sum the scores of items 10, 11, 18, 24, 27, and 31 and divide by 6.

The lower left corner of Figure 9.1 corresponds to the *avoiding style,* where you avoid conflict. To get a measure of that style, sum your scores of items 3, 7, 22, 23, 32, 33, and 34 and divide by 7. Finally, the compromise position of Figure 9.1 corresponds to the *compromising style* and can be measured by summing items 9, 20, 21, and 26 and dividing by 4.

The Rahim Conflict Resolution Style Inventory (Table 9.1) was developed by Rahim [1983], who validated the items and reported how various groups respond to it. For example, females were found to use more integrating, avoiding, and compromising and less obliging styles than males.

TABLE 9.1 The Rahim Conflict Resolution Style Inventory

1. I try to investigate an issue with my _____ to find a solution acceptable to us.
2. I generally try to satisfy the needs of my _____.
3. I attempt to avoid being "put on the spot" and try to keep my conflict with my _____ to myself.
4. I try to integrate my ideas with those of my _____ to come up with a decision jointly.
5. I give some to get some.
6. I try to work with my _____ to find solutions to a problem that satisfy our expectations.
7. I usually avoid open discussion of my differences with my _____.
8. I usually hold on to my solution to a problem.
9. I try to find a middle course to resolve an impasse.
10. I use my influence to get my ideas accepted.
11. I use my authority to make a decision in my favor.
12. I usually accommodate the wishes of my _____.
13. I give in to the wishes of my _____.
14. I win some and lose some.
15. I exchange accurate information with my _____ to solve a problem together.
16. I sometimes help my _____ to make a decision in his favor.
17. I usually allow concessions to my _____.
18. I argue my case with my _____ to show the merits of my position.
19. I try to play down our differences to reach a compromise.
20. I usually propose a middle ground for breaking deadlocks.
21. I negotiate with my _____ so that a compromise can be reached.
22. I try to stay away from disagreement with my _____.
23. I avoid an encounter with my _____.
24. I use my expertise to make a decision in my favor.
25. I often go along with the suggestions of my _____.
26. I use "give and take" so that a compromise can be made.
27. I am generally firm in pursuing my side of the issue.
28. I try to bring all our concerns out in the open so that the issue can be resolved in the best possible way.
29. I collaborate with my _____ to come up with decisions acceptable to us.
30. I try to satisfy the expectation of my _____.
31. I sometimes use my power to win a competitive situation.
32. I try to keep my disagreement with my _____ to myself in order to avoid hard feelings.
33. I try to avoid unpleasant exchanges with my _____.
34. I generally avoid an argument with my _____.
35. I try to work with my _____ for a proper understanding of a problem.

Source: M. A. Rahim, A measurement of style of handling interpersonal conflict, copyright 1983, *Academy of Management Journal*, 371–372, reprinted by permission.

9.6 UNIQUE ISSUES OF CONFLICT IN R&D ORGANIZATIONS

For a research organization, there are some ethical issues that either create special cases of conflict or provide a rather different framework for resolving conflicts. The following discussion of conflict within individuals, interpersonal conflict, and intergroup conflict focuses specifically on R&D organizations.

Conflict within Individuals. The need to find an intellectually challenging research environment, the need for research facilities, and, indeed, the simple need for employment forces many scientists to work in an organized environment. In addition, the needs of the organization and of society as they relate to a research project can be at variance with the moral beliefs or convictions of individual scientists. Some recent cases have involved scientists who are opposed to R&D related to the defense industry. However, when one looks at investment by defense organizations in R&D worldwide, it should come as no surprise that the majority of scientists are involved in activities related to the defense industry. When some prominent scientists at major research universities in the United States questioned programs such as the Strategic Defense Initiative, the so-called "Star Wars," they were perhaps responding to a conflict between their desire to make a contribution to science and their disapproval of the expenditure of resources for research programs that, from their perspective, served no meaningful human needs. In an open democratic society such differences should be expected.

Interpersonal Conflict. One scientist may be competing with another scientist within a research group for promotion, status (for example, principal investigator versus associate investigator), or other rewards (attending conferences, office space, etc.). Since many of these things are perceived by the individual as a zero sum game, the ethos of a scientific community, which emphasizes cooperation, universalism, and sharing of ideas as its underpinning, is often lacking. This, in turn, creates conflicts within the organization and also adversely affects the productivity of the organization.

Intergroup Conflicts. It is not unusual for one group in a research organization to compete with another for projects or resources. This inevitably creates conflicts. Again, the total resources and other amenities (such as laboratory space) that are available are finite. If one group gets a certain portion of these resources, then the other group may feel that they did not get their fair share. This inevitably leads to some conflicts and also may lead to a lack of cooperation between the groups. Some organizations have competing divisions undertake the same research project. This type of competition can, for some situations, speed up the innovation process by making participants work very hard and perhaps work very cooperatively within the division. Competition among different research groups in an R&D organization is inevitable and so is some of the resultant conflict. Some of this competition and conflict may in fact be beneficial. It may provide motivation to excel and thus positively affect performance. Benefits may exceed any adverse effect that may result from conflict and a lower level of cooperation among different groups.

Perhaps one should look at the fundamental issues involved with problems created when there is no sharing of knowledge or cooperation and conflict is viewed as a zero-sum game in which one group loses proportionately what the other one gains.

Zero-sum conflict needs to be avoided, and alternatives must be sought. One must take the position that most, if not all, conflicts can be translated into win–win situations provided that people have sufficient imagination and creativity. If the contestants see their confrontation as the result of a lack of creativity (rather than due to conflict), this lack of creativity becomes "our problem." One no longer sees the other side as the "enemy," but as a collaborator in the search for creative solutions that will be mutually satisfactory.

One of the problems one often faces in organizations is well described by the concept of the "tragedy of the commons." If every individual acts in a selfish way, the individual will obtain short-term benefits but the community will suffer. How can we persuade individuals to act less selfishly? Group discussions, in which norms of proper behavior in the laboratory are examined and shaped, can be beneficial. The individual can then be told, "You are not behaving the way we all agreed to behave." Some retaliation or "fines" for incorrect or norm-violating behavior may be needed in the case of certain individuals, but most people are sufficiently concerned about their good name to adapt their behavior to the norms.

Supervisors can also help by discussing with their subordinates the need for cooperation rather than competition. In doing this, they might use the results of a study that may or may not apply to all researchers, but that certainly seem convincing. In that study the researchers sampled all the social psychologists who were members of an elite organization. Membership in the organization required several refereed publications that were judged as significant by its membership committee. At the time of the study, the organization had a membership of 200. A random sample of 200 "ordinary" social psychologists was also obtained. These 400 individuals received a personality test in the mail. About 60% of each sample completed the test. One of the attributes measured by the test was "competitiveness." The data show that those who belonged to the elite organization were significantly *less* competitive than those who belonged to the random sample of social psychologists. One interpretation of these results is that the more competitive social psychologists were so concerned with whether or not they were doing well that they did less significant research than the elite social psychologists. In other words, some good advice to give to subordinates is, "Do your best, work hard, and do not worry about what other people are doing."

An important benefit of noncompetition with other scientists is the increased likelihood of joint publications. When researchers discuss their work with others and are not protective and secretive, it is more likely that joint publications will occur. Some of these publications may be of higher quality because they are joint. In short, noncompetitiveness, openness, and cooperation may have great advantages. Finally, the emphasis on joint publications is desirable because such publications constitute a superordinate goal (see above).

9.7 ETHICS

Two principles of ethics should be stressed in the laboratory: *reciprocity* and *benefiting the least powerful*.

Reciprocity is an old principle found in the ethical systems of both East and West. The dictum "Whatsoever ye would that men should do to you, do ye even so to them: for this is the law and the prophets" (*Matthew,* 7: 12) can be translated into "Do not do unto others what you do not want others to do unto you."

The principle of benefiting the least powerful is less well known and is based on the idea that in every social system some have more power than others. In a lab, the supervisors have more power than the subordinates; directors and principal investigators more than nondirectors and nonprincipal investigators; those with large research budgets have more power than those with small budgets, and so on. When a conflict develops between two people with unequal power, the more powerful has the duty to act generously *(noblesse oblige).* If there is any doubt, bend backward to benefit the less powerful. So, if there is doubt about who should be a co-author on a paper, the principal investigator should make sure that those without power are listed as co-authors.

9.8 SUMMARY

Conflict within individuals, in the form of role conflict or role ambiguity, is widespread. Individuals who understand this can do something to manage it better. Conflict between individuals is also common, and developing superordinate goals is very helpful in reducing it. Conflict between groups is commonplace. There are a number of ways of reducing conflict, including taking a win–win orientation, emphasizing problem-solving and creativity, finding superordinate goals, and developing a system of norms and laboratory ethics that will reduce it.

9.9 QUESTIONS FOR CLASS DISCUSSION

1 What forms of role conflict are likely to develop in R&D organizations?

2 What are some of the major ways to reduce intergroup conflict?

3 What are some of the major ways to avoid intercultural conflict?

9.10 FURTHER READINGS

Albert, R. (1983). The cultural sensitizer or culture assimilator. In D. Landis and R. Brislin (Eds.), *Handbook of Intercultural Training,* Vol. 2. New York: Pergamon.

de Laat, P. B. (1994). Matrix management of projects and power struggles: A case study of an R&D laboratory. *Human Relations,* **47**(9), 1089–1119 (September).

Ertel, D. (1991). How to design a conflict management procedure that fits your dispute. *Sloan Management Review,* **32**(4), 29–42 (Summer).

Kanfer, F. H. and B. K. Schefft (1987). Self-management therapy in clinical practice. In N. S. Jacobson (Ed.), *Psychotherapists in Clinical Practice.* New York: Guilford.

Walton, F. (1995). Spot the signs and stop the conflict. *Works Management,* **48**(2), 53–56 (February).

Wisinski, J. (1995). What to do about conflicts? *Supervisory Management,* **40**(3), 11 (March).

10
PERFORMANCE APPRAISAL—EMPLOYEE CONTRIBUTION—IN R&D ORGANIZATIONS

In this chapter we will examine (a) what researchers and managers in R&D organizations do and (b) the way we can tell how well they do it. We will also discuss the need for focusing less on "appraisal" (evaluation, judgment) and more on employee contribution to the organization.

Accepted wisdom would suggest that for an organization to function efficiently and effectively, the employees must work well toward meeting organizational goals and objectives. From a manager's point of view, it would seem prudent to reward those employees whose performance contributes to organizational success. Logically, performance appraisal systems need to be designed to motivate employees to improve performance and thus contribute to organizational productivity, effectiveness, and excellence.

In practice, there are many problems. Few management activities have challenged and intrigued executives as much as performance appraisal has. To some, appraisal suggests supervisors sitting in judgment as "Roman Emperors." To others, performance appraisal is thought of as a method of manipulating employees and intruding into their lives.

10.1 SOME NEGATIVE CONNOTATIONS OF PERFORMANCE APPRAISAL

The problem may lie in the negative connotations of the words "performance appraisal." Appraisal implies evaluation and making judgments as to the quality and quantity of an individual's productivity. To make such an evaluation or a judgment, a certain yardstick has to be available to ascertain whether the individual has measured up to the performance level envisioned by the evaluator. How is one to compare the performance of one individual who has clearly exceeded the low standards he set for himself versus another individual who failed to meet the rather difficult standards he set for himself? Dimensions associated with the yardstick are variable, and procedures available to evaluate many of these dimensions are subjective and often not well understood by the employee or the supervisor.

Would changing the focus from "performance appraisal" to employee "contribution" to the organization make a difference? Would this allow the supervisor to move away from evaluation in a negative sense and move to the concept of employee contribution to the organization? Would this allow the supervisor to say that, "Your contribution to the organization has been . . ."? The discussion can then move on to how the organization could provide the right environment, support, and resources for this contribution to be increased, and perhaps allow for achieving goal congruence between the employee and the organization objectives.

We often talk about the "success" of an individual and, in turn, the success of the organization in which the individual works. We find that we define success in terms of organization profitability, productivity, or effectiveness. Is it really possible to define "success" without defining "failure"? Has an organization failed if it is not profitable for one year? Has the employee failed if the organization does not stay profitable every quarter, every year? Temporal aspects of success and failure are often overemphasized in organizations. The performance review process is tied too closely to time periods, with the focus on achievements during 6 months or a year, while R&D organizations need to look at achievements over the long term, say 3–5 years.

Let us take the case of Employee "A," who attempts ten activities, succeeds at eight, but fails two, versus Employee "B," who attempts five activities and succeeds in all. Experience clearly shows that the weight given to failure is greater than the weight given to success. Consequently, the probability is very high that Employee "B" would be rated higher than Employee "A" by the supervisor. This could be partly due to the emphasis on making judgments and evaluations rather than focusing on employee contribution to the organization. Employee "A" may in fact have made a considerably higher contribution to the organization than Employee "B." In an R&D organization, where the nature of work would naturally include some failures, this example case would point to a fundamental problem in the employee appraisal system. Employees who are risk averse and low performers are likely to get better appraisals than innovative researchers who take initiative and make mistakes.

Recognizing that some of these issues are crucial to enhancing employee contribution to the organization, the first discussion that follows focuses on *difficulties with employee appraisal.* It is often stated that performance appraisal needs to be linked to the managerial activities and the management system and that performance appraisal should be tied to the stage of development of the organization. These items, along with performance appraisal and organizational productivity, are discussed. Most technology-based R&D organizations are staffed by engineers and scientists. Since there are differences in the goals and aspirations of engineers and scientists, as discussed in this chapter, the concept of employee contribution needs to be differentiated for engineers and scientists.

In an acquisitive and consumption-oriented modern society, monetary rewards could be thought of as the litmus tests for the level of contribution an employee makes to the organization. As seen in the discussions that follow, *monetary rewards* do not work out as well as one would initially assume. After discussing performance appraisal in practice and the university department case, we propose a *performance appraisal implementation strategy* that focuses on employee contributions to the organization.

10.2 DIFFICULTIES WITH EMPLOYEE APPRAISAL

When a supervisor appraises a subordinate, the process of appraisal can be analyzed as follows: First, the supervisor must have observed some performances. However, such observations in the case of R&D personnel are unlikely to be sufficiently coherent to be valid. If the supervisor were to observe a simple operation, he might be able to judge it. But R&D work is complex, and doing any one thing well is unlikely to provide a clue to the total performance. Thus, rather than observe an individual's specific performance, the supervisor is much more likely to observe large chunks of performance, such as the presentation of a research plan or the completion of a project. Usually these are products of groups rather than individuals. It then becomes difficult to know how much the particular scientist has contributed to the group product.

Second, the observations must be integrated into some sort of "schema." Unfortunately, there are several biases in the formation of such schemata. For example, research has shown that first impressions are extremely important. If the scientist has a good reputation, many acts that are ambiguous will be evaluated positively. Also, recent events tend to be given more weight in the formation of such schemata than events that occurred during the middle of the period of observation.

The fact that negative events are given more weight in such judgments than positive events creates a further bias. If the supervisor has observed ten events, and eight are positive and two are negative, the negative ones will be given more weight because they "stand out" as "figures" against the "background" of the eight positive events. This is because in our own lives we generally encounter few negative events, but when we do they are major negatives (e.g., loss of loved ones). On the other hand, although we encounter mostly positive events, they are seldom *major* positive events (e.g., getting married, winning a million dollars), and so we become especially vigilant about the negatives. Furthermore, a manager who overlooks the negative may be far sorrier than a manager who overlooks the positive. After all, if the manager does nothing about a major mistake, top management will blame the manager; but the manager is not likely to be blamed for failing to praise good performance.

Once the "schema" has been formed, regardless of the observations on which it is based, and it has been distorted by various biases, there is the problem of remembering it long enough to register it on paper. Here we have additional biases influencing what happens. Humans tend to remember positive events better than negative ones. Also, we are helped in our memories by previous schemata, such as stereotypes. That means we tend to distort what we remember in order to make it consistent with our stereotypes.

When making such judgments we often make major mistakes. For example, we generally do not give proper weight to the "base rates" of events. Suppose a department has a 15% failure rate for projects that are carried out. When judging a particular failure we usually do not take this into account.

Employee appraisal used in salary administration is affected by numerous other problems such as the "halo effect" (because an employee is good at one thing, he is seen as good at a lot of other things) and the "leniency effect" (everybody is first rate). Appraisal ratings tend to correlate with job difficulty, age, pay, and seniority, and they tend to increase each year. Such correlations suggest that they are contaminated by other factors.

In order to overcome these difficulties, psychologists have developed a number of strategies. These include forcing comparisons among subordinates [in a paired comparison format of N employees the supervisor makes $N(N-1)/2$ comparisons]. This may not really solve the problem because if all of them are excellent they should all get a high rating. Another approach is to ask supervisors to check specific statements using behaviorally anchored scales, or ask them to make forced choices between two statements describing the employee that are of equal social desirability but only one of which correlates highly with performance. In this way, only the selection of the "valid" alternative will ensure that the employee gets a high rating. These matters are beyond the scope of this book and are mentioned here only to sensitize managers to some of the methods that the personnel department can use that are slight improvements over the simpler rating scales.

There is a good deal of research that shows that observers attribute the other's behavior to internal factors (ability, attitude, personality, effort), while actors attribute their behavior to external factors (the situation, pressure from others, perceived consequences in the particular environment). Such biases are due, in part, to what observers and actors are looking at. The observer looks at the actor and naturally attributes the actor's behavior to internal causes. The actor looks at the environment and also knows that on other occasions he has acted differently. So, it is the particular environment on *this* occasion that is seen as determining the action.

The attributions we make give meaning to the behavior. If a failure is due to the difficulty of the task, it can be excused; if it is due to laziness, it cannot. People also tend to use their own behavior as a measure of how "typical" some action of the actor is. They underuse base rates (see above) and they see the actor as more responsible for acts leading to rewards than for acts that prevent failures. Furthermore, if they like the actor, they will attribute good actions to the person and bad actions to the environment; but if they dislike the actor, good actions are attributed to the environment and bad actions to the person.

If you keep these biases in attribution in mind, you might be able to improve your judgments. But in any case, a discussion with your subordinates of the way you look at their actions can increase the convergence between your attributions and the attributions made by your subordinates. If you two agree on the attributions, the subordinates will feel that the evaluation has been fair.

The performance appraisal, as usually carried out by an organization's management, is very often incompatible with the needs of research personnel. For most of the research and development personnel, these performance appraisals do more harm than good. Wilson [1994] proposes a two-tier approach. A management performance appraisal is normally conducted in most organizations, augmented by a professional development appraisal conducted by a highly respected mentor not directly involved in the management structure. The performance appraisal conducted by the mentor would focus on professional growth and contribution to the individual's field.

10.3 PERFORMANCE APPRAISAL AND THE MANAGEMENT SYSTEM

Performance appraisal needs to be linked to the managerial activities and the management system. Cunningham [1979, p. 657] has categorized the management system into two distinct areas: The *process* of management includes activities such as planning, organizing, controlling, budgeting, and staffing, and the key orientation of these processes focuses on integrating (work activities), making decisions, recording information, motivating, and negotiating. The *function* of management includes procurement, production, adaptation, and so on. The orientations of these functions are adaptability, productivity, efficiency, and bargaining.

The managerial processes are concerned with the administration of inputs, while the managerial function deals with the way inputs produce outputs (production) that are important and relevant to the organization. Managerial processes respond to day-to-day problems, and primarily involve problem-solving. The managerial functions, on the other hand, are concerned with prescribing specific operations, procedures, and standards for achieving a certain level of production or output.

Four major activities have been identified under managerial function: adaptability, productivity, efficiency, and bargaining. Twenty-one elements were listed under the managerial process. They included items such as integration, information recording, decision-making, and motivation. In evaluating managerial performance, the manager's effectiveness in accomplishing the specific process and function elements just mentioned can be considered.

10.4 PERFORMANCE APPRAISAL AND ORGANIZATIONAL STAGES

Some of the purposes of performance appraisal relate to management control and to achieving the congruence of organizational and individual goals and objectives. Management control and strategies for goal congruence also depend on (a) the stage of development of the organization and (b) several other factors such as the technology of the organization.

Salter [1971, p. 41] has defined four stages of corporate development in terms of "the structure of operating units" (dependent variable) and "product-market relationships" (independent variable). In a general sense, at Stage 1, the organization has a single operating unit, producing a single line of products on a small scale. At Stage 2, operating units increase and production becomes large scale, but the focus is still on a single line of product. At Stage 3, operating units may be at different locations and decentralized, each producing different or related products using multiple channels of distribution. At Stage 4, the number of autonomous units producing different products increases. Basically, as organizations move from Stage 1 to Stage 4, the number of autonomous operating units increases, these units become geographically decentralized, and the operating units produce technologically different product lines or research outputs for diverse markets using multiple channels of distribution. As this development progresses, the number of variables related to organizational products, operational centers, and market relationships increases. This would also point to the organization increasing in size (number of employees and volume of sales or the size of the research budget); it may, in some cases, lead to an increase in assets and profits.

Management control at each of these four development stages is different. At the earlier stages, the organization is small, and one owner or director can oversee most of the activities. At later stages the organization grows in size and in number of products and may also be geographically dispersed. As authority becomes decentralized, performance elements need to be designed differently, depending on the development stage of the organization. For example, performance elements could be less formal during the early stages of development and more structured and quantitative later. This approach was successfully used by Salter for four high-tech electronic firms. It should be used also by managers whenever they are setting up an appraisal system. Start by analyzing your situation. In what stage is your laboratory? Then design a system that fits your stage.

10.5 PERFORMANCE APPRAISAL AND ORGANIZATION PRODUCTIVITY

Organizational productivity can be defined as the ratio of outputs to inputs. Inputs can be determined by the level of resources invested. Outputs can be conceived as income minus costs. For a profit-making organization, profitability can provide a good measure of the organization's productivity. We must keep in mind that behavior is shaped by its consequences. If we want specific behaviors to occur, we need to use an appraisal system that rewards them when they occur.

Output measures for a research organization can be subjective or objective, quantitative or nonquantitative, and discrete or scalar and can include some measure of quality [Anthony and Herzlinger, 1984]. While the measurement of quality requires extra effort and, at times, human judgment, this dimension of output should not be ignored.

Since R&D organizations have multiple objectives and their outputs are often incommensurate, the output measures are usually nonquantitative and subjective. Quantitative measures for the output elements are usually in different units, thus defying precise comparison between different quantitative outputs. Anthony and Herzlinger [1984] suggest that it might be feasible to combine a multidimensional array of indicators into aggregate units, which could then provide trends, indicators, and patterns of the individual (and organizational) output measures.

One suggested categorization of output measures includes the following:

- Process measures (related to activities carried out in an organization; useful for the measurement of the current, short-run performance)
- Result measures (stated in measurable terms; end-oriented)
- Social indicators (stated in broad terms, related to overall objectives of the organization rather than specific activities; useful for strategic planning) [Anthony and Herzlinger, 1984]

Thus, based on these output measures one could construct requisite performance appraisal elements.

10.6 GOALS OF ENGINEERS VERSUS SCIENTISTS

Most technology-based research and development organizations are staffed by engineers and scientists. In developing performance appraisal systems, differences in the goals and aspirations of engineers and scientists need to be recognized.

In general, influence in an engineering-oriented organization is largely a function of position on the management ladder, while in a science-oriented organization, it is based on the scientist's reputation in the external scientific community. Terms such as "local" and "cosmopolitan" have been used to differentiate between an engineer and a scientist. "Locals," or engineers, are more interested in working on technology that is applicable to the business aims of the company; they pattern their behavior and measure their success against internal company standards. "Cosmopolitans," or scientists, are more interested in new concepts and basic research (the focus here still can be on the business aims of the organization); they interact more freely with the wider scientific community and pattern their behavior and measure their success against its standards [Ritti, 1982, p. 372].

These differences are not so rigid as to exclude one behavior pattern from the other; rather, they are described as typical examples of relative emphases by the two groups. Both groups still desire career development and advancement in their organizations. It is important to note that some scientists in research laboratories have had extensive engineering practice experience and are able to span the boundaries between these two categories.

Scientists and engineers working in technology-based research organizations make contributions to science, which in turn is reflected in the progress science makes. Commenting on scientific progress, Brooks [1973, p. 125] states:

> There are no simple objective measures of scientific progress external to the social processes of the scientific community which produces it. Thus, to evaluate scientific progress we are compelled to rely on a consensus of the scientific communities in each field . . . the highly structured system of mutual criticism; refereed publications; peer group evaluation of research projects; and personal recognition through prizes, fellowships, and academic appointments constitutes a kind of intellectual marketplace in which scientific contributions are valued in a more or less impersonal way.

Thus, in assessing the performance of many scientists and engineers, their contribution to science as reflected by refereed publications, personal recognition through awards and prizes, and contribution to scientific and professional society activities is quite proper. Since these contributions are external to the firm, "local" types will not do so well. In establishing performance standards, depending on the goals and objectives of the organization, appropriate emphasis can be given to these external contributions. And performance requirements can be established that accommodate the organizational goals and individual capabilities—whether "local" or "cosmopolitan."

Since most R&D organizations have both research and development activities in addition to their scientific contributions (reflected by publications, etc.), the development of marketable products and technical assistance for product improvement should be used as important aspects of performance. In practice, there are conflicts. Researchers who work on successful development projects often feel that their contribution is not as well recognized as researchers conducting mostly basic research. Since most research projects require a team effort, one way to overcome this conflict is to have a team share in the development and research activities based on individual capabilities and interests. Rewards and recognition can then be provided to the whole team. This approach would, in turn, suggest that the best way to structure teams in an R&D organization is to have a mix of locals and cosmopolitans.

10.7 PERFORMANCE APPRAISAL AND MONETARY REWARDS

Giving monetary rewards to those who perform well seems logical enough. In an acquisitive and consumption-oriented modern society, higher pay satisfies basic human needs and more. For an individual, receiving monetary remuneration above what is required for basic human needs can also provide security, autonomy, recognition, and esteem.

The motivation model, generally referred to as the *expectancy model,* suggests that high performance is likely to occur if the individual feels capable of achieving it, if pay is closely tied to performance level, and if the individual finds pay to be important (this would of course vary across individuals).

In a research and development organization, indeed in most complex professional organizations, a number of reasons make tying pay inexorably to performance appraisal an imprudent approach:

- Significant accomplishments in an R&D organization often require input by many individuals. Singling out one person for a monetary reward creates the problem of inequity for others.
- The purpose of performance appraisal shifts, on the part of the supervisor, to justifying the pay decision already made, and, on the part of the employee, to comparing himself or herself to others and shaping his or her performance data to outdo others.
- Cooperation among peers is reduced because of competition for pay for performance, where total monetary rewards are viewed as a zero-sum game.
- During performance appraisal the employee is likely to exaggerate (some might suggest falsify) his or her performance to gain higher monetary rewards. This would not create the proper environment for counseling and feedback—two of the important purposes of performance appraisal.

It is therefore desirable to hold performance appraisal discussions at one time and then, later, at a different time, discuss pay increases or monetary rewards, if any. Frequency of discussions can depend on factors such as duration of projects, the employee's tenure, and organization policies. If projects are of long duration, these discussions could be as far apart as 6 months or a year; for shorter-duration projects, discussions could be tied to completion of projects. New employees are most anxious to find out how well they are doing. Feedback after the first 30 days, however brief, is most reassuring to a new employee. Tenured employees are quite satisfied to have these discussions held roughly at a time when some pay decisions are going to be made.

Some of the pay decisions are rather subjective. While no system can provide complete equity in pay, it should be recognized that "relative deprivation" in terms of pay, or any other rewards, acts as a strong demotivator and, thus, adversely affects performance. External equity (by reviewing data from a professional society and other wage surveys) and internal equity (by reviewing the contribution of the individual vis-à-vis others in the organization) are essential. Any mechanism one can use to minimize inequities in pay is worth the effort. Academic institutions often use committees to review performance of faculty members. This tends to minimize individual biases. This practice is not so common in industry and government. Most managers find this committee approach an intrusion into their domain. Periodically seeking input from an independent board or committee about researcher performance can be worthwhile. A manager can always choose to use the outside review to justify a salary decision.

10.8 PERFORMANCE APPRAISAL IN PRACTICE

It is generally felt that a person should know where he/she stands, and, consequently, the supervisor should periodically discuss his/her performance with the employee. In practice, however, experience shows that neither the employee nor the supervisor is anxious to participate in the performance appraisal process.

To illustrate this problem, McGregor [1972, p. 134] cites an example. In one company with a well-planned and carefully administered appraisal program, an opinion poll included two questions regarding the performance appraisal system. More than 90% of those answering the questionnaire approved of the idea of an appraisal system, yet nearly 40% of the respondents indicated that they never had performance appraisals done at this company. The record, however, showed that over 80% of them had signed a performance appraisal form and that they had had more than one appraisal interview with their supervisors since they had been with the company. This is an interesting discrepency. The respondents had no reason to lie, nor was there any reason to believe that the supervisors had falsified the performance appraisal signatures of the employees. The most likely reason is that the supervisors were basically reluctant to undertake performance appraisal activities and, thus, had conducted the interviews in such a perfunctory manner that many subordinates did not realize or did not remember what had happened.

Could this be due to the fact that, in practice, the focus is on "evaluation" and "judgment" rather than on employee "contribution" to the organization?

In addition to the problem described above, it is very difficult to decide exactly what all the elements for performance appraisal ought to be so that measures used include all the behaviors expected, activities to be performed, and results to be achieved by the employee. When strictly quantifiable measures are used, this can introduce an intolerable rigidity into the system to the detriment of overall organizational goals. Subjective performance elements, though necessary, can be criticized because they have a built-in bias. Also, one could argue that in most organizations, the proper environment for mutual (the employee and the supervisor) goal setting and complete trust does not exist. Consequently, the ambivalence about performance appraisal and evaluation continues to be shared by both the supervisor and the employee. In practice, therefore, there are many underlying difficulties with the performance appraisal process. Yet, it is an important activity for organizational effectiveness and employee and organization goal congruence.

10.9 A UNIVERSITY DEPARTMENT CASE

In one university department, each faculty member writes an annual report that includes two pages on each of the following topics: teaching, research, impact, and service. Under teaching, there is information about the quantity and the quality (student comments, syllabi, textbooks written) of the professor's work. Under research, there is information about the particular problem the professor is attacking and the success so far; two reprints of papers from the current year are appended.

The impact section indicates what reviewers had to say about the work, who is working on the same topic, how often the professor has been quoted in the refereed literature and by whom, and the number of invitations for major lectures or participation at major conferences or colloquia. The service shows membership on the editorial boards of journals, on national and international committees, and on university committees. The information that is relevant to two categories is entered twice. For example, the invitation to become an editor of a journal is an index of impact, but doing that job is also recorded under service.

The reports for the whole faculty, based on these four categories, are read by a committee of nine elected from among the faculty. Having at least that many members on the committee ensures that the biases of a single judge will be avoided. Getting a substantial panel means balancing different perspectives and provides a range of expert opinion.

The ratings on each of the four criteria are converted into z-scores [$z = X - \bar{X}/\sigma$], which have a mean of zero and a standard deviation of 1.0. The sum of these four z-scores constitutes the professor's evaluation. However, one can use any number of criteria (see Chapter 3 for a discussion of criteria of effectiveness, especially the following section on implementation strategy). If in a given year the department of the particular university receives, let us say, a 5% raise, then those with a sum of z-scores of 0.000 get 5%. Those with z-scores around 1.0 get 10%, and those with -1.0 get nothing. Of course, the formula that translates z-scores to percentages should be arrived at participatively by the laboratory. The point of this example is that this model can be adapted in a number of ways for the use of a particular lab. Members of the lab should be encouraged to take part in the discussions. For example, one might introduce job evaluation as a factor by computing the z-scores only within the same range of job evaluation scores. One might introduce gainsharing by paying some bonus to all on the basis of the whole lab. The 5% figure in the example might be adjusted to take lab performance into account.

One might also provide flexible benefits since different employees have different needs for money, security, status, and so on, at different points of their careers. People should also be given some choice of fringe benefits, since not everyone wants the maximum of cash as opposed to vacation time, insurance policies, and so on.

10.10 IMPLEMENTATION STRATEGY WITH EMPHASIS ON EMPLOYEE CONTRIBUTION

The preceding discussion identified some of the underlying issues associated with the performance appraisal system. Commenting on performance appraisal, Dalton [1971, p. 1] states that "few things have been more baffling to managers than the results of their attempts to develop workable performance measures and controls, thus channeling the energies of their employees towards the firm's objectives."

Recognizing that many complex issues are involved in implementing a meaningful performance appraisal system, it is nevertheless useful to focus on three items:

- What does an individual's performance depend on?
- Why performance appraisal is needed.
- A suggested strategy.

Performance Dependency

One of the fundamental questions in an R&D organization that needs to be addressed is, What does the performance of an individual really depend on? For a technology-based R&D organization, Roberts [1978, p. 7] points out the following:

> Theory Y is lovely, and McGregor and Maslow are fine. But if you look for explicit measures of what affects a technical person's productivity, then all of the boss's [human] relations skills turn out not to matter very much at all. What matters far more, at the first-line technical organization level, where you're really talking about employing the scientist and engineer more productively, is the technical competence of this first-line supervision.

This is not to say that the supervisor should be insensitive to people and not have the ability to motivate and stimulate the researcher; rather it stresses the crucial role the *supervisor's technical competency plays in productivity and excellence.*

Studies indicate that in an R&D organization, the diversity in professional activities and the diversity in skills of American scientists both relate to enhanced levels of performance. Scientists who work on diverse R&D functions (such as basic research, applied research, and consultation) make greater scientific contributions and are more useful to their organizations than those who do not work in diverse areas. In addition, scientists who spend part of their time in teaching or doing administrative work outperform those whose sole activity is research. Diversity in terms of knowledge of several areas of specialization (rather than just one) and involvement in more than just a single research project are also related to higher levels of performance [Andrews, 1979, p. 269].

Lawler [1973, p. 9] suggests that an individual's performance depends not only on motivation but also on ability. This can be described as follows:

$$\text{Performance} = f(\text{Ability} \times \text{Motivation})$$

where

$$\text{Ability} = f(\text{Aptitude} \times [\text{Training} + \text{Experience}])$$

$$\text{Motivation} = f(\text{Extrinsic Rewards} + \text{Intrinsic Rewards})$$

The performance of a researcher then would depend on factors such as aptitude, training (academic and on-the-job training), experience (e.g., working with colleagues who impart knowledge through discussions and examples), extrinsic rewards (rewards given by the organizations—e.g., pay, promotion, job assignments, fringe benefits, and conference travel), and intrinsic rewards (rewards giving internal satisfaction, belonging to the essential nature or constitution of an individual). For intrinsic rewards, all the organization can do is to create conditions (e.g., a proper work environment) that make it possible for the individual to experience these rewards internally.

As we have seen in Chapter 6 on motivation, motivation can be broken down into social factors, the affect toward the behavior, and the perceived consequences of the behavior times the value of these consequences. The affect toward the behavior can be considered an intrinsic reward; the other two factors can be considered extrinsic rewards.

The performance of an individual thus depends on many factors. Organizations and their management can influence some factors—but not all. A person's aptitude, academic training, and previous experiences are givens, and only some aspects of these can be modified with time. Even when poor performance is caused by low motivation, management can work on extrinsic rewards and create conditions for the individual to experience the intrinsic rewards. In the final analysis, it is the individual who, by building on his or her abilities (aptitude, training, and experience) and by looking at the extrinsic rewards and the satisfactions provided by the work environment, puts in the necessary effort to perform effectively.

Regarding intrinsic rewards, it is important to note that if researchers are to be committed to their work, they must feel that their research efforts are inherently worthwhile and serve some useful social purpose. It would seem that the researchers' interaction with the user community and their participation in technology transfer would provide the necessary feedback concerning the utility of their research effort and would further reinforce the intrinsic reward structure. If researchers are fundamentally opposed to the overall goals and objectives of an R&D organization, the real commitment may be difficult to achieve. In such a case, it would be best for the individual to seek employment in a different organization.

Performance Appraisal Purposes

As suggested by McGregor [1972, p. 133], performance appraisal could have the following purposes:

Evaluation. To provide a systematic means of evaluating an employee's performance to back up salary increases, promotions, transfers, and sometimes demotions or other adverse personnel actions.

Feedback. The appraisal can serve as a means of telling the employee how he or she is doing and suggesting needed changes in employee behavior and attitude. In turn, the employee should be encouraged to bring out any concerns he has about the supervisor's behavior. That may not be easy for the employee to do. The supervisor could, however, ask the employee if he has any concerns about their mutual working relationship. This way, the employee does not have to attribute the problem only to the supervisor. The employee can also be asked to provide suggestions for making his work professionally more rewarding and for improving the organization's effectiveness.

Counseling. The performance appraisal system can also be used as a basis for counseling the employee and identifying training and other developmental needs.

Added to these are the following suggested purposes:

- Increasing the motivation of the employee
- Instituting organization control and goal congruence
- Enhancing organizational effectiveness and excellence

Suggested Strategy

No specific list of performance elements and no rigid step-by-step strategy would properly encompass the many complexities inherent in R&D or other professional organizations. The following categories of performance elements, along with suggestions that follow, might provide a framework or a general guide for developing a performance appraisal system at different levels of the organization. It is important to keep in mind (a) the manifold problems and issues related to implementing effectively a performance appraisal system and (b) understanding on what the performance of a researcher or a manager really depends.

Emphasizing employee contribution to the organization, the three categories of elements suggested are as follows:

Process Measures: Routine activities, short-run outputs

Result Measures: Ends-oriented activities, tangible, significant outputs

Strategic Indicators: Indicators focusing on recognition, awards, and reputation internal and external to the organization

It is important to note that these category titles are identical to the ones used in Chapter 3 ("Creating a Productive and Effective R&D Organization"). The items included under these categories, however, have different emphases; the emphasis is on employee contribution to organizational goals and objectives. Organizational effectiveness is ultimately based on the performance of the individuals that comprise it. In other words, in order to achieve a one-to-one correspondence of individual performance and organizational effectiveness, there should be a goal congruence between individuals and their organization.

Specific elements in the appraisal would also depend on the position the individual occupies in an R&D organization. For example, a first-line supervisor or a principal investigator may have "quality of leadership in project planning" as an important element for her appraisal, while an associate investigator or a research assistant may have "completion of assigned project tasks within allotted time and budget" as an important element of his appraisal. To focus effectively on this, employee position descriptions can be used.

Because research activities are a collaborative effort of many participants, considerable similarity in the performance elements of many participants, albeit at different levels of the organization, is quite understandable. Performance appraisal can be used to further strengthen this collaborative effort by giving some employees primary responsibility for certain elements and others supportive or secondary responsibility. For example, while "project planning" may be the primary responsibility of the principle investigator, associate investigators may also have project planning as a secondary responsibility. Since effective project planning is important to the ultimate success of the project, this would encourage associate investigators to provide input to project planning and facilitate the collaborative process.

The level of detail and actual appraisal elements would further depend on the organization itself. For a commercial, profit-making organization, ultimate profitability from research outputs may be given appropriate emphasis. Even for not-for-profit organizations, return on investment studies conducted by a third party can provide important information for individual employee performance and for organization effectiveness.

As an example, at the end of this chapter (Appendix 10.14), three documents used by the Argonne National Laboratory for performance review purposes are included. One document describes the main steps for effective performance review, and the second document is a typical position description. Measures of effectiveness included in the position description can form a basis for developing appraisal elements. The third document is the actual form used to document performance assessment.

Some other suggestions that might assist in the performance appraisal process include the following:

- The employee and the supervisor should attempt to look at the performance appraisal process beyond "judging" or evaluating the employee. The focus ideally ought to be on the employee contribution to the organization. In addition, focusing on other important aspects (such as feedback, counseling, motivation, and goal congruence) discussed earlier in this chapter would, to a great degree, overcome the problem of reluctance on the part of the employee and the supervisor to participate in the appraisal process.
- Generally, supervisors are very reluctant to rank order employees or give them overall ratings such as exceptional, excellent, good, fair, unsatisfactory. Before deciding to rank order employees, or giving them an overall rating, one should consider the purpose this would serve and what benefits or problems would result. Among the benefits might be an explicit message to the employee as to his performance vis-à-vis other employees. To do this, in practice, each performance element would have to be related quantitatively. In addition, performance elements themselves would have to be rank ordered or given relative weights, and all subjective elements would require some implicit quantifica-

tion. In a typical performance appraisal case, one is dealing with more than 20 such elements. Reducing the number of elements is a good idea, but experience shows that the number of elements actually grows over the years. Even if the number is small, this quantification is subjective and using this approach creates some problems.

A supervisor soon finds out that the employee who is ranked number 6 in a group of 15 feels his ranking should have been better than one who is ranked number 5 and certainly equal to the one who is ranked number 3. Number 3 does not understand why he is not ranked number 1, and number 1 has a hard time understanding how number 3 could have received that high a ranking, and so on. Giving overall ratings (such as exceptional, good, unsatisfactory) creates similar problems.

In organizations where performance elements can be explicitly stated and where reviews are conducted by a panel or a committee to minimize biases, such ratings or rankings may work well, as shown in our example for a university department. Normally, where possible, rank ordering or overall personnel ratings should be avoided. They do not serve as an effective means of communication between the employee and the supervisor. They also tend to affect adversely a mutually supportive environment so necessary within an R&D group. Assigning ratings to individual performance elements without assigning an overall rating to the individual is preferable.

Many R&D organizations, however, require managers to assign an overall rating to the employee. So, what is a manager to do? One approach might be to downplay the overall rating and focus more on the individual performance elements than on the overall rating or rank ordering when discussing performance appraisal with an employee. From our experience, this works quite well. The employee is more relaxed and there is more feedback and counseling than would otherwise be the case.

While the appraisal system can interact with the reward system, linking the appraisal system *tightly* to monetary awards is not desirable. This suggestion seems contrary to the accepted wisdom—higher pay for higher performance. So, let us examine the situation.

A higher-performing employee can indeed get a higher salary. The point here is that during performance appraisal reviews, discussion of pay and monetary rewards, as indicated earlier, would reduce the probability of any feedback or counseling. Organizational policies should encourage the utilization of a repertoire of items for rewards beyond the monetary aspects. If the premise is that pay will motivate an employee, one finds very quickly that pay has only limited ability to do that. The suggestion here is simply that it is not very useful to tie pay tightly to performance determined during the appraisal process. Of course, in situations where performance requirement can be specified explicitly and individual biases are minimized by using a committee or a panel, it may work quite well.

- In developing performance measures and goals, the supervisor needs to work with the employee so that the employee's needs can be positively related to the organizational goals. This can assist in establishing goal congruence. In fact, this means that the employee first establishes performance goals for himself/herself based on his/her knowledge of the goals and objectives of the organization. The supervisor then reviews performance elements and helps the employee relate career goals to the needs and, indeed, the realities of the organization.

- To the degree possible, performance appraisal elements should comprehensively cover all activities an employee is expected to perform and all results an employee is expected to achieve, as well as the total contribution an employee is likely to make to the organization beyond the performance appraisal period.
- Working with the employees, managers should set goals that are specific and difficult, but realistic. They should emphasize a code of ethics, such as the one we discussed in the chapter on managing conflict (Chapter 9). They should schedule special events to celebrate reaching goals, and they should give recognition to those who have been unusually successful. They should not hoard information. They should provide and communicate a *vision* of an organization that focuses on *excellence* and *productivity*.
- At times, a vexing problem in dealing with poor performers or problem individuals is not easily handled. Such a problem could result from employee ability, organization environment, or conflict between the employee's personality and that of the supervisor and/or the organization culture. At times, separating such an employee from the organization is best for all concerned. When that is not easily done—for example, in the case of a tenured employee—a different strategy is needed.

In one case, a new laboratory director found that the research organization environment had been confrontational rather than collaborative. Serious turf problems were endemic. Researchers had to compete for research funding, special-purpose equipment, and laboratory and office space in such a way that all problems were viewed as zero-sum games. Poor performance and low organization effectiveness were the direct cause of the organization culture.

In another case, a new laboratory director found that one research division director who had been viewed as a "star performer" actually focused mainly on process indicators. He did what the previous laboratory director asked him to do and never took any initiative or provided any technical or management leadership. Division productivity over the years was mediocre.

Such organizational culture and individual behavior problems cannot be changed precipitously or rashly. Methodical change, while reinforcing the positive behavior, would indeed take some time. During the performance review session, the supervisor may focus on the following:

1. Positive aspects of the employee contribution.
2. The behavior and its adverse effects on the organization's effectiveness and the employee's contribution to the organization.
3. Seek employee input as to the source of the problem and ways to overcome it.

4. Use peer pressure and rewards rather than "sticks" or exhortation as instruments of change. For example, demonstrating to the employee how his peers are able to do what he has not been able to or is reluctant to pursue can be quite effective.
5. Finally, the supervisor should be firm and reach a mutually agreed-upon strategy and a timetable for change. If the change is handled methodically and incrementally, this mutual agreement should not be too difficult to achieve.

10.11 SUMMARY

In summary, regardless of the many problems associated with implementing a meaningful performance appraisal system in an R&D organization, a system for performance appraisal that focuses on employee "contribution" to the organization is indeed needed to enhance organizational productivity and effectiveness. A performance appraisal system can be used to achieve organizational and individual goal congruence and organizational excellence. The vexing problem of dealing with poor performers or problem individuals, sometimes caused by a lack of employee ability or by the organization structure, is not easily handled.

Discussions here have focused on providing a basic understanding of problems and issues related to performance appraisal in R&D organizations, where possible ideas to overcome these problems were presented. The chapter culminated with a discussion of individual researcher performance dependency, performance appraisal purpose, and performance appraisal strategy that focuses on researcher contribution to the organization.

10.12 QUESTIONS FOR CLASS DISCUSSION

1 What is a good system for rewarding creativity in a R&D lab?

2 What principles should we keep in mind when we design performance appraisal systems?

3 Take the case of an R&D lab. Take position descriptions of a principal investigator and his supervisor. Develop main elements of the performance appraisal for both positions.

10.13 FURTHER READINGS

DeBandt, J. (1995). Research and innovation: Evaluation problems and procedures at different levels. *International Journal of Technology Management,* **10**(4–6), 365–377.

Greene, R. J. (1992). Effective compensation strategies for professional and scientific personnel. *Compensation & Benefits Management,* **8**(4), 57–65 (Autumn).

Performance Appraisal, Harvard Business Review (1988). Special Collection of Papers, No. 90070.

Wilson, D. K. (1994). New look at performance appraisal for scientists and engineers. *Research-Technology Management,* **37**(4), 51–55 (July–August).

Yalowitz, J. S. (1993). Application of Deming's principles to evaluating professional staff members in R&D services companies. *Engineering Management Journal,* **5**(1), 27–31 (March).

10.14 APPENDIX: ARGONNE NATIONAL LABORATORY PERFORMANCE REVIEW INFORMATION*

POSITION DESCRIPTIONS, WITH THEIR AGREED-UPON MEASURES OF EFFECTIVENESS, FORM THE ESSENTIAL FOUNDATION FOR PERFORMANCE REVIEWS.

Each employee should have a current position description. These descriptions should be prepared using a single-page standard format and should include the following sections:

BASIC PURPOSE – A brief explanation of the essential function of the job.

TYPICAL ACTIVITIES – A listing of six to eight key activities of the job. Each activity written in the format: what the employee does, how the employee does it, and why the employee does it. Usually, these statements cover 85% of the workday and tend to be recurring, typical, or very important facets of the job.

WORK ENVIRONMENT – The working conditions under which the job is performed. This might include such elements as extended hours, deadlines, health hazards, and travel.

KNOWLEDGE, SKILLS, AND EXPERIENCE – A summary of the knowledge, skills, and experience necessary to perform the job.

MEASURES OF EFFECTIVENESS – Criteria by which the employee's performance is evaluated. These are the key measures upon which merit is assessed.

A NOTE ABOUT MEASURES OF EFFECTIVENESS

Since these are the yardsticks by which performance is measured, they must be specific! A measure such as "maintain good quality" is too general. A specific measure of quality should be expressed. In writing measures of effectiveness, the supervisor should ask: When this responsibility of the individual is performed properly, what will take place?

A primary responsibility of each supervisor is the preparation **and revision** of position descriptions for the jobs of each subordinate. **It is extremely important that these descriptions be current, reflecting present expectations. They should be routinely reviewed and revised as job changes occur**. Prior to each performance review, supervisors should examine position descriptions to determine if the measures of effectiveness are still relevant and then choose those measures appropriate for inclusion on the performance assessment form. The annual performance review should also produce position descriptions that apply to the forthcoming review period.

Position descriptions should not be written by supervisors alone. Input from the Division office and the subordinate should always be sought. All parties should agree on the measures of effectiveness written on the position description. Position description revisions should be coordinated with the Compensation section of Personnel.

---

*From: Argonne National Laboratory (undated), "Three Steps to an Effective Performance Review: An Instructional Guide for Supervisors at the Time of The Annual Performance Assessment," author, Argonne, Illinois. Reprinted by Permission of Argonne National Laboratory.

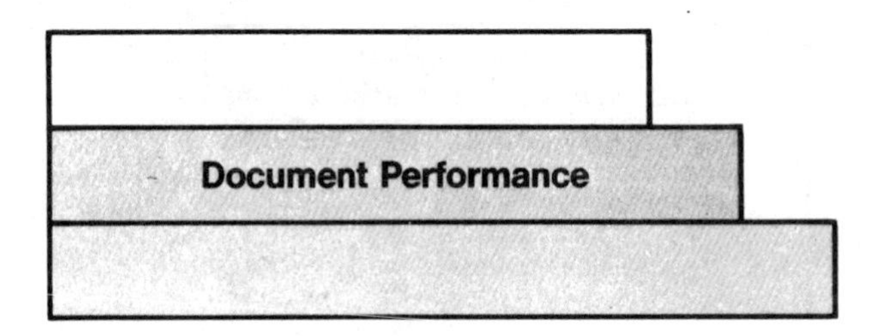

FILLING OUT A PERFORMANCE ASSESSMENT FORM can be a formidable task. However, this task can be made much simpler if the supervisor has done his or her homework by establishing and maintaining personal contacts with the employee and by evaluating the employee's performance throughout the review period. A requisite, therefore, is the noting, preferably in writing, of critical incidents in the subordinate's performance during the review period, including positive or negative comments expressed by others. This provides a resource for the documentation of performance required on the assessment form.

Using information gathered throughout the period reduces the possibility of violating a cardinal principle: **BE OBJECTIVE!** Nothing can undermine a supervisor/subordinate relationship faster than a performance review based on sheer opinion regarding the employee's traits or personality. Of importance is what the individual has accomplished in the past year.

THE PERFORMANCE ASSESSMENT FORM (AS SHOWN BELOW) REQUIRES THAT THE DEGREE TO WHICH EACH MEASURE OF EFFECTIVENESS HAS BEEN ACHIEVED BE RATED, FOLLOWED BY STATEMENTS WHICH SUBSTANTIATE AND DOCUMENT THE REASONS BEHIND THAT RATING.

MEASURE:

WEIGHT ☐ RATING: Low |____|____|____|____|____| High

PERFORMANCE DOCUMENTATION TO SUPPORT RATING:

There are spaces provided on the form for listing up to six measures of effectiveness. These should be taken directly from the position description. Attaching a weight to each measure is optional. Required is a rating of performance regarding that measure along a scale from "Low" to "High." In the section "Performance Documentation to Support Rating," the supervisor is required to document specific instances occurring since the last Performance Assessment that will substantiate the rating given. **A RATING WITHOUT DOCUMENTATION IS INCOMPLETE AND DOES NOT THOROUGHLY COMMUNICATE THE REASONS BEHIND THE ASSESSMENT**. This is particularly important in cases where documentation is necessary to substantiate future management actions.

The "Overall Relative Assessment of Performance," at the end of the form, should summarize the ratings for each measure of effectiveness and should correlate with the merit increase received by the employee.

EMPLOYEE ACCOMPLISHMENT STATEMENTS

Each employee should prepare a statement of accomplishments during the review period. This action is a vital link in the communications necessary for effective performance reviews. Though not specifically required by the Argonne performance assessment procedure, accomplishment statements give supervisors a summary of achievements during the review period as the subordinate sees them.

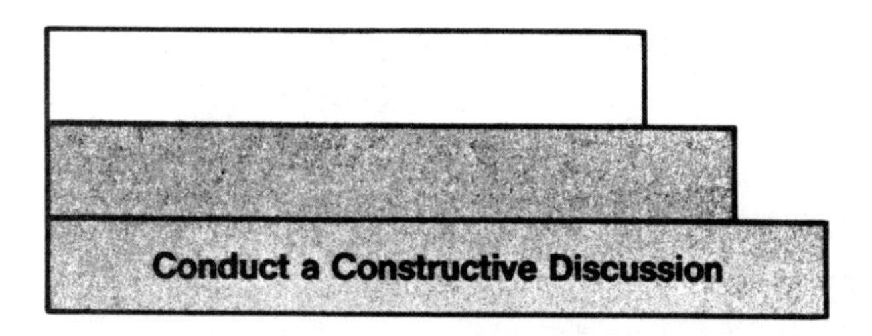

THE PERFORMANCE REVIEW DISCUSSION. This discussion should be built upon a foundation of free-flowing, timely communications during the review period. If it goes well, it can honestly summarize the employee's contribution, substantiate a merit increase or lack thereof, and provide the basis for improvements in performance. If done poorly, it can undermine relationships, discourage both supervisor and subordinate, and diminish the credibility of documentation of personnel actions.

The following checklist for the performance review discussion should be referenced before each talk takes place.

PREPARE – Good discussions don't just happen, they are planned! Inform the employee in advance. Review background material. Again review the position description and the completed assessment form. Set objectives for the discussion. Think it out. Jot down key questions and topics.

PICK A GOOD TIME AND PLACE – Hold the discussion where it will not be interrupted, where physical facilities and time do not present limitations.

FACILITATE SUPPORTIVE COMMUNICATIONS – Open the discussion by explaining the purpose of the talk, putting the employee at ease, and describing the review process, timing, and format. Stress again the importance of position descriptions and measures of effectiveness.

The goal is to keep defensiveness to a minimum. **Defensive** behavior occurs when an individual perceives or anticipates threat. People who become defensive, devote an appreciable portion of energy to defending their capabilities rather than listening to the supervisor. Communications are **supportive** when both parties engage in a mutually beneficial discussion of the work performed, the ratings, and the documentation of measures of effectiveness.

ENCOURAGE DISCUSSION – Ask open questions that stimulate response. Above all, **listen** and don't respond until you understand what the other person has said.

REVIEW THE PERFORMANCE ASSESSMENT FORM – Go over it together in detail. Allow the employee to insert legitimate additions or corrections. Have the employee sign the form, indicating that he or she has read it, not necessarily that he or she agrees with its contents. Allow the employee to make a copy if desired.

EMPLOYEE REBUTTALS

The employee who strongly disagrees with the performance assessment should be allowed, and encouraged, to write a rebuttal. This should be submitted within a week of the performance review discussion, should be attached to the performance assessment form, and should be made a permanent part of the performance assessment record.

ARGONNE NATIONAL LABORATORY
9700 South Cass Avenue
Argonne, Illinois 60439

EES 0515
JOB CODE

POSITION DESCRIPTION

COST CODE

TITLE: Oceanographer/Physical (Descriptive)	DATE: 01/22/85

BASIC PURPOSE:

To support energy resources and technology development by describing, quantifying, and evaluating physical oceanographic (and other geophysical) processes, including functioning as a project leader when so designated.

TYPICAL ACTIVITIES:

This position will gather and analyze data regarding the physical oceanographic and/or geophysical aspects of resource and technology development projects. Such data may come from existing sources; but in some cases, the position requires the oversight of and participation in the acquisition of physical data on transport and mixing water. Interaction with others in the Geophysics and Engineering Section (both modelers and data acquisition personnel) is required. This position may require interaction with those involved with environmental assessment or compliance activities. Interfaces with sponsors and/or investigators from other agencies may be necessary in field activities. Preparation of reports on data acquisitions and analysis is a function for this position.

WORK ENVIRONMENT:

Typical office setting. Field work and extended hours often required. Occasional travel for program reviews and technical conferences. As a project team member reports to project leader and as a project leader reports to the Section Leader for Geophysics and Engineering.

KNOWLEDGE, SKILLS, AND EXPERIENCE:

This position requires:

1. Considerable knowledge of oceanographic problems in coastal waters;
2. Considerable knowledge of physical hydrologic processes;
3. Considerable knowledge of measurement techniques for tracers in water;
4. Good knowledge of measurement and analysis of data pertaining to geophysical, particularly hydrologic, processes;
5. Good knowledge of oceanographic/hydrologic equipment and devices;
6. Good knowledge of environmental concerns posed by energy resource and technology development;
7. Considerable skill in oral and written communication;
8. Considerable skill in establishing and maintaining effective interpersonal relationships;
9. Considerable skill in performing qualitative and quantitative data analysis;
10. Considerable skill in conducting physical measurements;
11. Working skills in using oceanographic/hydrologic measuring devices;
12. Working skill in interpreting oceanographic data.

This level of knowledge and skill is usually attained through advanced formal education in oceanography (masters level) and several years of experience in physical oceanography/liminology/hydrology, including working with tracers.

Oceanographer/Physical (Descriptive) (Cont'd)

MEASURES OF EFFECTIVENESS:

1. Quality, quantity, and originality of publications with consideration given to the magnitude and complexity of the work, and emphasis on refereed publications in books, journals, and formal Argonne reports.
2.a. For research activities: quality and quantity of work, including: approaches taken, techniques applied or developed (including patents), and interpretations of results in general technical and project-specific terms.
2.b. For review activities: quality and quantity of work including: thoroughness in dealing with relevant literature, and clarity and tactfulness of critiques.
2.c. For assessment activities: quality and quantity of work, including: relevance of approaches adopted, professional community.
*3. Completion of project tasks within allotted time and budget to meet project goals and to provide sponsor satisfaction.
4. Initiative and resourcefulness in creating new or improved approaches to ongoing work, to provide better ways to pursue existing work or expand sponsor support.
5. Effective interaction with project and section members to accomplish project goals and enhance group research capabilities.
6. Communication of research, review, or assessment activities and results, in oral and written form, to other project personnel, the sponsor, and relevant professional community; organization of technical sessions and related outside professional activities, where applicable.
7. When designated a project leader, quality of leadership in planning project goals and implementing plans within project resources, and success in developing new or expanded sponsor support.

ARGONNE NATIONAL LABORATORY

MERIT REVIEW PERFORMANCE ASSESSMENT

STAFF AND SALARY EMPLOYEES

Employee's Name	Payroll No.	Job Classification

Division/Department	Evaluator	Review Period

1. MEASURES OF EFFECTIVENESS (from current position description)

A. MEASURE:

WEIGHT ☐ RATING: Low |____|____|____|____|____| High

PERFORMANCE DOCUMENTATION TO SUPPORT RATING:

List Ways Employee Can Strengthen Performance in This Area:

B. MEASURE:

WEIGHT ☐ RATING: Low |____|____|____|____|____| High

PERFORMANCE DOCUMENTATION TO SUPPORT RATING:

List Ways Employee Can Strengthen Performance in This Area:

C. MEASURE:

WEIGHT ☐ RATING: Low |____|____|____|____|____| High

PERFORMANCE DOCUMENTATION TO SUPPORT RATING:

List Ways Employee Can Strengthen Performance in This Area:

D. MEASURE:

WEIGHT ☐ RATING: Low |____|____|____|____|____| High

PERFORMANCE DOCUMENTATION TO SUPPORT RATING:

List Ways Employee Can Strengthen Performance in This Area:

E. MEASURE:

WEIGHT ☐ RATING: Low |____|____|____|____|____| High

PERFORMANCE DOCUMENTATION TO SUPPORT RATING:

List Ways Employee Can Strengthen Performance in This Area:

F. MEASURE:

WEIGHT ☐ RATING: Low |____|____|____|____|____| High

PERFORMANCE DOCUMENTATION TO SUPPORT RATING:

List Ways Employee Can Strengthen Performance in This Area:

2. SAFETY PERFORMANCE (Training of subordinates, corrective actions, use of personal protective equipment, etc.) Comment on how safety principles were applied in work area:

3. AFFIRMATIVE ACTION (for supervisory positions) Comment on how supervisor supported Argonne's Affirmative Action Program:

4. DEVELOPMENT PLANS: To assist employee in strengthening performance or to prepare for promotion:

5. OVERALL RELATIVE ASSESSMENT OF PERFORMANCE:

Low ____|__|__|__|__|__|__|__|__|__|__|__|__ High

ADDITIONAL COMMENTS AND REACTIONS:

Evaluator's Signature

Date

Employee's Signature
(indicating only having seen completed form)

11
TECHNOLOGY TRANSFER

This chapter focuses on technology transfer for a mission-oriented research organization. A mission-oriented research organization has its objectives defined in specific organizational goals rather than in technical terms and is vertically integrated. In other words, these organizations span activities covering basic research, applied research, development, and even technical support for operational activities. Non-mission-oriented research organizations have their objectives defined primarily in scientific terms—for example, the study of high-energy physics, nuclear energy, toxic substances, atmospheric physics, and bioacoustics. Academic research is generally non-mission-oriented and is usually small-scale research carried out in academic departments of universities. Much of the technology transfer from non-mission-oriented research organizations to application in real-life situations is likely to occur via a buffer organization similar to a mission-oriented R&D organization, and hence the focus on such organizations.

For R&D organizations, technology transfer, or tech transfer, may be defined as the process by which science and technology are transferred from one individual or group to another that incorporates this new knowledge into its way of doing things.

A new technology has to have considerable *relative advantage* and has to provide significant value to the customer before it is embraced by the wider user community. The new technology can be more expensive than the older technology, but the value in terms of quality, flexibility, and responsiveness it provides motivates the user to take the necessary steps in adopting this technology.

In utilizing new technology, there are numerous management challenges. Continuous improvement is the basis of future competitive advantage for a firm. Howard and Guile [1992] suggest some rules of thumb for a manager responsible for adopting new technology. These are:

- Do not accept performance as it is and focus on continual improvement.
- Do not just do the same thing a bit faster (or cheaper, or automatically). Careful reexamination of product and process designs is essential to make significant improvements.
- Recognize and learn to deal with people's natural reluctance to accept change that is necessary to incorporate innovation in the firm. Some examples given are:

  "There is nothing wrong with our manufacturing processes; the problem is the product designers."

  "We can't afford to spend time and money on incorporating changes and this will only slow us down."

  "We simply need to get labor rates down."

  "If only our foreign competitors were not allowed to dump their lower quality products in our market, then there is nothing we could not handle internally."

11.1 TECHNOLOGY TRANSFER HYPOTHESES

The following general hypotheses are related to technology transfer:

- Technology transfer of research results is essential if a mission-oriented research organization is to be effective in fulfilling its task.
- The effectiveness of technology transfer provides the essential measure of productivity of a mission-oriented R&D organization.
- Effective technology transfer increases user involvement in the innovation process, which, in turn, positively affects R&D productivity and has long-term benefits in terms of funding support from the sponsor groups.
- Institutional and organizational constraints, as well as improper planning for technology transfer, impede the process.
- Technology transfer techniques and approaches can be developed to facilitate the process.

11.2 STAGES OF TECHNOLOGY TRANSFER

Transferring technology from an R&D lab to manufacturing, marketing, and the ultimate user is an important function. Different organizational elements can play useful roles in successfully reaching this goal. To see how different roles and functions can best be organized, it would help to examine the stages or steps of technology transfer.

Rogers [1983, 1995] suggests five main steps leading to the adoption of technology:

- Knowledge
- Persuasion
- Decision
- Implementation
- Confirmation

Knowledge occurs when a potential user learns about the new technology and gains some understanding of its capabilities and usefulness. At this stage the user wants to know what the innovation is, what its capabilities are, and how it works. *Persuasion* occurs when the user forms a favorable or an unfavorable attitude toward the innovation. Here the user is looking at comparative advantages and disadvantages of the innovation. *Decision* occurs when the user engages in activities that lead to adoption or rejection of the innovation. *Implementation* occurs when the user incorporates the innovation into the way of doing things. *Confirmation* occurs when the user seeks to confirm the implementation decision and continues to use the innovation. This step is not always well understood, which is why many innovations first implemented are later discontinued. Certain activities to reinforce user acceptance of the innovation need to continue after implementation.

Adoption of innovation involves considerable uncertainty and thus some risk since it is not always clear what benefits will follow. Operational problems can often occur during the implementation stage, thus increasing costs and reducing benefits. Some of this uncertainty can be reduced by demonstration projects and by implementing the innovation on a partial basis. Organizations that do not reward prudent risk-taking are less likely to adopt innovations.

Innovation adoption typically follows an S-curve. Rogers [1983, p. 247] describes five categories of adopters. In general, early adopters are prudent risk-takers, are better informed and educated, and act as opinion leaders for the organization. The role of early adopters is to decrease the uncertainty about an innovation by adopting it and by adjusting it to fit the organization's needs. Early adopters then communicate this information to other potential users within the organization and to peers outside the organization.

Adoption of innovation normally requires resources (people, funds, and time), some training in using the innovation, and, at times, some changes in the way organizations operate. This involves commitment to and acceptance of the innovation at both the individual and organizational levels. Organizational structure and its routine functioning provide stability and continuity to an organization. The adoption of innovation may seem to threaten this stability and continuity, and thus it is understandable that there often is some resistance to innovation.

Some innovations may require manufacturing before they can be utilized by the ultimate user. For example, if the innovation involves a longer-lasting light bulb or a complex instrument to monitor toxic wastes, the device must first be manufactured. Some innovations, such as computer systems, improved analysis procedures, or improved design criteria can be transferred to the user without major intermediate steps. In both cases, before the innovation is implemented, the manufacturing department or the user has to become aware of the innovation and be persuaded to go on to the next steps: decision and implementation.

During the early steps—knowledge and presentation—marketing people can play an important role. Marketing people may, for example, develop information brochures or demonstrations that capture the imagination of the users, motivating them to seek further information. As users move to the decision stage and beyond, the R&D group and other individuals intimately familiar with the innovation need to play the pivotal role.

11.3 APPROACHES AND FACTORS AFFECTING TECHNOLOGY TRANSFER

Roberts and Frohman [1978, p. 36] describe three general approaches used by industrial research organizations to facilitate research utilization: These are the personnel approach, the organizational link-pins approach, and the procedural approach.

The Personnel Approach

The personnel approach involves movement of people, joint teams, and intensive person-to-person contact between the generator and the user of the research. Suppose an R&D group develops an intelligent and stand-alone air-pollution monitoring device that has a built-in microprocessor capable of real-time analysis. The innovation is complex, requiring some modifications or debugging during manufacturing. Some key members of the R&D group may be transferred to manufacturing to facilitate the process. The enthusiasm and keen insight of the R&D group can thus be transferred to manufacturing, increasing the probability of effective technology transfer.

The Organizational Link-Pins Approach

This encompasses specialized transfer groups that contain engineering, marketing, and financial skills, use of integrators who act as third-party transfer coordinators, and new venture groups.

Some organizations may find that the movement of people creates other unacceptable personnel problems or is not economical. A special "technology transfer group" is formed to specialize in moving innovations from R&D to demonstration, to manufacturing, and to the ultimate user. It is important to recognize that a technology transfer group cannot consist of just a sales or public affairs office (PAO). In one case we studied, the PAO was driving the train and results, predictably, were disappointing. After the initial knowledge and presentation stages, further activities quickly faded away. The PAO group did not have the technical understanding to successfully carry out other tech transfer activities. Even at the knowledge and persuasion stages, misleading and at times erroneous information was provided to the user groups. This further reduced the probability of success for the follow-up stages. For a technology-based innovation, it is essential that knowledgeable engineers and scientists play a leading role in the technology transfer group at all stages. As the technology moves to the decision stage and beyond, the PAO group's role is minimal.

The Procedural Approach

This includes joint planning, joint funding, and joint appraisal of research projects using research and user groups from manufacturing and marketing.

This procedural approach, which involves joint planning and participation in the innovation process by the user community, can be utilized quite effectively. User groups that include personnel from manufacturing, marketing, field users, corporate funding sponsors, and the research community can be organized for major R&D products. It is important to note that participants in these user groups still continue their normal duties. Their participation in the user group is an added responsibility. Researchers often comment on how many new ideas are generated as a result of their interaction with this user group. Such approaches require considerable organizational support, but the effort is worth the cost. In many cases, movement of people or formation of specialized technology transfer groups is simply not feasible due to organizational or cost considerations. Procedural approaches such as formation of user groups can serve as a tool for effective technology transfer without requiring movement of people or extra resources for establishing technology transfer groups. Procedural approaches can also be used to complement the other two approaches.

Factors Affecting Technology Transfer

In a study involving 26 companies, Bosomworth [1995] found that the central research efforts of large corporations vary widely in their organization, objective, and strategic approach to research investment and technology transfer. One of the most important findings was that a formal technology transfer process tends to shorten the time required to move the technology from research to commercialization.

A study of high-performance computer development companies and projects found that the differences in performance are correlated with skills and routines aimed at technology integration [Iansiti, 1995]. Furthermore, high project performance is linked to a broad approach to resolving critical problems, merging deep technical knowledge with a detailed understanding of the specific environment in which the new technologies would be applied.

Cetron [1973, p. 11] describes a number of factors affecting technology transfer:

- National policies, laws, and regulations (e.g., taxes and tax credits, tariffs, and health and safety regulations)
- Corporate policies
- Market demand

- Scientific base of the nation and industry
- Level of R&D effort
- Education level
- Availability of capital

11.4 ROLE OF THE USER

Von Hippel [1978, p. 31] makes a strong case for the role the user plays in the innovation process and in technology transfer:

> We have found that 60–80% of the products sampled in those industries [manufacturing process equipment or scientific instruments] were invented, prototyped, and utilized in the field by innovative users before they were offered commercially by equipment or instrument manufacturing firms.

In the case of scientific instruments, Von Hippel [1978, p. 12] states the following:

> In 81% of all the innovation cases studied, we found that it was the user who perceived that an advance in instrumentation was required; invented the instrument; built a prototype; improved the prototype's value by applying it; and diffused detailed information on the value of the invention and how the prototype device might be replicated. Only when all these steps were completed did the manufacturer of the first commercially available instrument enter the innovation process. Typically, the manufacturer's contribution was to perform product engineering work which, while leaving the basic design and operating principles intact, improved reliability, convenience of operation, etc., and then to manufacture, market, and sell the improved product.

The implications of these findings are significant for some industries. Management strategies should be set up to discover and utilize user-developed innovations. The manufacturer's contribution to product engineering and to setting up the necessary facilities to manufacture and market the product are also significant. Issues of innovation ownership in terms of patent rights or trademarks need to be investigated before committing resources for manufacturing and marketing.

11.5 CHARACTERISTICS OF INNOVATION

Rogers [1983, 1995] describes five different characteristics of innovation, as perceived by the potential adopter, that affect its rate of adoption:

Relative Advantage. The degree to which the innovation is superior to ideas it supersedes.

Compatibility. The degree to which the innovation is consistent with existing values, past experiences, and needs of the user.

Complexity. The degree to which the innovation is relatively difficult to understand and use.

Trialability. The degree to which an innovation may be tried on a limited basis (in other words, without committing to full-scale, total operational change).

Observability. The degree to which the results from the use of an innovation are visible and easily communicated to users and other decision-makers.

Clearly, characteristics of an innovation play an important role in technology transfer. For example, before the user adopts new technology, the user has to weigh the extra effort and investment in adopting new technology against the *relative advantages* presented by the new technology. Since existing technologies can be modified and can "stretch" to be more efficient, the new technology has to represent considerable advantages over existing ones before the extra effort involved in adopting this new technology would be considered a worthwhile undertaking.

Relative advantages relate to such items as reduced cost, increased profitability, increased convenience, reduced time, enhanced capability, and associated social status. While cost factors may stay the same or even increase, some innovations could provide relative advantage by reducing the time required to accomplish a mission or by markedly increasing product performance. For example, for military hardware such capabilities could provide a strategic or tactical advantage and thus facilitate adoption.

An innovation that is *compatible* with existing values and past experiences of the user is more likely to be adopted. For example, if the user has had a positive experience with innovations from a particular research laboratory, user adoption in the future will naturally be higher. The felt needs of the user can also play an important role. Sometimes external forces can create this need. For instance, regulatory requirements could create a strong need for adoption of advanced wastewater treatment technologies.

Some innovations are complex because their capabilities are difficult to understand and may require specialized training, equipment, and user capabilities. For such innovations, efforts need to be made to communicate capabilities simply and to provide the necessary training and equipment to increase the adoption rate.

Users are often willing to try new technology but are not willing to make full-scale and total operational changes, for obvious reasons. The risks outweigh the benefit to be derived. Therefore, when technology can be tried on a limited basis and if the changes can be made incrementally *(trialability),* the probability of its acceptance increases. Many innovations in office automation have followed this pattern.

If the benefits from the adoption of an innovation can be readily seen and easily communicated to potential users *(observability),* the rate of adoption is naturally greater. Hardware items fall in this category. Benefits from the adoption of software items (procedures, methodologies, and computer systems), however, are not as observable and not as easily communicated to potential users and thus have relatively slower rates of adoption.

11.6 ROLE OF MARKET AND PEOPLE

According to Roberts and Frohman [1978, p. 38], several studies provide persuasive evidence that market needs, rather than technological opportunities, provide the main pull and motivation for the use of research project outputs. They found that 75% of the innovations judged most important by the company originated in response to perceived needs in the marketplace rather than any technological opportunities. Market factors and user needs are therefore important considerations in facilitating technology transfer.

The role of people in technology transfer has been well recognized. The existence of a *technology gatekeeper,* a person who links the organization to the outside world of scientific and technical knowledge, has been documented by Allen [1977, p. 141]. Roberts and Frohman [1978, p. 37] describe two other gatekeepers—market gatekeeper and manufacturing gatekeeper—who have relevance to technology transfer.

The market gatekeeper is a communicator who understands what competitors are doing, what regulators might be up to, and what is happening with regard to the marketplace. This type of a gatekeeper brings vital information to the R&D organization and keeps the R&D research focus on target and toward the kinds of activities that are likely to be accepted and implemented successfully.

The manufacturing or operations gatekeeper understands enough of the practical and constrained environment of manufacturing and of the operations of the user community to keep the R&D personnel well informed about the manufacturing and operations requirements. This individual makes sure that the concepts developed by R&D can either be manufactured profitably or be made a part of the operation procedures of the user community.

As discussed previously, a strong case has been made for the crucial role people play in technology transfer. Roberts and Frohman [1978, p. 37] have stated: "Nothing transfers enthusiasm so well as working with or watching a person who has faith, conviction, and excitement about an idea." At times, moving research project personnel so they can work more closely with manufacturing and marketing can help overcome many unforeseen transfer difficulties. This point is made by many authors; but moving people presents several practical, financial, and organizational problems.

When moving personnel, it is essential to have an effective plan for replacing the research personnel with others who can continue needed research activities. Sufficient investment inherent in movement and replacement of people is required. The goals and objectives of the affected individuals need to be considered. The effects of movement on employee motivation and future job opportunities and promotion potential (within the organization and outside the organization) need to be analyzed. In summary, the strategy recommended would be to move some research project personnel selectively for a short period to act as catalysts and mitigate other concerns associated with moving people.

11.7 BOUNDARY SPANNING

One main and perhaps crucial ingredient in technology transfer is *boundary spanning*. This requires some elaboration.

Engineers should be able to communicate effectively with other engineers, economists with other economists, and so on. Furthermore, an engineer with a specialization in air pollution control presumably can communicate better with another engineer in air pollution control than with an engineer specializing in groundwater hydrology. One could continue this argument further. The point is that with the increase in the complexity of science and technology, specialization has become necessary, and this increased specialization can inhibit communication. As discussed previously, an increase in communication among scientists and engineers is essential for the innovation process. The added problem in tech transfer is that the communication network needs to go beyond the research community to include the user community, the marketing people, and the manufacturing groups. This would require boundary spanning—going beyond the immediate boundary of one's discipline.

To be sure, it is not enough to just increase communication beyond the immediate boundary or group of R&D personnel. For the communication to be effective, it must result in the understanding of "user needs" and creative collaboration among the various groups (researchers, marketing, manufacturing, and ultimate users) to facilitate significant tech transfer.

Sole responsibility for boundary spanning or gatekeeping cannot be formally assigned to an individual. In an R&D organization, this will have to be one of the responsibilities that R&D personnel undertake as an integral part of the innovation process, although some will do it more effectively than others. When utilizing any of the approaches for technology transfer (e.g., personnel, organization link-pins, procedural), participation of personnel who are able to span the boundary is likely to produce the best results.

11.8 ORGANIZATIONAL ISSUES IN TECHNOLOGY TRANSFER

In an article titled "Implementing New Technology," Leonard-Barton and Kraus [1985, p. 102] identify a number of organizational issues that affect technology transfer. Some of these are discussed and summarized below.

The technology transfer manager (or implementer of new technology) has to integrate the perspectives and needs of both the R&D personnel and users. The focus in technology transfer needs to be on *marketing* the product rather than *selling* it, with user needs and preferences given proper consideration. Like many other authors, Leonard-Barton and Kraus have identified involvement of opinion leaders in the user organization as a critical element for technology transfer success.

They point out that for implementing new technology, enthusiasm alone is not enough. New technology implementation usually requires a supportive infrastructure and the allocation of scarce resources for its implementation. Implementation teams for this purpose should include [Leonard-Barton and Kraus, 1985, p. 107]:

- A sponsor (a high-level person who can help provide adequate financial and manpower resources)
- A champion (salesperson, problem solver)
- An integrator (manages conflicting priorities)
- A project manager (oversees the administrative details)

In addition, they suggested that sufficient authority needs to be vested in one member of this team with enough power in the organization to mobilize the necessary resources (both in the R&D and user organizations) to make things happen.

There is a natural tendency on the part of humans to resist change. Leonard-Barton and Kraus [1985, p. 108] state that "tacit resistance does not disappear but ferments, grows into sabotage, or surfaces later when resources are depleted." At times, the reasons for resistance relate to real or perceived problems that the new technology may cause:

- Loss of jobs
- Loss of control
- Loss of autonomy or authority
- Many benefits to the organization, but not many to the individuals involved

In addition to resisters, another group that can adversely affect technology transfer efforts is referred to as the "hedgers" [Leonard-Barton and Kraus, 1985, p. 109]. Hedgers refuse to take a stand against an innovation so that others can address their objections, but they can affect the future of a new technology when they are a key link in the implementation plan. To overcome this, top managers, presumably in the user organization, should take some kind of symbolic action (memo, speeches, etc.) providing full support to the technology transfer effort [Leonard-Barton and Kraus, 1985, p. 109].

One of the key factors affecting technology transfer is the perceived risk and uncertainty associated with new technology. Adopters have to weigh expected benefits or rewards (if adoption is successful) against perceived risks (if the adoption is unsuccessful). In organizations where initiative is not valued highly and failures are severely criticized, adoption of new technology is not likely to be pervasive.

In one organization an employee commented that the way to get ahead was "don't rock the boat, keep a low profile, and don't make any mistakes." Naturally, in such organizations there will be resistance to change, and adoption of new technology is going to be less likely. On the other hand, adoption of new technology is likely to flourish in organizations where employees are eager to understand and learn about new technology, where management recognizes that mistakes are likely during any change, and where personnel are rewarded for prudent risk-taking.

11.9 THE AGRICULTURAL EXTENSION MODEL

One government agency that has been extremely successful in technology transfer of research results in the past is the Agricultural Extension Service. The approach used by the Agricultural Extension Service is commonly called "The Agricultural Extension Model," which consists of three main components [Rogers, 1983, p. 159]:

1. *A research subsystem,* which conducted agricultural research. This was a cooperative effort of the U.S. Department of Agriculture and the 50 state agricultural experiment stations.
2. *State extension specialists* who were stationed in land-grant universities. They linked the research group to the county extension agents.
3. *County extension agents,* who worked as change agents with farmers, and other rural people at the local level.

Once an agricultural innovation reached an individual, via the county extension agent, horizontal transfer of the innovation took place through the peer networks.

The Agricultural Extension Service was established in 1914 "to aid in diffusing among the people of the United States useful and practical information on subjects relating to agriculture and home economics, and to encourage the application of same." Thus, the Agricultural Extension Service had a long and successful history of technology transfer. Unique among its characteristics was the fact that the budget for the extension service came from federal, state, and county governments [Rogers, 1983, p. 160].

Since federal agencies and industry are concerned about the lack of adequate mechanisms for transferring technology from the R&D groups to the users, it has been suggested that the agricultural extension model be replicated. However, the application of this model for other purposes has not been very successful.

What are the reasons for this? One has to look at the unique characteristics of agricultural research and this model. In this case, the farmer is an identifiable and unique target of research, sustained level of funding was available from the federal, state, and county governments, and there were over 70 years of history and experience in implementing this model. Many of these characteristics are not present in other technology transfer situations.

11.10 NASA TECHNOLOGY TRANSFER ACTIVITIES

The National Aeronautics and Space Administration (NASA) has established a network of industrial application centers at six universities that provides technical assistance and literature retrieval. The network is perhaps the largest information storehouse of technical information in the world [Ruzic, 1978]. Another method of transferring technology is the NASA-funded biomedical and technology applications teams. These teams try to apply technology to problems encountered in medical facilities and public sector agencies.

A study of factors affecting adoption of NASA-developed technology indicates that a number of technical and economic factors impact the rate of technology adoption [Chakrabarti and Rubenstein, 1976]. This study indicates that top management support is of primary importance in the success of innovation adoption; it also shows how technical and economic variables, organizational climate, and involvement of the innovator in implementation of the technology affect the rate of technology adoption.

Recently, NASA has reshaped its technology transfer program by making commercial technology transfer an integral part of the agency mission. The objective of this mission is to involve proactively private sector participation in every NASA program from the outset. This way, technology developed in the course of each aeronautics and space program is likely to have immediate commercial linkage and a good potential for transfer. NASA has created an Aerospace Industry Technology Program (AITP) which was funded at approximately the $20 million level in 1994 [Scott, 1994]. AITP intends to establish a new paradigm for the conduct of aerospace R&D and commercialization of technologies through industry-led partnerships [Scott, 1994].

11.11 IBM CASE STUDIES

Cohen et al. [1979, p. 11], focusing on the transfer of technology from research to a profitable commercial enterprise, describe a study of 18 IBM projects; some of them were successful, while others failed. They produced valuable guidelines for moving technology from research to project development. This study can form an archetype for the development of guidelines for technology transfer that are responsive to the unique requirements of a given organization.

As a result of this study, factors identified that affect technology transfer are discussed in the order of their relative importance:

Technical Understanding

- It is necessary that research personnel fully understand the main technology before passing it on. Though this may seem obvious, it is not always the case.
- It is necessary to evaluate the benefits of new technology in comparison to what is already available and to other competitive advancements.
- One must identify where it will fit in the product line and what requirements must be met to reach the fit.
- One possible means of manufacturing needs to be exhibited.

Feasibility

- Both the research and the receiver unit must reach an agreement on what constitutes feasibility and then what should be established.

- Some estimate of cost effectiveness should be made.
- In some cases, feasibility implies acceptability by the end user. This would recognize some kind of joint study with real users to establish feasibility.

Advanced Development Overlap

- For projects being transferred out, some overlap of research activities may be needed either to support development or to explore advanced or related technologies.
- For systems work (computer software), creation of a special advanced development effort is often the answer to problems of scaling-up or is helpful in answering questions of economic feasibility.

Growth Potential

- When projects are narrowly focused on a specific need and do not have paths to technical growth and product applicability, technology transfer may suffer. This is because existing technologies "stretch" themselves and the limited advantage offered by the new technology may not be sufficient to warrant change.

Existence of an Advocate

- A strong proponent activity is needed to help overcome many hurdles during the technology transfer process.

Advanced Technology Activities in a Development Laboratory

- In moving technology from research to manufacturing, advanced technology programs in the development laboratories are often necessary. (For some research organizations, research and advanced development units may work in the same group.)

External Pressures

- In some cases, parallel activity by a competitor may help provide the push for technology transfer; in others, regulatory requirements may necessitate adoption of new technologies—for example, advanced waste treatment technologies.

Joint Programs

- Although joint programs with receiver groups are good to have, they do not ensure success.

Other secondary factors affecting technology transfer relate to timeliness, internal users, government contracts, high-level involvement, individual corporate responsibility, and proximity. For the IBM projects studied, however, in no case was the proximity of a development laboratory to a research laboratory an important factor for technology transfer. Being close was convenient and saved money, but no transfer failed because of distance [Cohen et al., 1979, p. 15].

In thinking about the transfer of technology we must be careful not to give sole weight to technical and rational criteria. The following true story makes the point. In India, an agricultural team convinced a farmer to use some new seeds. The results were dramatic. Production was 10 times as great. In evaluating the event, the farmer was asked for comments. To the amazement of the questioner, he indicated he was not planning to use the seed again. "Why?" asked the city-raised Indian agricultural engineer. "Because I have no room to store that much extra production, my cows can't eat the plants that are left on the field after the crop is harvested, and I have no way to get that much production to market." In other words, the engineer had used productivity as the *only* criterion, not taking into account social and collateral activities associated with the crop.

11.12 TECHNOLOGY TRANSFER STRATEGY

After reviewing all that has been written about the subject, one may feel a little overwhelmed by the many requirements necessary to transfer technology effectively. Trying to get several individuals to do some specific tasks for tech transfer, trying to get support from top management, trying to get the necessary financial resources, and then anticipating all the problems (for example, "hedgers", etc.) make tech transfer seem like a difficult, if not an impossible, task. Because of the uncertainty associated with each step and because of the difficulties in finding the necessary people and resources on a timely basis, successful tech transfer may seem like an elusive dream. In practice, rarely does anyone have all the resources available for tech transfer except for those special projects that are necessary for the survival of the organization.

More often than not, technology moves from research and development to the user in small increments. The size of the transfer effort varies. For some large projects, resources required for effective tech transfer may indeed be extensive. For most projects, the tech transfer effort may have to be accomplished within existing resources, by fending off the skeptics who are opposed or simply reluctant to accept new technology and without all the support mechanisms at high levels of the organization.

No grand scheme or all-encompassing formula for tech transfer is offered and none seems obvious. Based on the considerable practical experience of the authors in moving technology from a research organization to the user community, and after reviewing some of the insightful suggestions made by others, the following approach may allow one to develop a strategy in response to the unique requirements of the organization—its history, its culture, and its technology. A generalized tech transfer strategy development plan is depicted in Figure 11.1, and a description of major activities of this plan follows. To understand this approach clearly and to

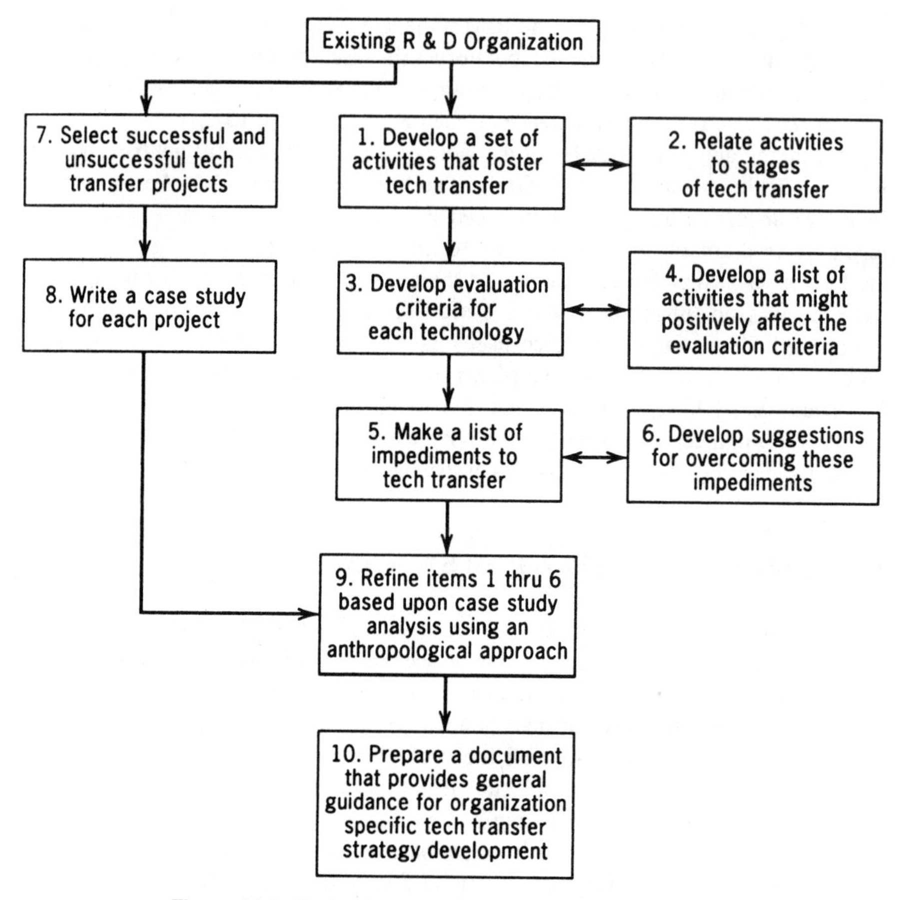

Figure 11.1. Technology transfer strategy development plan.

operationalize the concept, real research project execution and actual organizational experiences are needed. Hypothetical examples cannot easily convey the organizational and individual behavior contexts that effect tech transfer. Where possible, some examples are presented, although these examples may not be applicable to all cases.

1 Tech Transfer Activities and Documents

Based upon the knowledge of the R&D staff and the user community, prepare a preliminary list of tech transfer activities and documents that foster or enhance technology transfer. For larger and more complex projects, this list preparation may require extensive interviews; for smaller projects, telephonic information from selected participants would be sufficient. The general approach of allocating resources based on the size and complexity of the project and, indeed, based on the availability of resources is the prudent course of action. (Discussion material presented earlier in this chapter could be used as a guide for this and other activities that follow.)

Some examples of tech transfer activities and documents are:

- User involvement in research project identification
- User involvement in research program execution
- Sponsor at high levels
- Effective information brochures, audiovisuals, etc.
- User manual
- Design criteria
- Patents
- Licensing for manufacture
- Operation and maintenance document
- Support center for training, etc.
- Hot line to respond to questions
- Demonstration projects
- Successful implementation for selected users

2 Activities and Tech Transfer Stages

Relate these activities to the five tech transfer stages or steps (knowledge, persuasion, decision, implementation, confirmation) discussed earlier. Use some relevance scale. As an example, using a matrix, activities having the most relevance to a given stage could be rated A, those with least relevance could be rated C, and others without relevance could be left blank.

3 Technology Evaluation Criteria

It must be recognized that, fundamentally, the ability to transfer a new technology is limited by its utility. Utility encompasses such items as relative advantage, marketability, economic feasibility, and user acceptability. Trying to push a new technology that is marginally utilitarian will result in failure in the end. At times one must deal with a technology in which considerable R&D resources have been invested and which seems utilitarian to the R&D community, but which the user community judges to be of marginal utility. It is a poor strategy for the R&D community and its top management to zealously push marginally useful technology without making a genuine effort to understand and to overcome the user community's objections. Not only is the effort likely to fail, but it could adversely affect future worthwhile efforts. In such situations, it would be prudent to recall that the focus of tech transfer is supposed to be on marketing the product rather than selling it. Thus, user needs and preferences should be given proper consideration. Emphasis then should be on activities (discussed earlier) that enhance user acceptance, though this is not always an easy course of action.

A list of evaluation criteria should be developed. Based on the discussion of characteristics of innovation and key issues in technology transfer, a suggested list follows:

Relative Advantage. To what degree is the innovation more advantageous to existing technologies? Does it save cost, time, or improve quality?

Compatibility. To what degree is the innovation compatible with existing values, experiences, capabilities, felt needs, and organizational and cultural settings?

Complexity. To what degree is the innovation complex and difficult to adopt by the users? What degree of specialized training is required before the innovation can be adopted? What specialized equipment is needed?

Trialability. To what degree can the innovation be tried on a limited basis?

Observability. To what degree can the advantages of the innovation be easily communicated to decision-makers and users?

Technical Understanding. To what degree does the research personnel fully understand the main technology?

Resource Requirement. What level of resources is required to implement the new technology? Is this resource requirement compatible with previous user experiences? Is the capital needed for the new technology available?

Advanced Development Concepts. Are the research activities going to continue to debug problems and further supplement the technology?

Growth Potential. Does the technology have a potential for growth and product applicability? Will the new technology overcome "stretching" of existing technology capabilities?

Advocate. Are there advocates at higher levels and at user level?

Market Pull. To what degree is there a market pull?

External Pressures. To what degree are there external pressures (such as regulations, competitor development, etc.)?

A numerical or relative importance scale rating scheme can be developed to evaluate a new technology vis-à-vis such criteria.

4 Activities to Enhance Evaluation Criteria

The purpose of the evaluation criteria is not necessarily to determine whether the project should be tech transferred or not, but rather to see how the viability of the tech transfer can be improved. It would seem that if the evaluation criteria are used to make a "go or no-go" decision, there may be a natural, though unfortunate, tendency on the part of the researchers to be less objective about the criteria. In actual practice this is a crucial point and needs to be emphasized. If, for example, the evaluation criteria show that the product (a new concept, process, system, or design) is not compatible with existing and past experiences, then the emphasis should be on how the innovation can be improved so that it becomes compatible.

5 Impediments to Tech Transfer

Make a list of impediments to tech transfer. This list should include organizational items, resource requirements, and general behavioral-related items. In the earlier part of this chapter, much information is provided on this topic and can be used to generate a list that can be supplemented, depending on organization experience and project characteristics.

6 Suggestions to Overcome Tech Transfer Impediments

For every impediment to tech transfer, it would be useful initially to develop suggestions for overcoming the impediment. As an example, if at the higher management level there is a tepid response concerning the immediate benefits from the new technology, showing evidence of tangible, intangible, and unexpected benefits should help. Some examples of such benefits that might accrue from tech transfer include:

- Improved quality of the product
- Increased market share due to improved quality
- Flexibility to use the new technology for purposes other than those intended at this time
- Strategic advantage over competitors if the new technology can provide the flexibility for necessary product lines
- Reduced time required to do the job; even though savings in time have already been used as a savings in cost, reduced time can often provide a crucial advantage (for instance, in a military tactical situation)

7 Selection of Successful and Unsuccessful Projects

Based on the past experience of the R&D organization, select a number of projects in which tech transfer was or was not successful. The number of projects selected would depend on the total domain, diversity, and resource availability of the study. A minimum of three of each type of project is recommended, with a higher number of successful ones when more than three projects are selected.

8 Case History

Prepare a case history for each project. The examples provided in the referenced IBM projects [Cohen et al., 1979] should prove useful.

9 Refine Items 1 through 6

After analyzing the cases, items 1 through 6 should be modified. Considerable effort would be involved in executing this activity. Input from R&D and user community personnel is needed. Emphasis here should be on doing qualitative, anthropological analysis. Quantitative analysis should be avoided unless sufficient quantifiable data are available.

10 Guidance Document for Tech Transfer

A guidance document should be prepared based on analysis performed to provide information for R&D performers. The focus of the document should be on flexible general guidance. Rigid, mandatory requirements will only be counterproductive. The document can provide a framework that allows R&D managers to develop policy and implement strategies that foster technology transfer. The format of the document and level of detail will depend on the nature of technology, characteristics of the R&D organization, and the user community among other things.

It needs to be recognized that not all research outputs can or should be pushed to tech transfer. This is because need and technology may change during the R&D process, and R&D may not be able to produce what was thought possible during the planning stages. This type of uncertainty in R&D results should be accepted if an organization is ever going to undertake challenging R&D projects. Not being able to transfer technology successfully should not be viewed as a loss or necessarily a poor investment in the unsuccessful R&D project. Projects with unsuccessful tech transfer records can be useful as building blocks for future related research activities. There could be other unintended benefits to ongoing research efforts, though the links may seem less obvious at the time.

11.13 SUMMARY

In developing an effective management strategy for technology transfer, it is important to understand stages of technology transfer and fundamental issues and factors affecting adoption or rejection of new technologies. All these issues and factors that relate to the innovation itself, the adopter, and the organization are discussed in this chapter.

This chapter briefly describes tech transfer activities of the Agricultural Extension Service and NASA. The Agricultural Extension Service has been most successful in technology transfer of agricultural research to the farmer. Its work started in 1914. Recently, NASA has undertaken many activities in order to facilitate adoption of space technology for industrial and biomedical applications.

In the high-tech area, effective and timely technology transfer from research to manufacturing is essential to maintain a competitive position in this industry. Eighteen IBM projects were studied by Cohen et al. [1979]; some were successful projects and others were not. Factors that affected technology transfer are discussed in detail.

Finally, this chapter presents an approach that should allow one to develop an effective strategy to meet the unique requirement of a given R&D organization.

Technology transfer is an issue on which many R&D managers and others involved in research have some interesting, useful, and insightful comments. Some of these comments include:

"People by nature are conservative; therefore, reluctance to adoption of new technology is understandable."

"Because of some uncertainties involved in adopting new things, people are just not going to easily accept innovation."

"Changes and innovation are going to put some people to trouble so there will be resistance."

"Tech transfer becomes a question of personalities. Some people resist innovation and change so you really need a champion at a high enough level to push new technologies through."

"Organizations profess that they encourage adoption of new technologies, but, in practice, initiative and risk-taking are not valued much or rewarded; mistakes are penalized quickly, and promotions are given to managers who are caretakers and take little or no initiative or risk."

"Some of the new technologies sound and look good, but, when implemented, benefits end up being marginal and costs higher than expected. There just are too many uncertainties and hidden costs."

"Nothing encourages adoption of innovation more than external pressure from a competitor. If you don't change you may not survive."

"Time is a premium commodity, so new technology has to be convenient to adopt and benefits have to be transparent to the user."

11.14 QUESTIONS FOR CLASS DISCUSSION

1 From your experience as a consumer or a user of new technology, what kinds of factors persuaded you to adopt new technology? What factors worked against adopting new technology?

2 Now that you know about problems and issues related to technology transfer, take an R&D organization and develop a guidance document for facilitating technology transfer.

11.15 FURTHER READINGS

Betz, F. (1994). Basic research and technology transfer. *International Journal of Technology Management*, **9**(5–7), 784–796.

Dorf, R. C. and K. K. F. Worthington. (1987). *Study of Technology Transfer Arrangements for Natural Laboratories*. Lawrence Livermore National Laboratory, Report No. UCRL-15967. NTIS 64-88-ISS24.

Dubarle, P. (1994). The coalescence of technology. *OECD Observer*, **185**, 4–8 (December 1993/January).

Greis, N. P. (1995). Technology adoption, product design, and process change: A case study in the machine tool industry. *IEEE Transactions on Engineering Management*, **42**(3), 192–202 (August).

Howard, W. G. and B. R. Guile (Eds.) (1992). *Profiting from Innovation*. National Academy of Engineering Report. New York: Free Press.

Interagency Study of Federal Laboratory Technology Transfer Organization and Operation (1985). Federal Laboratory Consortium and Industry Working Group. Department of Energy Report No. DOE/METC-85/6019.

Rebentisch, E. S. (1995). A knowledge asset-based view of technology transfer in international joint ventures. *Journal of Engineering & Technology Management*, **12**(1, 2), 1–25 (July).

Rogers, E. M. (1995). *Diffusion of Innovations*, 4th ed. New York: Free Press.

Scott, W. B. (1993). U.S. labs embrace technology transfer. *Aviation Week & Space Technology*, **139**(8), 64–66.

12

ORGANIZATIONAL CHANGE IN R&D SETTINGS

This chapter is for those who are thinking of introducing change in their organization and is designed to give the manager a general, overall idea of what goes on in organizational change and how to evaluate it. When the change is substantial, the manager should get help from consultants either inside or outside the organization.

Organization change can focus on individuals, dyads (e.g., supervisor–subordinate), teams, or the whole organization. One can focus on cognitive skills (e.g., how to analyze a problem), affective changes (e.g., how to feel about one's competitors), or behaviors (e.g., how to behave correctly in particular situations). Thus, potentially, there are $4 \times 3 = 12$ kinds of organizational changes. However, to simplify this chapter we will discuss only some of these: changing individuals (cognitive, affective, behavioral), teams, or the whole organization.

Before deciding what aspect of the organization to improve, it is important to analyze the strengths and weaknesses of the organization. A needs assessment is highly desirable. This can be done by interviewing at all levels of the organization, focusing on what is done well and what is done badly and needs improvement in the views of the participants.

In fact, such needs assessment should be done routinely every few years because few R&D organizations are doing today what they did 5 years ago. Research projects, technology, and customer needs keep changing. To be responsive to such changes, the organization needs to change.

The first step is to determine specifically what needs to change. Some behaviors can be changed directly; in some cases, attitudes and values linked to many behaviors need to change. The strategic planning of the organization must be coordinated with the activities of each part of the organization. Then a needs assessment and a plan for change can be developed that can indicate whether people, teams, or the organization needs to change, and whether the change is mostly cognitive, affective, or behavioral. Standard operating procedures may need to be developed that will respond to the changed environment of the laboratory.

Organization change is particularly difficult to implement in an academic institution. Cole et al. [1994, p. 9] has suggested that at universities ". . . we have neither the rules that permit for orderly governance of choice nor the conceptual frameworks to guide those choices." In nonacademic research organizations, whether they be industrial or governmental, change can be achieved by two means: the ability to replace people and the ability to make personnel accountable to management for performance and for achieving mutually agreed-upon goals. In universities, neither of these tools are readily available [Kennedy, 1994]. Each academic unit within a university has its own history and tradition. Thus, each unit has to develop means to implement change, especially the ability to redistribute resources (faculty, laboratory space, student support, etc.), to support new areas of knowledge, and to deemphasize areas no longer relevant.

12.1 WHY ORGANIZATIONAL CHANGE?

It is sometimes necessary to change the organization. The need for this change may be due to the following:

Change in the Stage of the Development of the Organization. The organization may have matured, may have become too large, or may have become too static. Such changes may require new teams, work groups, different perspectives, or new managerial structures.

Program Fluctuation. In a dynamic R&D organization, it is not uncommon for the program to change. Considerable increase or decrease in a program may necessitate major change.

New Program Emphasis. Again, no research organization working on different programs at present can anticipate the emphases of future programs. As in any dynamic organization, management must respond to changing needs; and as the program emphasis changes, it becomes necessary to change the structure of the organization in response. In addition, it may be necessary to make structural changes and move people in the organization in order to provide visibility for new programs and a focal point for emerging requirements.

Customer Interface. When problems and issues with the sponsor interface exist, it is not uncommon for an R&D organization to restructure the organization to eliminate these problems.

Personnel Changes. Even if the total program and the focus of the research organization remain the same, some personnel changes (for example, loss of key personnel) could necessitate restructuring the organization.

Performance Problems. If some individuals or units are not performing well, some structural changes in the organization may be needed.

The Relationship of a Work Group to the Organization. If an R&D group within a research laboratory needs to relate differently to the main organization, major changes within the research group may be needed.

12.2 STEPS IN ORGANIZATIONAL CHANGE

Organizational change may involve implementation and understanding of the following steps:

Diagnosis. What is the problem? Can it be solved?

Resistance. Who has a vested interest in the status quo? If we change things, what are we going to change, and who will object?

Transfer. Introducing training, attitude change, or other new procedures.

Evaluation. Empirical determination of whether the change has been successful.

Institutionalizing. Establishing new norms and procedures, restructuring work, and changing schedules to fit the new norms.

Diffusion. Telling others that the change was successful. Developing opinion leaders to increase the use of the successful methods.

12.3 PROBLEMS AND ACTION STEPS

There are a number of problems associated with implementing major organizational changes. Managers who are not sensitive to these problems and attempt to implement organizational changes in an autocratic manner will find themselves imposing an enormous cost on the organization, as well as to the individuals involved and, indeed, to the very objectives that they had hoped to accomplish.

Management should provide an analysis indicating the need for organizational change and articulating the objectives and goals that need to be achieved as a result of this change. Implementation of major changes in an organization can present significant problems and must be dealt with in such a way as to minimize their adverse impact on organization effectiveness.

Nadler [1982, p. 449] identified three major problems that occur when implementing major organizational changes:

Resistance to Change. Since change represents some uncertainty, it has an impact on the stability and security of the individuals affected. For the individual, this could mean finding new ways of coping with new situations and new environments.

Organizational Control. Organizational change may alter the existing system of management control and may change some of the existing power distribution. Organizational change may also make it difficult to monitor performance and make corrections during the transitions.

Power. Since an organization is in a way a political system consisting of different individuals, groups, and coalitions holding and competing for power, an organizational change that alters the power distribution is likely to cause some political activity.

Implications and specific action steps to overcome the problems are identified in Table 12.1. Each situation is different and each organization is unique with regard to its history, individuals involved, and specific problems being addressed by the organizational change. Consequently, the general action steps suggested can serve only as a guide for developing an organizational change plan and strategy.

12.4 INDIVIDUAL CHANGE

Discussed under individual change are cognitive, affective, and behavioral changes. Some R&D organization-related examples are also provided for these types of changes.

Cognitive Change

Individuals often need new ways of thinking, or new cognitive skills—for example, how to analyze particular technical problems. In research settings this is particularly common, since new techniques in laboratory work, or new developments in mathematics, or the scientist's discipline may require learning. Seminars, courses, and workshops are often used for such training. The lecture is one of the main methods of delivering cognitive information, but guided reading, working with a model that can demonstrate the new skills, discussions, doing problems, or presenting seminar papers are also quite effective. One must consider carefully the best mix of such activities, for the particular need.

In this age of computers, a particularly useful way is to present information in programmed learning format. For example, a problem situation may be presented, and five possible courses of action may be suggested. The trainee must select one of these courses. After the course is selected the trainee receives feedback, indicating the strengths and weaknesses of the choice. If the most optimal choice was not selected, the trainee is asked to try another of the suggested options and again receives feedback, until the optimal option has been selected, and then the most extensive feedback is offered. This approach can be made to look like a game. For example, if the trainee has selected a bad course of action, the feedback can be provided in an interesting and challenging manner without discouraging the trainee.

A computer does not have to be used. The same programmed learning format can be presented in a book. Even learning how to lead has been put in this format, in the so-called Leader-Match Leadership Training Program [Fiedler et al., 1977]. The program teaches managers how to analyze their leadership style and how to change the work environment to match their style. It has been evaluated and found to be useful (improved mine safety and productivity; see Fiedler et al., 1984). This

TABLE 12.1 Organizational Change Management and Action Steps

Problem: Resistance. Implication: Need to motivate change

ACTION STEPS

- Identify problems and issues that necessitate the need for change. Individuals affected need to be jolted out of the present complacency and stability.
- Affirmatively seek participation of the affected staff in the change. Focus on building ownership of the change among the participants so that they would be genuinely motivated to not only accept the change but make it work smoothly and effectively.
- Build in rewards for the behavior that is needed during the transition stage and after implementation. Restructure awards (e.g., pay, promotion, monetary rewards, job assignments, other recognitions) to support the direction of the transition.
- Allow time for dealing with the feeling of loss caused by the change. Show understanding and provide necessary information to help overcome the problems it might create.

Problem: Control. Implication: Need to manage the transition

ACTION STEPS

- A critical step is to develop and communicate a clear image of the future. Recognizing that the transition and the implemented states are still dynamic systems, written information, which explains the reasons for the change, what the new organization would be like, how the transition will occur, and how individuals will be affected, should be provided.
- Include all changes needed to accommodate the new organization. This may include structural change, task change, change in the social environment, etc.
- Organizational management for the transition needs to include items such as a transition plan, a transition manager, and resources for the transition. At times, other transition management structures such as task forces or experimental units may be needed during the transition period.

Problem: Power. Implication: Need to shape the political dynamics of change

ACTION STEPS

- Develop multiple and effective mechanisms (e.g., surveys, sensing groups, consultant interviews, formal reports on key milestones, informal channels) for generating feedback about the transition to management.
- Assemble and mobilize the key power groups in support of the change. Work with individuals adversely affected and develop ways to investigate any adverse effects caused by the change.
- Use leader behavior to generate enthusiasm and energy in support of change.
- Use positive symbols and language to generate support and enthusiasm for change.
- To build in some stability and minimize anxiety, provide some sources of stability (e.g., physical location, people, some programs, etc.) that stay unchanged. These should be identified and communicated to the organization members.

Source: Adapted from Nadler [1982, p. 446].

technique was also used to (a) teach managers going overseas to work in other cultures and (b) learn something about the local customs and behaviors so as to improve their performance abroad [Triandis, 1977].

The in-basket technique is also useful in cognitive training. A manager is given a case and a series of communications—for example, a letter from a research program sponsor complaining about responsiveness, a request for information regarding an ongoing project, procurement documents, programming documents, information about upcoming meetings and visitors requiring immediate action, personnel documents, and telephone messages. The manager must decide which of the documents must be handled first, and how to deal expeditiously with them. This requires analysis of each item, decisions to delegate some items, and the development of procedures for effective follow-through on all items.

Much of the written communication sent to a manager or principal investigator usually is deposited in the in-basket. Managers who do not respond effectively to this information cause inefficiencies and delays. Organizations also require effective information dissemination, which may require the manager to make phone calls or write letters. Such actions can constitute correct responses to the in-basket training. It is particularly useful to teach managers which of the in-basket items must be handled immediately, which require more information, which can be delegated, and which require setting up follow-through procedures to be checked in the future. A common problem in organizations is that no effective follow-through mechanism exists and some items, while being acted on at different levels of an organization, are simply forgotten (i.e., fall in the cracks). This often causes a burden on the organization and dissatisfaction with the manager's performance. Failure to follow-through and respond to action items often increases the manager's own workload. New managers may not realize this, and thus the in-basket is an opportunity to teach them about the importance of follow-through.

The in-basket method has the advantage that it is very similar to the actual work of the manager. So, the training has validity and can be accepted easily by the manager. Experienced managers, who are known for the excellence of their management methods, may be used as instructors.

Affective Change

Changing the way people feel about their particular activities, co-workers, or projects involves attitude change. Chapter 5 presented much information about this subject. One procedure that can be used to change emotions, not discussed so far, is sensitivity training. Sensitivity groups encourage people to provide candid assessments of the way they perceive others. They usually meet for 2 or 3 days, or a weekend, and provide an intensive experience. They are popular among those who wish to escape from loneliness, wish to receive warmth and support from others, and are supposed to teach people to tolerate anxiety, understand themselves better, change interpersonal behavior, and resolve conflict. Evaluations of these groups have generally supported the expectation that people will change for the better. Institutions have well-established norms, and people have strong habits, so spending a weekend on this kind of training will not overcome habits that were developed over a long time. To change the norms, one needs to work on groups or even departments rather than individuals and to provide training for everyone. Also, the consultants who do lead these workshops are not always well trained. When clients break down in the course of discovering something unpleasant about themselves, workshop leaders are often unable to help them. Thus the casualty rate can be high and the objective benefits are often questionable.

Behavioral Change

Clinicians have developed very effective procedures for changing clients. However, these changes require quite a lot of one-on-one work and are quite expensive for organizations. For example, Kanfer [1988] has developed a "self-regulation" approach that begins with the clinician working on the client's motivation to want to change. Next, the client is helped to see that certain settings, stimuli, and people are the causes of the undesirable behaviors. The client has to learn to avoid such situations and thus master a particular technology that keeps the undesirable behaviors at a low probability.

Clients take responsibility for arranging contingencies—that is, placing themselves in situations where the undesirable cues are not present, so the behavior that needs to be suppressed is less likely to occur, and placing themselves in situations that contain the desirable cues, so that the desirable behavior increases in probability. Once the desirable behavior takes place, the clients learn to reward themselves. The clinician helps the client to set goals and to define the desired responses and the specific rewards to be administered after each goal has been reached. Thus clients learn to monitor their own performance and to compare it to established criteria, as well as to self-administer rewards. Also, the client has to learn to resist temptation—that is, to avoid behaviors that are immediately enjoyable but detrimental in the long run. This requires learning to recall the distant undesirable effect at the moment a behavior is chosen. The amount of time required for an important change using this kind of approach is approximately 40 hours of clinician time. Thus, the cost is in the thousands of dollars. However, if the effectiveness of an important manager is reduced by the undesirable behavior, the organization may well be able to justify such expense.

12.5 GROUP CHANGE: TEAM BUILDING

One of the reforms frequently used in organizations is called *team building*. The team definition is simply "a group of people who must relate to each other in order to accomplish some task." Team building is the process of encouraging effective working relationships among members of the team and, also, reducing barriers that exist in effective cooperation of members of the team. There is no doubt that many teams do not work very well. Some people are not well integrated, and team members do not plan together. They do not use their resources to achieve needed communication, and people misunderstand or do not trust each other. A team facilitator can help members talk to one another so they can discover how much they have in common. For example, superordinate goals (goals desired by both teams that neither team can reach without the help of the other) might become salient, and members of the organization may realize that some of the mutual distrust they feel is not justified.

Since a lot of team building involves interactions with colleagues, it is a very sensitive matter. It is true that if reform does not work out as planned, the relationship may become worse than before. Team building can only be effective if the total culture of the organization supports it. In fact, the emphasis on team building used today contrasts with the emphasis on sensitivity groups that was more prevalent about 20 years ago. This is a result of the realization that one has to work with the whole organization rather than with just a few groups if one is to be effective in changing the organization.

One of the problems management faces when it attempts change is that the change may affect only isolated individuals. In the past, individuals from different parts of the organization were selected by management and sent for training. When these trainees were returned to the organization, they found that the organization did not respond to the change that they had experienced. As a result, they went back to their old habits and the effect of the training was wiped out. In contrast to training specific individuals, team building involves training everybody who is part of the team; it also ensures that other groups, particularly other teams that are part of the organization, receive the training.

An effective approach is to allow members of the team to talk to one another or send messages to each other concerning behaviors that they find desirable, objectionable, or neutral. One such technique is called "From Me to You." Each team member writes a message on a sheet of paper aimed at another team member. The message specifies behaviors that should be kept up ("keep doing that"), stopped ("stop doing that"), or started ("it would be nice if you did that"). Some of these behaviors can be job-specific. Other behaviors might be social. For example, a subordinate may ask a boss for more frequent comment on the boss's perception of the subordinate's work performance or for an invitation to the boss's house once a year. These messages may be sent anonymously, or not, depending on the extent the team members are ready for the exchange of intimacies. A contract might be negotiated as a result of such exchanges. For example, "if you do more of this, I will do more of that" could be part of the agreement.

A related technique is role clarification. The trainer asks each member to identify the four or five people the member interacts with most frequently when on the job. The member then "visits" each of them and asks them to describe how they perceive the member's job. "What do you think I am supposed to do? With what frequency? When, where, how?" Such information defines the "emitted job role" as perceived by that person. By going on to all the relevant others (superiors, subordinates, and peers), the person can identify varieties of the "emitted role." Quite frequently there will be a discovery that the various important others define his/her role differently. Then it is possible to discuss the discrepancies with them as a group. The role can then be clarified. This can be very helpful in improving job relationships. The team member can negotiate a different role and align the subjective and emitted roles. The subjective role is thus more salient and clearer to everyone, and the enacted role can become a much closer version of the subjective role.

A usual approach to team building is to begin with a diagnostic phase, in which team members answer a questionnaire that indicates team problems and difficulties. Usually about 10 questions are asked about each of the following areas [Francis and Young, 1979]:

1. *Effective Leadership.* For example, "Team members are uncertain about their individual roles in relation to the team."
2. *Suitable Team Membership.* For example, "We need an input of new knowledge and skills to make the team complete."
3. *Team Commitment.* For example, "No one is trying hard to make this a winning team."
4. *Team Climate.* For example, "There is much stress placed on conformity."
5. *Team Achievement.* For example, "In practice the team rarely achieves its objectives."
6. *Relative Corporate Role.* For example, "We do not work within clear strategic guidelines."
7. *Effective Work Methods.* For example, "Team members rarely plan or prepare for meetings."
8. *Team Organization.* For example, "We do not examine how the team spends its time and energy."
9. *Critiquing.* For example, "The team is not good at learning from its mistakes."
10. *Individual Development.* For example, "The team does not take steps to develop its members."
11. *Creative Capacity.* For example, "Good ideas seem to get lost."
12. *Intergroup Relationships.* For example, "Conflicts between our team and other groups are quite common."

If a team agrees with many negative statements in a particular area, it indicates that some team building work is needed. Special exercises are available corresponding to each of these 12 areas. For example, the "From Me to You" exercise described above can be used to improve team climate.

Team building also involves communication exercises, reviewing the progress that is made in each of the areas that has been targeted for change, and, finally, taking a second measure with the questionnaire mentioned above. The second time, if the team checks fewer negative items that correspond to the 12 dimensions of team building, one can assume that some positive change has occurred. However, a multimethod approach to team change evaluation is recommended and should include a wide range of measurements.

The different phases of team building can include cognitive, affective, and behavioral changes. Specifically, after the administration of an instrumentation, such as the one previously described, there is bound to be some cognitive change—for example, "our team has a problem learning from its mistakes." The various team-building exercises change both affect (how people feel about themselves and their team colleagues) and behavior (how people respond to each other). Decisions by the team to institute new standard operating procedures result in behavior change. Of course, some behaviors are easier to change than others. If the behaviors are automatic and determined by habits, there needs to be a substantial interference with the cue-behavior sequence to modify the behavior. Other behaviors are easy to change. For example, discovering that a task one finds unpleasant to do is undesirable from a wide variety of perspectives can easily lead to change.

12.6 ORGANIZATIONAL CHANGE

Two techniques, survey feedback and grid organizational development, can, at times, be useful in understanding organizational problems and ways to overcome them. These techniques require administering a questionnaire and working with different levels of management. This means an experienced consultant would be needed to help implement these techniques.

Survey Feedback

In this case a consultant distributes a questionnaire throughout the organization, the data collected are aggregated in a variety of ways and fed back to the organization, and the organization discusses them in several sessions. Much of the discussion turns out to be technical: It criticizes specific questions and indicates how they can be interpreted in different ways. The organization can obtain some benefit from this exercise, particularly when top management realizes that certain departments do have low morale, that communication upward is poor, that supervisors are really unaware of problems faced by subordinates, and that researcher ideas are not sought for organizational goal setting. Management can use survey results to focus on the problem and to introduce change using techniques such as team building.

Grid Organizational Development

This approach starts with top management and works through to the bottom of the managerial structure in an attempt to sensitize managers to the importance of the human factor and healthy relationships. Generally, managers are task-oriented (otherwise they would not have gotten there), so the emphasis on human relationships sensitizes them to a dimension that they tend to underuse. The approach also provides training in conflict management and is an opportunity to review policies and objectives, make plans, and evaluate changes that have already been made. In the hands of a good consultant this approach is quite helpful.

12.7 EVALUATING ORGANIZATIONAL CHANGE

The several kinds of change outlined above need to be evaluated. The organization needs to know whether the cost of particular changes can be justified. It needs to assess what techniques of change work and what techniques are ineffective. Since each organization is to some extent unique and with its own culture, the fact that one technique worked in another organization does not mean it will be effective in your own.

The most important attitude that managers need to develop about organizational change is that every change is an experiment. When we do an experiment we modify some independent variables and measure some dependent variables. Similarly when we undertake an organizational change (our independent variable) we must measure its effects on the organization (our dependent variable). The reason we do experiments is to unlock the secrets of nature. Organizations are also part of nature. In fact, they are a very complex part of nature. Designing changes for organizations that will improve them is much more difficult than doing a chemistry experiment. So, we must be modest in our expectations and not anticipate miracles. But with systematic change, and careful measurement, we should be able to sort the changes that are effective from those that are ineffective or hurt the organization.

Unfortunately, administrators often have the wrong attitudes about organizational change. Since they are the ones who approved the change, they feel ego-involved. Since they want the change to succeed, they are unable to take an objective, open, experimental approach. Yet, that is exactly the approach that is needed. It is important that managers train themselves to see organizational change supportively, but also critically. If it fails, try again. In other words, the correct attitude is to view change as an experiment, which may or may not work. If all our experiments came out the way we expected, there would be no point in doing them! It is exactly because we get unanticipated results that we keep experimenting. Similarly, we should not assume that every innovation will benefit the organization, and we should not put down our peers or subordinates who fail to introduce successful innovations. In fact, if one never fails, this may be a sign of too low a level of risk. The bold innovator has more failures than successes, but the few successes often change the world.

In sum, we must not expect that our first idea about organizational change is going to work. Try, and try again. Keep measuring what happens and you will gradually find out what works.

When doing evaluation research following organizational change, we need to use many and very different methods. For example, one may wish to measure job satisfaction, turnover rates, productivity rates, quality of publications, and many other dependent variables before deciding that a particular reform has or has not been successful.

In evaluating reforms, one must also consider that any particular reform continues over time and thus cannot always be evaluated at only one point. In fact, people who are specialists in evaluation research have distinguished between *formative* and *summative* evaluation. Formative evaluation examines the effects of the change as it happens. Thus one modifies the organizational change to take into account the results of the evaluation. In the case of summative evaluation, on the other hand, one waits until the change has occurred, and has been in place for some time, before making the evaluation.

Specialists in evaluation have used a variety of ways to make their evaluations. For example, some people advocate self-study as a means of getting the group affected by the change to assess how the particular reform has worked for them. Another approach consists of forming a blue-ribbon committee, usually consisting of people outside of the administrative unit in which the change has occurred, who come in, ask a lot of questions, and make a judgment as to whether the particular reform has been effective. Still another method is to look at particular data sets as criteria for an effective reform. These data sets may include the dollar amounts of grants and contracts obtained to support the research, the number of publications, judgments about the quality of the publications, the extent to which the persons or groups who were part of the changes are being quoted, or the reputation of the group. Still other possibilities include bringing in a specialist with an adversarial view, whose role is to discover that the reform is ineffective. This specialist is usually a critic of the reform and may often uncover problems that the participants may not see. Still another approach is goal-free evaluation, in which the evaluator simply tries to find out what "really" is happening or has happened. The idea in this case is that the evaluator is an unbiased spectator who can evaluate the change most appropriately. There are also classical evaluation specialists, who utilize "experimental" and "control" groups or look at the results of change over time. They use the particular group's performance prior to the introduction of a reform as a control for the evaluation of the change that has occurred since the reform was introduced. Still another approach is the one used by anthropologists who look at what is happening and describe it as well as they can from the perspective of the "natives" (the members of the organization). This is done without any idea about the antecedents or the correlates of the observations they make.

It is often the case that a combination of these approaches may be optimal in order to gain a really good understanding of the effects of the reform.

Another issue in evaluation research is the question of whose perspective to take more seriously. For example, in a department in a particular organization, the members of the department may represent one perspective, top management may represent a different perspective, the supervisor of the department may have a third perspective, and the peers in other departments may have a fourth perspective. Who is to say whose perspective should be taken more seriously? Should one weigh the various perspectives to get a single index that reflects the particular reform?

12.8 SUMMARY

Organizations often need to be changed. One needs to understand problems associated with organizational change and the action steps that might help overcome these problems. Some individual changes can be accomplished through training, whereas others require extensive one-on-one clinical work. Some group change can be accomplished by team building techniques; other techniques for organizational change involve survey feedback and grid organizational development. These aspects of individual and organizational change are discussed in this chapter. Evaluation of the change is necessary to ascertain that desired objectives are reached. Different procedures for evaluating change are outlined in the chapter. A combination of these approaches needs to be used to cover many perspectives and to reduce bias.

12.9 QUESTIONS FOR CLASS DISCUSSION

1 Suppose you were going to do some "team building" in your R&D lab. What would be the steps?

2 What would be some considerations that may militate against team building?

3 What criteria for effective team building would you use?

4 Review the methods of evaluation of an organizational change. Which method should be used for which condition?

12.10 FURTHER READINGS

Stone, F. (1995). Overcoming opposition to organizational change. *Supervisory Management,* **40**(10), 9–10 (October).

Hutt, M. D. (1995). Hurdle the cross-functional barriers to strategic change. *Sloan Management Review,* **36**(3), 22–30 (Spring).

Nystrom, P. C. and W. H. Starbuck (1981). Remodeling organizations and their environments. *Handbook of Organizational Design,* Vol. 2. New York: Oxford University Press.

Peak, M. H. Coming to grips with change (1994). *Management Review,* Vol. **83**(7), 40–44 (July).

13 THE UNIVERSITY RESEARCH ENTERPRISE

Vannevar Bush [*Science the Endless Frontier,* 1945] articulated national science policy after World War II. He described the relationship between basic research and application and how basic research leads to new knowledge. He suggested that basic research is scientific capital, a common fund from which practical applications can be derived and which provides a rationale for public support of basic research. The academic research enterprise has benefited and flourished as a result of the public support of the basic research policy articulated by Vannevar Bush.

While the basic research enterprise in the U.S. is composed of governmental, industry and academic institution research labs, the largest and probably the most creative segment in the U.S. resides primarily within the academic institutions (National Academy of Engineering, 1993, p. 63). Some of the unique characteristics of the U.S. academic research enterprise relate to free flow of ideas and talent from throughout the world, a long history of public support of this enterprise, opportunities it provides for initiative of individual scientists at relatively early stages in their career, and the history of preeminence of the academic institutions in the scientific accomplishments (National Academy of Engineering, 1993).

Much of the research funding at academic institutions comes from governmental and nonprofit organizations as discussed in more detail later in this chapter. Industry support has been increasing during the last decade, but it still represents less than 10% of the total funding. University–industry linkage is important, however, for reasons other than absolute numbers in terms of research funding. To understand the reasons for this and to understand other issues related to the University research enterprise, we discuss the following topics in this chapter:

- Basis for University Research Activities
- Federal Support of University Research: An Entitlement or a Means to Achieve National Goals?
- Basic Research—Who Needs It?
- University–Industry Linkage
- University–Industry–Government Interaction
- Rethinking Investment in Basic Research

13.1 BASIS FOR UNIVERSITY RESEARCH ACTIVITIES

There is considerable debate about the role and purpose of research and scholarly activities in research-intensive universities. It is suggested that the research grant at a university has become the true badge of honor and that the faculty member who devotes primary energy and creativity to teaching is in fact penalized and often looked down upon by peers [Griffiths, 1993]. It is well known that promotion and tenure decisions at universities are greatly affected by a faculty member's accomplishments in research, and these accomplishments are to a great degree affected by research grants. Thus, research grants become the driving forces for faculty recognition and promotion. As an example, one Dean refused to promote a biologist who did not have outside support on the grounds that tenure commits $1 million to a professor's career, and making such a commitment is too risky if the federal government is not participating financially.

A comprehensive university is more than a teaching college. A university is a repository of new knowledge. Scholars at a university generate new knowledge and participate in disseminating this knowledge through teaching and research publications. The university community also participates in many outreach and service activities.

Quality graduate education is undoubtedly linked to research and scholarship. The American higher education system is unique in that the education of scientists and engineers takes place at academic institutions where basic research is undertaken and thus new knowledge is generated. Research in academia, therefore, serves the most crucial purpose of educating the next generation of scientists and engineers.

A clear connection between academic research and commercial use of technology and innovation is difficult to establish. In a recent study Mansfield [1995, p. 64] concluded that ". . . a substantial proportion of industrial innovation in high-technology industries like drugs, instruments, and information processing have been based directly on recent academic research. . . ."

To clearly understand the importance of research to a university, imagine the implications of a precipitous decline in external research funding at a university.

At the outset, there would be an immediate impact on all budgets. Why? Simply because, at most research-intensive universities, external research funding represents more than 50% of the budgets of science-and-technology-based departments and colleges within the university. Much of the research infrastructure at a university is supported by external research funding. The effect on such research infrastructure would be considerable. This would, in turn, affect the quality of the graduate program. Since many of the graduate students in science and engineering are supported from external research funding, the number of graduate students would decline.

Many faculty members draw supplemental compensation for summer research activities from research grants. This is an important issue. First-rate faculty members are naturally drawn to research-intensive universities where they can work on significant research programs and are usually able to augment their normal academic salaries with funds drawn from research grants. Thus, a decline in external funding would inevitably affect the faculty quality at a given university and, in turn, affect the program quality and reputation, especially at the graduate level.

One could argue that undergraduate programs benefit from high-quality faculty and that a decline in the research funding would also affect the quality and reputation of the undergraduate program. Finally, many outreach programs and service through innovation would be adversely affected if the research funding were to decline.

Many crucial decisions affecting faculty promotion and tenure, program quality, and the reputation of a university are all inextricably tied to external research funding. For universities that are not research-intensive, arguments presented here may not be as relevant.

Notwithstanding all of these points, arguably, *research in academia is best justified to support the institutions' primary mission of education, generating new knowledge, and service.*

Walter Massey, former Director of the National Science Foundation [1994, p. 201], has stated that: "Commitment to scholarship and research is what distinguishes the research university from other types of institutions, but it does not define them." He further suggests that the primary mission of a research university still is to educate and it is this mission that an academic institution needs to focus on. Research and scholarship in an academic institution should be undertaken to support excellence in teaching and service.

13.2 FEDERAL SUPPORT OF UNIVERSITY RESEARCH: AN ENTITLEMENT OR A MEANS TO ACHIEVE NATIONAL GOALS?

Vannevar Bush's assertion that benefits from basic research would be too far in the future for industrial support and that the cost is too great for philanthropy alone has historically proven to be correct. Over the years, industry has accounted for less than 10% of the total research funding at academic institutions in the United States (as discussed later in this chapter on sources of academic research funding). This then points to the need for the government to assume this responsibility and to nurture basic research efforts. It was further argued that such basic research would take place primarily in the universities and that major decisions regarding the allocation of funds for basic research are best left to the scientific establishment [Bush, *Science, The Endless Frontier,* 1945].

As discussed earlier in the chapter, U.S. academic institutions' unique characteristics are related to (a) free flow of ideas and talent from throughout the world, (b) a long history of public support of this enterprise, (c) opportunities it provides for the initiative of individual scientists at relatively early stages in their career, and (d) the history of preeminence of U.S. academic institutions in scientific accomplishments [National Academy of Engineering, 1993].

Neal Lane, Director, National Science Foundation, has stated that the Federal government's support of basic research at academic institutions is important because it fulfills crucial national objectives of generating new knowledge and educating scientists and engineers. Furthermore, many authoritative studies have shown that support for basic research is a prudent investment that pays substantial dividends, exceeding 20% in real economic terms [Lane, 1996].

U.S. industry leaders, normally critical of government spending, recognize the importance of Federal government support of academic research. Fifteen chief executive officers of the largest U.S. corporations, in letters to the Senate Majority Leader and Speaker of the House [March 13, 1995], vigorously supported Federal investment in academic research and they stated:

> We . . . urge your continued support of a robust federally supported university research program. We recognize you face some tough choices as you deliberate and ultimately decide which federal programs merit continued support. We understand that priorities must be established. We strongly recommend, however, that you maintain a high priority on supporting the research efforts being carried out in our universities.
>
> . . . America's leadership position in an ever-increasing globally competitive economy has been fueled by our technological prowess. Our universities, and the research programs pursued therein, have played a pivotal role.
>
> Our message is simple. Our university system and its research programs play a central and critical role in advancing our state of knowledge. Without adequate federal support, university research efforts will quickly erode. American industry will then cease to have access to the basic technologies and well-educated scientists and engineers that have served American interests so well.

Investment in basic research is clearly needed to achieve national goals of economic growth, job development, and international competitiveness and to address crucial societal problems. Decisions about the level of investment in basic research is an open question. This is further elaborated in the following sections. Investment in basic research at universities, as discussed earlier, is best justified to support the institutions' primary mission of education, generating new knowledge, and service. Thus, it is not an entitlement in any sense. Considerable criticism, however, has been leveled against the university research enterprise. Some argue that it has increasingly become an entitlement rather than a means to achieve national goals. Questions have been raised about the utility, effectiveness, and efficiency of this investment. Some examples of concerns and criticisms related to the university research enterprise follow.

In the engineering colleges in the United States, there is an emphasis on pursuit of excellence and leadership in basic research; this, in turn, leads to undervaluation of other types of technical activity. By ". . . overvaluing the pursuit of original knowledge relative to excellence in execution, many engineering schools have helped to create, or at least to sustain, dysfunctional walls between research and other downstream technological activities in American industry" [National Academy of Engineering, 1993, p. 64].

There is a mismatch between the university research agenda and the problems and needs of society, especially those perceived by industry [Brooks, 1994a]. Academic research is inspired by the culture of the discipline. Problems and needs of society arise in a complex technological and social context, thus requiring cross-disciplinary collaboration. Though, recently, at universities, considerable emphasis is being placed on interdisciplinary research collaboration, there is little evidence that this is either rewarded or increasing in any appreciable manner.

The manner in which Congress earmarks academic research is another reason for serious criticism of the university research enterprise. Hanson [1994] and Long [1992] discuss the practice of appropriating federal funds by the use of academic earmarks. Universities use political clout to get funding, and this political process is not reflective of the merit of or need for a particular research project. This process, very likely, would gradually lower the quality of scientific research by rewarding political strength rather than true scientific merit.

13.3 BASIC RESEARCH—WHO NEEDS IT?

The conduct of basic research that does not have immediate commercial benefit raises questions such as "Who needs it?" and "Why do it?"

The genius of Vannevar Bush in articulating the importance of basic research over half a century ago, when it was an untested concept, is quite remarkable. In commenting on the importance of investment in basic research, which is responsible for scientific progress, he stated [Bush, 1945, pp. vi and 1]:

> Scientific progress is one essential key to our security as a nation, to our better health, to more jobs, to a higher standard of living, and to our cultural progress. . . . Without scientific progress no amount of achievement in other directions can insure our health, prosperity, and security as a nation in the modern world.

Benefits from basic research are hard to predict. Even the scientists are unable to see how their discoveries could someday address crucial human needs. J. Michael Bishop (Nobel Prize winner in Physiology) describes his experience to illustrate this point [1995, p. 63]:

> In 1911, Peyton Rous at the Rockefeller Institute in New York City discovered a virus that causes cancer in chickens, a seemingly obscure observation. Yet 65 years later, that chicken virus was the vehicle by which Harold Varmus and I, and our colleagues, were able to uncover genes that are involved in the genesis of human cancer. The lesson of history is clear, the lines of inquiry that may prove most fruitful to science are generally unpredictable.

As discussed in the chapter on R&D organizations and research categories (Chapter 1), a great majority (approximately 67%) of the research conducted at universities falls into the category of basic research. However, basic research conducted at U.S. government and industry laboratories is a much smaller proportion of their overall research activities: 16% and 6%, respectively. Thus, investment in basic research by the nation is of special concern to the university community.

In the public sector, as in the industrial sector, decision-makers are often concerned with current problems and issues. Since basic research involves discovering fundamental mechanisms rather than achieving practical applications, and there is considerable uncertainty and risk, it is not difficult to understand why support for basic research is not always as strong as one might expect.

Personnel involved in basic research need resources to undertake such activities. In addition, scientists want freedom to investigate the topics they deem worthwhile. Basic research requires a great deal of time, but decision-makers tend to be impatient; when faced with the problem of distributing scarce resources among competing requirements, decision-makers may find it difficult to support basic research.

One could argue that if resources are not invested in basic research, the foundation necessary for technological innovation (infrastructure for training scientists and engineers at universities, trained personnel, and new inventions) would be missing. On the other hand, without technological innovation and investment in technology, the increase in productivity and the general economic well-being of society would be missing. After all, commercialization of technology is the engine that produces resources for use in basic research and for investment in the future. While the scientific and the university community may want the decision-makers to understand the importance of basic research for the long-term viability of an industrialized society, it is equally important that scientists understand the importance of the innovation process, which turns outputs from research and development into useful commercial products so that the extra resources needed for basic research can become available.

Basic research focuses on the development of new knowledge, much of which is embodied in scientific information that cannot be turned into a marketable private property. As Merton [1973, p. 273] has suggested, the findings of science are a product of collaboration within the scientific community. Discoveries are the property of the commons, and the rights to these properties are assigned to the wider scientific community. This implies that in most cases the output from basic research is not directly marketable. This raises questions such as:

- Who should fund basic research?
- How should the resources devoted to basic research be determined?
- How should efficiency in the use of these resources be achieved?

Funds for Basic Research

There seem to be three major possible funding sources; private enterprise through the free market economy, governmental agencies, and nonprofit foundations. It is useful to examine which sources are most likely to fund activities that will improve society.

Arrow [1974, pp. 144–163], discussing the allocation of resources for invention, has treated this subject very elegantly and thoroughly. He suggests that the possible failure of perfect competition to achieve optimal resource allocation through the free market system is related to several factors. His discussion is too technical for review here, but the conclusion is clear: The free market system is not going to be able to allocate the necessary resources to basic research activities. In addition, no matter what the demand for the output of basic research might be, it would be less than optimal in a free market system, for two reasons: "(1) since the price is positive and not at its optimal value of zero, the demand is bound to be below the optimal; (2) . . . at any given price, the very nature of information will lead to a lower demand than would be optimal" [Arrow, 1974, p. 154]. What this means is that even if the free market system could provide the necessary basic research funds, this would not be in the best interest of society since the basic research output would be utilized at a level that is less than optimal for society. Noncommercial sources have historically been the major funding sources of basic research. And this not only makes sense, but is desirable from a societal viewpoint.

It is important to note that sometimes the output from basic research is needed to conduct applied research and that science (the output from basic research) and technology interact with the market or human needs to foster the innovation process. Logically, the only basic research supported by industry then would be in areas where investment in basic research is essential to complete product developments that have present or potential commercial benefits.

In discussing this issue of basic research, Professor Andrew Schofield of Cambridge University pointed out that in pursuing new knowledge, in addition to the wages paid to a researcher and the capital invested in laboratory facilities, there is the fundamental issue of the individual scientist's motivation and drive to discover. Many scientists invest considerable personal time, effort, and emotional energy in pursuing new knowledge regardless of its marketability or immediate utility. This investment by the individual scientist may be far more important than the institutional investment in basic research. Management practices that recognize this and foster an innovative environment—via job design, organization development, and reward system—may result in increased effectiveness of institutional investment in basic research.

Figures 13.1 and 13.2 show historical trends for research funding sources for academic institutions in the United States. As these figures show, federal government support for academic R&D has been declining and nonfederal funding sources are increasing. Also, the rate of increase for total research funding has declined, and this funding is barely keeping pace with inflation. This has obvious implications for research-intensive universities. The type of research (for selected fields) undertaken by universities is shown in Figure 13.3.

Since Federal government and nonprofit organizations are, and are likely to continue to be, the main supporters of university research in the future, any decline in such support will inevitably reduce the total support available to academic institutions. As the figures show, Federal government support for university research has declined from 70% of all research funding in the 1970s to an incredible low of 57% in 1993. The focus of research is changing from defense to dual use (defense and civilian use), to strategic knowledge generation for international competitiveness, and to industrial collaboration. Universities need to consider the implications of this trend for research and graduate training.

Level of Resource Allocation

Because of the reasons cited above, private enterprise is not able to allocate the necessary resources for basic research using the efficient competitive market pro-

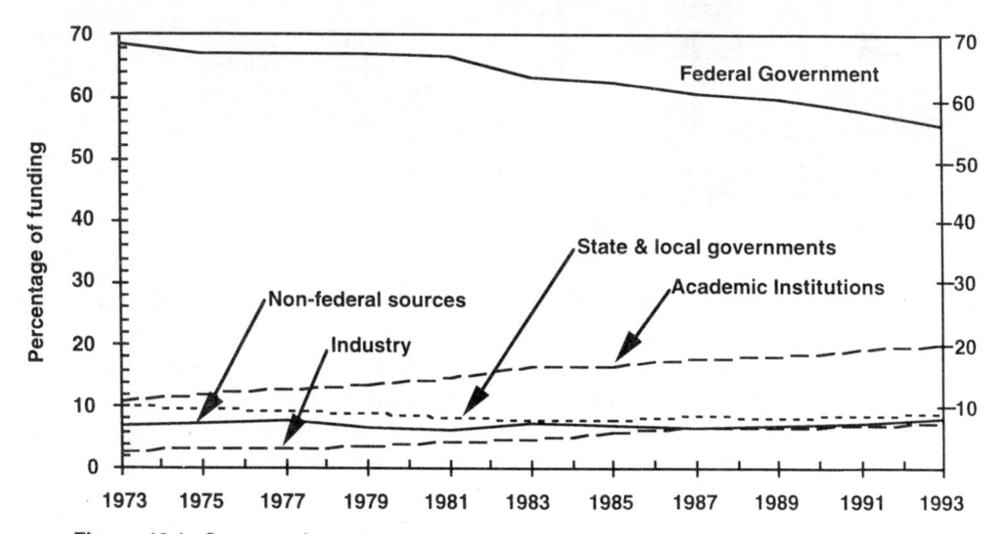

Figure 13.1. Sources of academic R&D funding, by sector. Data for 1993 are estimates. (Source: *Science and Engineering Indicators,* 1993, p. 369.)

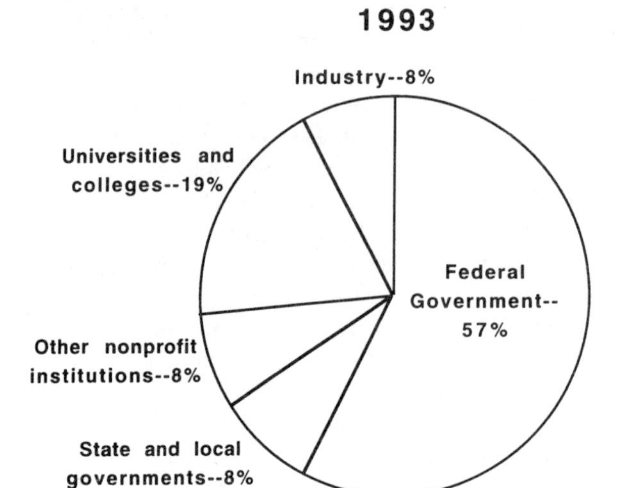

(Millions of dollars)						
Year	Total	Federal Govt.	State and local govt.s	Industry	Univ./ Colleges	Other nonprofit institutions
1984	8,620	5,430	690	475	1,411	614
1985	9,686	6,063	752	560	1,617	694
1986	10,927	6,710	915	700	1,869	733
1987	12,152	7,342	1,023	790	2,169	828
1988	13,463	8,192	1,106	872	2,357	936
1989	15,019	8,997	1,234	998	2,714	1,076
1990	16,344	9,637	1,339	1,130	3,033	1,195
1991	17,638	10,226	1,484	1,210	3,404	1,313
1992	18,880	11,087	1,506	1,302	3,576	1,409
1993 (est.)	19,950	11,500	1,600	1,550	3,750	1,550

Figure 13.2. Academic R&D expenditures, by source of funds. (Source: National Science Foundation, NSF 94-323, 1994.)

cesses, except where basic research is needed for commercial product development. Any time government or other nonprofit organizations find themselves engaging in such activities, the level of resource allocation becomes a problem. Other surrogate measures and social needs have to be used, recognizing that the level of resource allocation in such cases cannot be as efficient or as self-correcting as the market mechanism. Discussion of some of these measures and needs follows:

Economic Considerations. Presumably, resources should be allocated so that the expected marginal social benefit exceeds or equals the marginal social benefit

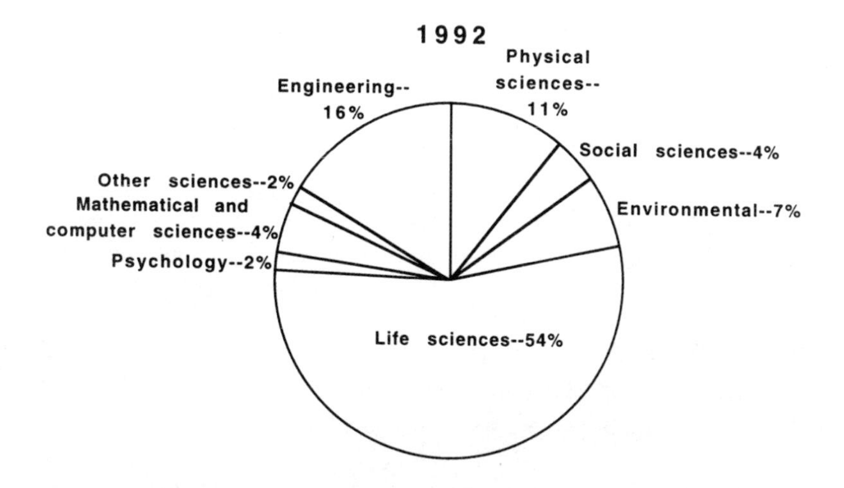

Field	1985	1986	1987	1988	1989	1990	1991	1992
	(Millions of dollars)							
Total	9,686	10,927	12,152	13,463	15,019	16,344	17,638	18,880
Physical sciences	1,148	1,286	1,398	1,544	1,649	1,809	1,945	2,058
Mathematics	128	152	177	199	214	221	230	247
Computer sciences	281	321	372	409	472	514	554	556
Environ. sciences	705	776	839	894	1,014	1,080	1,130	1,249
Life sciences	5,279	5,890	6,528	7,257	8,082	8,748	9,496	10,228
Psychology	158	170	187	213	237	258	291	336
Social sciences	383	462	502	552	637	706	755	817
Other sciences	186	228	256	289	316	335	330	307
Engineering	1,418	1,641	1,892	2,096	2,398	2,663	2,907	3,082

Figure 13.3. Academic R&D expenditures, by field. (Source: National Science Foundation, NSF 94-323, 1994.)

of competing usages. Because of considerable uncertainty and other complexities, these computations may simply be based on a preference index, which might itself be determined by the expected economic benefits. As mentioned earlier, authoritative studies have shown that support for basic research is a prudent investment that is estimated to pay dividends to society exceeding 20% in real economic terms [Lane, 1996].

Its Contribution to Social Welfare. As Freeman [1982, p. 201] has stated, "The advance of science and technology must find its support and its justification, not merely in the expectation of competitive advantage, whether national or private, military or civil, but far more in its contribution to social welfare, conceived in a wider sense." One of the colleagues at Cambridge University stated that basic research should be looked at as "an important cultural activity. If one is looking for direct returns, one is not likely to see a penny back." Therefore, decisions on allocating resources for basic research have to consider its enormous societal benefits.

Based on Comparative Needs. By looking at the historical record, one could gather percentages of gross national product devoted to basic research by other nations and compare these data to the impact of varying percentages of investment in basic research on economic productivity and the social welfare of society.

Other Considerations. In discussing this with some eminent R&D managers (identified in the Preface), a number of other practical considerations emerged, as reflected in the following questions. Basic research investment should run 10–15% of total R&D and depends on:

"Do we have the people who are interested?"
"Do we have the people who have the time?"
"Do we have the people who are able to conduct basic research?

This really determines the level of basic research funding in most cases.

As discussed in Chapter 1, a scientist who engages in a mix of basic and applied research is likely to be more productive than one who does not. Therefore, even for a profit-oriented industrial organization, allowing the scientist to undertake some basic research, regardless of its profitability or immediate utility, is best for all concerned. This will keep the scientist at the cutting edge of the discipline, provide a higher level of motivation, and, inevitably, result in greater productivity in the applied aspects of the research as well.

Efficient Usage of Basic Research Investment

At the macro level, efficient usage is partially achieved by establishing an appropriate science policy. For basic research, by and large, payment is independent of the results achieved. So, the traditional market mechanism is irrelevant. There are, however, other mechanisms that ensure efficiency at the macro and micro levels:

- The peer review process to determine the merits of proposed basic research projects.
- Awarding research contracts to those who have previously performed research successfully.
- The ethos of the scientific community (universalism, communalism, disinterestedness, and organized skepticism) that provides a vigorous review and analysis mechanism contributes to efficiency.
- Suggestions presented in this book, for creating a productive and efficient R&D organization, might be helpful.

Peer review and other processes mentioned here have historically worked well, but they tend to be a bit conservative and lack flexibility. Major breakthroughs are unlikely to be anticipated by a committee. In an R&D organization a portion of total R&D funding can be provided to managers at different levels as discretionary funds to be used for high risk research projects. Our experience indicates that this approach has proved very fruitful. It reduces the lead time required for the normal funding cycle for a project and provides the flexibility so necessary to undertaking high-risk, exploratory research requiring relatively low investment.

13.4 UNIVERSITY–INDUSTRY LINKAGE

As participants in a Technological Innovation Symposium savored the food and drink in the Fellows' Dining Room at Churchill College (Cambridge University), Maurice Goldsmith [1970, p. xiii] stated that "it was easier to accept why the 'educational purists' of the past, in such a cloistered, cultivated atmosphere, had been able to contribute to Britain's slow decline in industrial efficiency by insisting on the separation between the university, technology and industry."

If a university can contribute to the decline of a major world industrial power like Great Britain, presumably the same university has the potential to contribute to the rise of a nation to a world industrial power. This would further point to the importance of the university community contribution to innovation and industrialization. Cambridge University provides an example of university–industry cooperation that is worth noting.

The Cambridge Phenomenon

Notwithstanding Goldsmith's comments, the fact is that few universities, if any, have contributed so much to Britain's industrial efficacy, economic well-being, and technological leadership as has Cambridge University. During World War II, Cambridge scientists helped develop technologies such as radar, telecommunications, and nuclear physics that helped provide the allies with a winning edge! At present, the well-known "Cambridge phenomenon," which has helped foster development of many science parks around Cambridge, has resulted from the collaboration between the Cambridge University community and industry. Capital investment for the infrastructure (land, buildings, and laboratory facilities) for one of the largest science parks around Cambridge was provided by one of the colleges of Cambridge University [Wicksteed, 1985]. In fostering university–industry links, Cambridge University consciously avoided a structured and detailed policy governing these ties. On the issues of intellectual property rights, risk and liability, and industrial liaison, the university position evolved as follows.

Cambridge believes that the ownership of intellectual properties rests with the individual University member, unless the contract governing the work in which the know-how is acquired specifies otherwise. The University does not exercise any control over, or have a financial interest in, the exploitation of an academic's know-how, unless the academic asks the University to play such a role. It is felt that successful exploitation must ultimately depend on the motivation and skill of the academic.

Cambridge University has consistently taken a relaxed and liberal attitude toward the time spent by faculty on outside work. It is presumed that outside activities will be beneficial to teaching and research activities. The University does not accept any legal liability for work done by faculty members for outside organizations. Naturally, if the rewards accrue to the academics, the latter must accept the risks too and make their own arrangements for professional liability, and so on.

The Cambridge experience shows that a relaxed attitude and the simplicity of this industrial liaison arrangement have helped nurture a culture that encourages and is supportive of university links with industry [Wicksteed, 1985, p. 77]. As exemplified by the Cambridge phenomenon, the university-industry link positively affects the institution's responsibilities for teaching, research and public service. In addition to direct economic benefits to the nation, this link enhances the faculty's intellectual and technical capabilities.

Science Parks

In the United States, high-technology enclaves include "Silicon Valley" in California, Route 128 in Boston, the Research Triangle Park in North Carolina, and others. In 1940 Santa Clara County, where Silicon Valley is located, was a peaceful agricultural region with a population of only 175,000. It is now one of the densest concentrations of high-technology enterprises in the world and has reached a population of over 1 million. A 1984 estimate put the number of high-technology firms in Silicon Valley at a little over 1200, employing around 190,000 persons [*Washington Post,* December 3, 1984]. This development has been fostered by the presence of major universities such as Stanford and the University of California, Berkeley. Similarly, the Route 128 phenomenon has been helped by the proximity of MIT and other universities in the Boston area. The size and scope of spin-off industries in both Silicon Valley and Route 128 are far more extensive than the industrial parks around Cambridge. In the United States there are other university locations as well where there is a concentration of high-technology industry, though none on the scale of Silicon Valley or Route 128.

Discussing the lessons from establishing science parks, Saxenian [1994] writes that California's Silicon Valley and Boston's Route 128 attracted international acclaim in the 1970s as the world's leading centers of innovation in electronics. As traditional manufacturing sectors and regions fell into crisis, policy-makers and planners around the world looked to these fast-growing regions as models of industrial revitalization, seeking to replicate their success by building science parks, funding new enterprises, and promoting links between industry and universities. In the years following the high-tech manufacturing slump of the 1980s, Silicon Valley rebounded strongly while Route 128 did not. Silicon Valley's experience shows that, paradoxically, some regions offer an important source of competitive advantage even as production and markets become more global.

A case study of the Connecticut Technology Park near the University of Connecticut was used to examine the question of whether such university-related high-technology parks represent an effective development tool for the university and the region [Lewis and Tenzer, 1992]. Findings indicated that:

1. These ventures require time and money to succeed.
2. The choice of methodology for implementation is critical.
3. Poor relations with governments at the national, state, and local levels inevitably cause bureaucratic delays and political dissention.
4. There must be some reasonable expectations that market demand for the project exists or will exist in the society and economy.
5. Projects should have clearly defined objectives and goals.

Increasingly, industry is looking to outside sources for conducting research and for technology development. Maintaining in-house high-quality personnel and laboratory facilities is becoming increasingly expensive. Leveraging industry investment in research by collaborating with academic institutions is likely to become more prevalent than ever before.

In a study conducted by Roberts [1995], he looked at university–industry linkages in the United States, Europe, and Japan as related to four areas: collaborative research efforts, obtaining innovative ideas, determining technology trends, and training company personnel. In all of these categories, European and Japanese firms displayed far greater commitment to university–industry linkages than did the U.S. firms. For example, university collaborative research activities were 2.5 times greater for Japanese firms than for U.S. firms [Roberts, 1995].

13.5 UNIVERSITY–INDUSTRY–GOVERNMENT INTERACTION

As discussed in *Science Indicators* [1985, p. 108, 1993], the American system of university–industry research connections is without parallel in the world. The system is complex, involving individual, institutional, and corporate responses to many needs and opportunities. This interaction may consist of general research support to universities in the form of monetary gifts, equipment donations, endowment funds, construction of research facilities, or exchange of personnel. It may also take the form of cooperative research support through research consortia, cooperative research centers, and university-based institutes serving industrial needs.

The American university and government relationship is unique. As discussed by Donald Kennedy [1985, p. 480], the former President of Stanford University, the Federal government could have established a set of quasi-independent laboratories in cooperation with the industrial sector, like the Max Planck Institutes of Germany, or it could have established large government laboratories. Instead, for publicly supported fundamental research, the government has historically relied on the university community. This approach guaranteed that new discoveries in science and the training of such scientists could take place at the same locations—the great American universities.

The former President of Harvard University, Derek Bok [1982, p. 153], has suggested several mechanisms for improving university–industry linkages. Some examples are: industrial scientists teaching as adjunct professors and thereby contributing to the academic program while providing a channel of communication between the university and an industrial laboratory; arranging postdoctoral programs for industrial scientists at universities; university scientists working as consultants to industry, thus bringing recent developments and critical judgments from the university to industry, or performing contract research for industry; combined university–industry–government research programs; or the establishment of high technology research enterprises by faculty members themselves as exemplified by the Cambridge phenomenon discussed earlier.

Science and technology, as we have seen, contribute to national economic efficiency and productivity. The expediency and efficiency with which we are able to translate scientific knowledge into innovation—commercially useful products and processes—form the mechanism for this increase in national productivity and economic well-being. The universities employ outstanding scientists and engineers who perform the research that forms a basis for the innovation process. An industry or a government research laboratory that fosters interaction between its scientists and the university community is bound to improve its research productivity and quality. In addition, unique laboratory research facilities at the universities, and the personnel needed to operate such facilities, would be quite expensive if not impossible to duplicate at most commercial or governmental laboratories. In many cases, therefore, especially for basic and related applied research projects, sponsors can leverage their investment by providing research grants to a university.

According to *Science and Engineering Indicators* [1993], there has been a considerable increase in industry's interaction with the university community. As one would expect, by supporting academia, industry gains access to both cutting-edge research and highly talented personnel trained at the academic institutions. On the other hand, the university researchers benefit from funding, industry research facilities, and a clear understanding of user and societal needs. Many industry–university interactions have benefited from a variety of federal and state government programs set in place explicitly to encourage such collaboration. In many of these cases, programs require participation and contribution from industry and state. This sort of an interaction has also helped in undertaking research activities which cover the full spectrum from basic research to application and commercialization.

During the last 10 years, industrial R&D support to academic institutions has grown more rapidly than support from other sources. Though still a very small part of the total research investment, the industrial share of academic support has grown from 3.9% to 7.3% of total research funding between 1980 and 1993 [*Science and Engineering Indicators,* 1993, p. 132]. While research funds from industry are distributed to academic institutions through various means, it seems that the most used mechanism by far is via industry funding of university affiliated research technology centers. Figure 13.4 shows how these centers have grown over the years and also shows the different levels of funding provided by each of the partners [*Science and Engineering Indicators,* 1993, p. 121].

In summary, there are many ways industrial and governmental research labora-

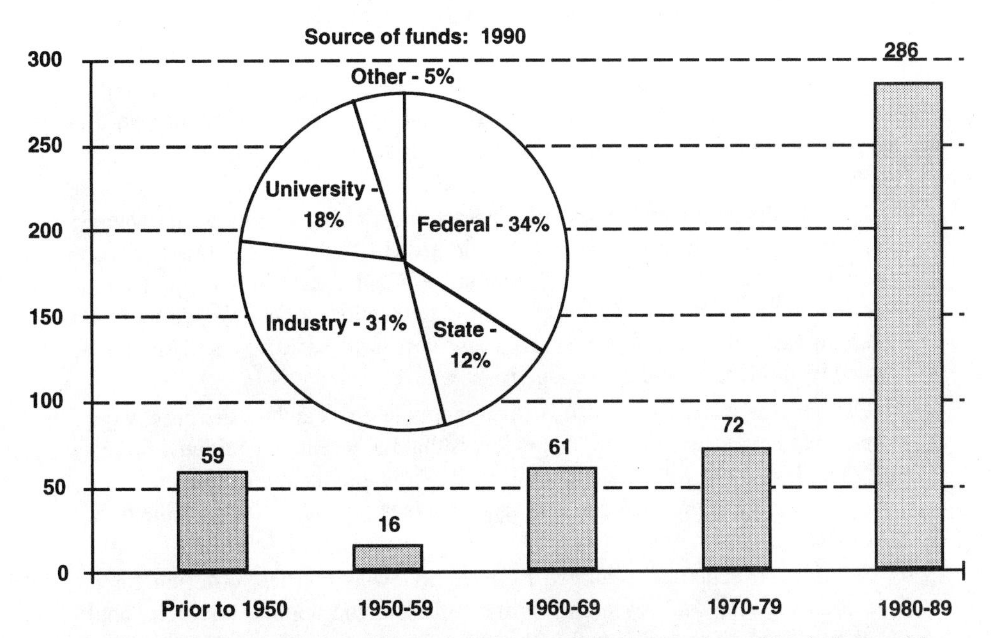

Figure 13.4. Growth in university–industry research centers, and source of funds. Data are for centers existing in 1990. Of an estimated 1058 centers, 458 provided funding data and 494 provided founding data. (Source: W. Cohen, R. Florida, and W. R. Goe, *University Industry Research Centers in the United States* [Pittsburgh: Carnegie Mellon University, 1993], as reported in *Science and Engineering Indicators,* 1993, p. 121.)

tories can leverage their investment in research by developing partnerships with academic institutions. For example:

- By providing seed money to a university researcher to explore an area of interest to an industrial consortium. If research results are promising, other agencies such as the National Science Foundation (NSF) may provide funding to continue the effort.
- By providing support to carry out, beyond the basic research stage, research activities originally funded by NSF or other noncommercial organizations.
- By providing graduate student support for research projects. This engages not only the student but a significant effort of faculty in a research area at a very nominal cost. Experience shows this has other benefits as well, such as the availability of trained scientists in areas of interest to the sponsor.
- By jointly building and operating some aspects of a research facility and thus sharing equipment, personnel, and costs. There are numerous successful examples of such endeavors.
- By creating research technology centers at universities.
- By exchange of research personnel.

13.6 RETHINKING INVESTMENT IN BASIC RESEARCH

The debate about benefits from investment in basic research and how such activities should be managed is likely to continue. Some of the general comments and differing views are summarized in the following paragraphs.

1. Basic research is a source of fundamental knowledge, leading to innovation, technology development, and economic growth. Basic research is best conducted primarily in universities without political constraints or application-oriented management. Overemphasis on application orientation would be an unmitigated disaster. The contrary view is that all basic research should be tied to application and so managed.
2. One view is that individual-investigator-initiated research is the only way to produce new ideas. The opposite view is that the researcher-initiated projects would lack a clear understanding of technology and its application and, thus, are not likely to produce good science or robust new ideas for possible utility to society.
3. Scientific excellence as judged by peer review is the best criterion for research funding, and research funding for basic and applied research should be managed by an individual with technical expertise. The opposing view is that research funding for basic and applied research should not be judged only by scientists. Individuals knowledgeable about application and marketing should have a major role in these decisions. Peer review process is too conservative and is not likely to produce major breakthroughs.
4. Applied research is likely to restrict creativity and serendipitous discovery. The connection between research results and societal benefits is indirect; this means that high-quality basic research will naturally translate into more benefits for society at large. The opposing view is that investment in basic research can only be justified by the direct benefits to the nation in terms of economic well-being and quality of life. This connection needs to be explicit and not indirect.
5. Science and technology have served society well. The high standard of living enjoyed in the industrialized countries is a tangible example of this. Investment in basic research, mostly at universities, has contributed greatly to this. The contrary view is that science and technology have not served society well; that is, they are considered to be a Faustian bargain. All benefits have had a darker side even more devastating. Examples provided are chemicals, drugs, pesticides, and herbicides that are having long-term adverse effects on human health and the environment.
6. Scientific research conducted at academic institutions has not contributed significantly to international economic competitiveness and national economic well being. Economic benefits as evidenced by commercialization and industrial success from such research have not been fully realized. The other view is that scientific research conducted at universities is only a small part of

commercialization and industrial success; the scientific community is being held hostage for the rest of the system failure [Brooks, 1994b].

13.7 SUMMARY AND CONCLUDING COMMENTS

What follows are our opinions about investment in basic research. Readers may find them debatable, but we hope to stimulate their thinking on this subject.

- In the United States, the largest and the most creative segment of the basic research enterprise resides at the academic institutions [National Academy of Engineering, 1993].
- We believe that both the public and Congress need to be educated about the importance of basic research and about the need for increased funding of this activity. In a democracy there is always some pressure to fund projects that would meet constituents' current needs. Basic research often has its payoffs years later and, at times, after the present constituents have died. It is important to understand this time lag.
- The National Research Council of the National Academy of Sciences did a very thorough review of the evidence and concluded that it is very clear that basic research gets initiated by individuals. It recommended support of investigator-initiated, merit-reviewed, university-based research to a much greater extent than support of particular universities in particular states (funding favored by some members of Congress, for obvious reasons). (For details see *Allocating Federal Funds for Science and Technology* [1995], National Academy Press, Washington, D.C.).
- There is a need for research on science and society. Some scientific developments have adverse effects on health, social well-being, and the environment. A full understanding of these relationships is important and can be used to influence some applied research. In our opinion, basic research rarely can be blamed for the adverse effects of science on society. It is usually the applications of basic research that lead to societal problems.
- The amount spent on basic research is minuscule relative to the amount spent on applied research. It is not appropriate to blame basic research for not solving the problems of international competitiveness. The blame can often be placed on the poor decisions of corporate Boards of Directors. In Japan the Chief Science Officer of the Corporation is a member of the Board of Directors in about 95% of the corporations; in the United States, that is the case for only 20% of the corporations [Roberts, 1995].

In sum, we believe that basic research must receive considerably more emphasis in the future if U.S. competitiveness is to be maintained in the international market place.

13.8 QUESTIONS FOR CLASS DISCUSSION

1 What problems do you see with the university research enterprise? And what strategies do you propose to address those problems?

2 What policies and strategies would you propose to ensure that research and scholarship in academia support the institutions' main mission of education?

3 Who should provide funds for academic research and why?

4 Should university–industry–government research laboratory linkages be strengthened? If so, how?

13.9 FURTHER READINGS

Cole, J. R., E. G. Barber, and S. R. Graubard (Eds.) (1994). *The Research University in a Time of Discontent.* Baltimore: Johns Hopkins University Press.

Holden, C. (1995). Careers '95: The future of the Ph.D. *Science,* **270,** 121–122 (October).

Lederman, L. M. (1991). Science: the end of the frontier? *Science (Supplement).* Report to the Board of Directors of the American Association for the Advancement of Science, Washington, D.C.

Lederman, L. M. (1992). The advancement of science. *Science,* **256,** 1119–1124 (May).

14 STRATEGIC PLANNING FOR R&D ORGANIZATIONS

This chapter covers strategy in various contexts (e.g., corporate, military), technology strategy, strategic planning, strategic thinking, and strategic management, as well as environmental scanning and external trends affecting strategic planning. Discussion of these topics is in the context of technology and R&D organizations and should provide a basis for developing a strategic plan for such organizations. To further assist the reader, the major elements of the strategic plan for a research-intensive college in a university is included at the end of the chapter.

14.1 "STRATEGY" IN CORPORATE CONTEXT

One commonly used corporate strategy model, the Harvard Business School model, known by its acronym SWOT, encompasses organizational strengths, weaknesses, opportunities, and threats. By understanding an organization's strengths and weaknesses, according to this model, a strategic plan can be developed and it can assist in exploiting opportunities and minimizing risks due to threats and uncertainties.

Strategy in the corporate context is used as a means of establishing the organizational purpose, long-term objectives, action plans, and resource allocation priorities. Some definitions of strategy include external threats, opportunities, internal strengths and weaknesses, and the stakeholders. Examples of some of these definitions are as follows:

> Strategy is the determination of the basic long-term goals of an enterprise and the adoption of courses of actions and the allocation of resources necessary to carry out these goals [Chandler, 1962, p. 13].

> Strategy formulation and implementation include identifying opportunities and threats in the organization's environment, evaluating the strengths and weaknesses of the organization, designing structures, defining roles, hiring appropriate people, and developing appropriate rewards to keep those people motivated to make contributions [Argyris, 1985, p. 1].
>
> Corporate strategy is the pattern of decisions in a company that determines and reveals its objectives, purposes, or goals, produces the principal policies and plans for achieving those goals, and defines the range of businesses the company is to pursue, the kind of economic and human organization it is or intends to be, and the nature of the economic and non-economic contribution it intends to make to its shareholders, employees, customers, and communities [Andrews, 1980, p. 18].

Hax and Majluf [1988, p. 102] combined many of these into a single, more comprehensive definition of strategy:

> Strategy
> —Is a coherent, unifying, and integrative pattern of decisions;
> —Determines and reveals the organizational purpose in terms of long-term objectives, action programs, and resource allocation priorities;
> —Selects the businesses the organization is in or is to be in;
> —Defines the kind of economic and human organization the company is or intends to be;
> —Attempts to achieve a long-term sustainable advantage in each of its businesses by responding properly to the opportunities and threats in the firm's environment and the strengths and weaknesses of the organization;
> —Engages all the hierarchical levels of the firm (corporate, business, functional); and
> —Defines the nature of the economic and noneconomic contributions it intends to make to its stakeholders.

According to a study by Rheem [1995] investigating the relationship between productivity growth and corporate activities, highly productive companies had long-term strategic planning whereas less productive ones did not. High-performing companies planned four to six years ahead and reviewed their plans every two to three years, whereas low-performing companies did not have any long-term strategic plans. If long-term thinking and strategic planning are important for corporations, then strategic planning for R&D organizations is even more important because of the future orientation of the enterprise.

14.2 "STRATEGY" IN MILITARY CONTEXT

Historically, the military has used strategic planning concepts to establish and achieve its goals. We thought it might be useful to take a brief look at this. According to the U.S. Joint Chiefs of Staff, military strategy is ". . . the art and science of employing the armed forces of a nation to secure the objectives of national policy by the application of force or the threat of force." General Maxwell

Taylor characterized strategy in terms of ends, ways, and means [Lykke, 1989] as depicted below:

Strategy =	**Ends**	**+ Ways**	**+ Means**
	Objectives	Concepts	Resources
	(What? Why? When?)	(How? Where?)	(With what? With whom?)

14.3 TECHNOLOGY STRATEGY

In a comprehensive study related to technology strategy, which has significant implications for R&D strategy, Roberts [1995] describes the importance of the linkage of technology strategy to the business strategy. His study found that over 70% of Japanese firms have strongly linked their technology and overall corporate strategy, while in the United States only 50% did so. Some of the underlying reasons for this are lack of understanding and knowledge of research and technology on the part of the company's board of directors and the lack of presence of Chief Technology Officers (CTO) (i.e., engineers or scientists primarily responsible for R&D and technology policies of a corporation) on the company boards.

One of the most significant findings of a study by Edward Roberts [1995] was that, incredible as it may seem, in Japan, 95% of CTOs are members of main boards or boards of directors while in the United States this figure is only 20%. Roberts [1995, p. 48] states that ". . . this figure presents the strongest damnation of U.S. senior executive practice and prioritization." Results of this study also show that chief executives in Japan are far more engaged in technology strategy development and implementation than their American counterparts. Japanese chief executives spend considerable time in assessing both internal and external technological investment opportunities, while U.S. CEOs involve themselves primarily in controlling R&D budgets [Roberts, 1995].

Most of the decisions about investment in R&D to develop technologies are made in the corporate board rooms. Not having CTOs involved in setting priorities for resource allocation and making tradeoffs among competing demands is very likely to result in underinvestment in R&D. There is ample evidence that this is happening in U.S. industry. For example, Japan and Germany spend approximately 3.0% and 2.7%, respectively, of their GDP on nondefense R&D while the U.S. in comparison spends only 1.9 percent. Technology perspectives in industry thus can increasingly become short-term, making U.S. industry less competitive in a technologically intensive world.

Engineers and scientists, who benefit from increased support for R&D, can naturally be viewed as making self-serving arguments for enhanced research support. For an R&D unit in an organization, the arguments need not focus on increased R&D investment. Instead, the R&D units should involve top corporate managers in the research strategic plan development and in ensuring that technology strategy is integrated with corporate strategy. In the long run, we hope, in the United States as in other highly successful industrialized nations like Japan, the CTOs will be represented in the great majority of the U.S. corporate boards, especially in those corporations that are technology-intensive.

14.4 STRATEGIC PLANNING

A strategic plan may be viewed as a well-thought-out and disciplined effort to establish a vision, define long-term goals, and develop an action plan for *achieving those goals*. A strategic plan can serve many useful purposes. It can:

- Help provide a unity of direction and a cohesive framework for action
- Assist in coordinating disparate organizational goals and resources from a total organizational vantage point
- Provide an assessment of both the internal and external organizational context
- Provide a mechanism for focusing on future needs and staying in tune with fundamental organizational priorities

Overemphasis on numerical goals and rigid deadlines is likely to undermine the very objectives the strategic plan is to achieve. This could easily lead to the establishment of trivial goals and exaggeration or even falsification of results achieved.

Strategy Levels and Perspectives

Strategies are often developed at different levels with different perspectives. For example, in corporations they could be at the corporate, business, and functional levels. The corporate strategy would deal with decisions that address broad corporate interests, business strategy would aim at obtaining superior financial performance and would attempt to have a sustainable advantage over competitors, and functional strategies would consolidate the functional requirements demanded by corporate and business strategies [Hax and Majluf, 1996, p. 24]. At the functional level, presumably, goals, objectives, and specific actions will be formulated.

In a research organization or in academia, perspectives similar to a corporation are relevant. A research unit would correspond to a functional level and it would need to understand and respond to requirements postulated at higher levels that are consistent with overall university and college educational missions.

Horizontal Strategic Integration

In a research organization, horizontal strategic integration is not well done due to the disciplinary interests of individual scientists and due to the further segmentation of research programs and projects within a research unit. In academia, this is a very real problem. Faculty members are recognized and rewarded for their disciplinary work which requires little or no horizontal integration. Interdisciplinary research, so necessary for addressing crucial societal problems, does not get the support it needs. Thus, by and large, the culture of research organizations and the reward systems of research organizations work against horizontal strategic integration.

Take the example of a research-intensive college in a university. It would be prudent to identify potential opportunities to undertake interdisciplinary research and related educational programs. By establishing superordinate goals for the departments of the college, this synergism can be fostered. This would clearly add value beyond the simple sum of independent departments or units. Hax and Majluf [1996] suggest that appropriate pursuit of horizontal strategy may be one of the most critical ways to establish a superior competitive position. While many management approaches could help bring about horizontal strategic integration, rewards and organization culture that foster such integration are the key to shaping such behaviors.

Steps for Strategic Plan Development

Various authors and consultants use different steps for developing strategic plans. Most of these steps have common elements. Based upon several publications [Hax and Majluf, 1991; Hensey, 1991], some suggested steps for developing a strategic plan for an R&D organization are as follows:

- *Develop the Vision and Mission Statement.* Include vision, mission, corporate philosophy, and crucial strategic issues.
- *Identify Trends.* Include information about external trends based upon environmental scanning at the corporate level (as discussed earlier).
- *Identify Strengths and Weaknesses.* Delineate inherent competencies and emerging deficiencies.
- *Identify Potential Opportunities and Threats.* Based upon the environmental scan and trends, identify major potential opportunities and threats that are likely to adversely affect the activities and the mission of the organization.
- *Elements of a Successful Strategy.* This would include identification of those critical elements that are necessary for making the organization successful.
- *Strategic Goals and Objectives.* This would include specific goals and objectives that the organization would like to achieve within a certain time horizon.
- *Specific Actions.* This would include specific activities that the organization needs to undertake to reach each of the goals and objectives.
- *Resource Allocation.* This would include allocation of resources and the determination of performance metrics for management control.

14.5 STRATEGIC THINKING AND STRATEGIC MANAGEMENT

Nadler [1994] suggests that in most organizations too much effort is spent on strategic planning and not enough on strategic thinking. Inordinate amounts of time are spent on analysis and drawing up strategic plans that influence very little behavior and that cause no real changes to occur. The real value of strategic planning should be in the learning and development of a common frame of reference and the shared context for many decisions that are made in an organization over time [Nadler, 1994]. Strategic thinking and strategic management, thus, are important concepts.

Developing a strategic plan and implementing it involves strategic thinking and strategic management. *Strategic thinking* is a dynamic process of:

- Developing a vision of what an organization can be
- Understanding its history, strengths, resources and weaknesses
- Recognizing opportunities that may lie ahead
- Appreciating potential threats

Strategic management, then, is the art and science of implementing the strategic plan and managing strategic change. It serves to channel managerial tasks at different functional levels within the organization to achieve organizational goals established in the strategic plan. It acknowledges uncertainties and divergent points of view and provides the necessary mechanism to coordinate activities of various participants [Hax and Majluf, 1988].

14.6 ENVIRONMENTAL SCANNING AND EXTERNAL TRENDS

To ground the plan in reality, a strategic planning process must include an assessment of external trends based upon environmental scanning at the corporate level. A summary of these trends and their anticipated impact on the organization should be included in the plan. Some trend areas to examine may include:

- National and international security
- Changing world order
- National policies
- Customer dynamics
- Economics
- Investment in science and technology
- Organization mission and activities
- Organizational culture and leadership values
- Population, family structure, and other demographics
- Environment
- Regulations
- Human health and welfare
- Education
- Workforce demographics

14.7 MAJOR ELEMENTS OF THE STRATEGIC PLAN OF A RESEARCH-INTENSIVE COLLEGE IN A UNIVERSITY

Strategic Plan Outline

- Background
  - History
  - Staff
  - Responsibilities
- College research overview
- Importance of research
- Trends and driving forces
- Strengths
- Weaknesses
- Opportunities
- Threats
- Elements of a successful research strategy
- Vision/mission statement
- Strategic goals and objectives
- Example of specific actions

Background

History. The College of Engineering Research Office at the University of Trinity was established in 1970. Professor Jonathan Moore served as Associate Dean for Research from 1970 to 1986. Professor John Randolph, the current Associate Dean, assumed his responsibilities in 1987. The Associate Dean for Research serves as the focal point for all research activities within the college and is responsible for providing effective leadership and integration of effort within the college.

Perhaps the genesis of research activities in the college could be traced to 1948 when the College of Engineering Advisory Committee recommended that a survey be made of the research potential of the faculty. Based on the results of this survey, the committee recommended that faculty should be encouraged and rewarded to pursue research in order to improve the quality of the graduate and undergraduate education.

Staff. In addition to the Associate Dean, John Randolph, the College Research Office has one staff assistant to discharge research-related responsibilities. To undertake additional significant activities, the Associate Dean can engage temporary help on an as-needed basis.

Responsibilities. Responsibilities assigned to the Office of the Associate Dean for Research (OADR) are outlined below:

- Development of liaison with federal, state, and local agencies and private industry
- Coordination of major interdisciplinary research activities and programs
- Training and assisting faculty in grantsmanship and contract administration
- Liaison with the university's Office of Research and Advanced Studies
- Promotion of college research activities
- Coordinating and directing college-level Interdisciplinary Research Centers
- Managing a $1.4 million budget derived from overhead return and university and state research incentive awards.

College Research Overview

The table below summarizes College of Engineering contracts and grants for 1995–1996:

Federal	$13,793,081	71%
State and city	1,350,317	7%
Industry	2,383,088	12%
Internal, Foundation, and Other	1,793,509	9%
Total	*$19,319,995*[a]	*100%*

[a]University internal research support is not included.

Research performers for the college include:

- College of Engineering faculty 150
- Graduate students 850
- Research associates 35
- Technicians 15
- Collaboration with other colleges within the university
- Collaboration with other universities—over 25 nationally and internationally
- Collaboration with industry, state agencies, and other research institutes

Importance of Research

The college position on this is articulated in the following:

> Research in academia is best justified to support the institution's PRIMARY MISSION OF EDUCATION, generating and disseminating NEW KNOWLEDGE, and SERVICE.*

* Service to the profession, the college, university community, the state, and the nation.

One way to consider the importance of research would be to see how a decline in research funding would impact the college. The following is a suggested list:

- Budgets (immediate impact on all college initiatives since external research funding represents about 60% of the total college budget)
- Research infrastructure
- Graduate program quality
- Number of graduate students (an impact on college and University budgets)
- Faculty quality
- Program reputation
- Undergraduate programs
- Service through innovation
- Technology transfer

Trends and Driving Forces

It is important to understand international, national, and state trends and driving forces. Summary of relevant trends generated during the strategy development meetings follows:

- New world order—end of cold war!
- International economic competitiveness
- Paradigm shift in research
    - In the past, federal support for academic research doubled every 10 years. This trend is not likely to continue during the next 10 years.
    - How academia gets research funding is changing.
    - Shifting focus of funding
    - Changing technology emphasis
    - More applied research
    - Focus on enhancing technologies
    - State and university support stable or declining
- Tremendous growth in competition for research
- Emphasis on more collaborative and multidisciplinary efforts
- More opportunities for university–industry linkages

Strengths

Over the years the college faculty has distinguished itself and earned a national and international reputation in the following research areas:

- High-performance computing
- Information processing/intelligent systems

- Bioengineering
- Electronic materials and devices
- Infrastructure
- Bioremediation
- Air pollution
- Water and wastewater treatment
- Hazardous waste
- Chemical separation
- Pollution prevention
- Composites and polymers
- Biomaterials
- Ceramics
- Aerospace engineering

The research infrastructure that includes well-equipped laboratories, excellent technicians and support staff, and high-quality graduate students has resulted in one of the best research settings in the nation.

There is a history of research collaboration with other units of the university—for example, the Medical College and the Biological Sciences, Chemistry, and Physics Departments. This would allow the college to undertake interdisciplinary research programs, and this is expected to be the growth area for research.

Weaknesses

- The percentage of faculty who are research active is declining.
- University support for research is expected to be flat or decline.
- State support for research is declining.
- Research infrastructure in some areas is deteriorating.
- Budgets for federal agencies that provide the majority of external funding are declining.

Opportunities

- The creation of major interdisciplinary research centers with collaboration with other units within the university
- Establishment of university/industry research centers that focus on a full spectrum of research activities
- New areas such as biomedical engineering, dual-use (military and civilian) technologies, environmental engineering and sciences, health-risk assessment, environmental policy, and material sciences

Threats

- A precipitous drop in federal support which accounts for over 70% of current research funding
- Graduate student enrollment decline due to national economic factors
- A major decline in university and state support of the college research infrastructure

Elements of a Successful Research Strategy

- Associate Dean for Research must understand and learn about the capabilities and research interests of all faculty members.
- Recruit and support graduate students and research personnel of top quality, support new faculty, and suitably recognize and reward research active faculty.
- Recognize and focus on unique strengths and key areas of college interest by identifying research leaders in these areas and emphasizing high-quality research.
- Provide seed funding for necessary facilities, labs, equipment, and emerging research areas.
- Provide support for young faculty while they develop and establish their research programs.
- Help create an environment and a system conducive to collaboration and successful acquisition of large grants.
- Identify, organize, prepare, and submit multidisciplinary college-level research proposals.
- Encourage and support faculty/scholar programs.

Vision/Mission Statement

Foster research and external funding to support excellence in *education* and *scholarship*.

Strategic Goals and Objectives

1. Generate significant *interdisciplinary research* programs.
2. Facilitate the development of productive and active *research centers*.
3. Provide *recognition for quality research*-related accomplishments and efforts of faculty, research assistants and students, with a balance across departments.
4. Increase funding support for *research infrastructure*.
5. Integrate research programs with college goals of *excellence in education and scholarship*.

6. Support individual faculty *disciplinary basic research* efforts.
7. Increase *externally funded research program to $X/year* per faculty.

Example of Specific Actions

The following is an example of specific actions under one goal.

Generate significant *interdisciplinary research* programs:

- Identify research leaders.
- Form interdisciplinary research teams.
- Develop interdisciplinary research proposals.
- Provide seed money to develop ideas and working relationships among researchers that may lead to interdisciplinary research programs.
- The Office of the Associate Dean for Research actively assists in preparing a minimum of five major interdisciplinary research proposals per year, with a goal of successfully acquiring $*Y* million or more for interdisciplinary research per year.

14.8 SUMMARY

Strategic planning is particularly important for research organizations due to many uncertainties and due to the need to coordinate disparate activities to meet organizational goals and objectives. A strategic plan can provide a mechanism for focusing on future needs and staying in tune with fundamental organizational priorities and goals.

Specific goals and related action plans established in a strategic plan can be detrimental if misused. For example, overemphasis on numerical goals and targets often proves to be too rigid and tends to undermine the dynamic flexibility and creative vision that a strategic plan should strive to achieve. General introduction to topics relevant to strategic planning and the strategy example that follows are presented in the context of R&D organizations and their unique requirements.

14.9 QUESTIONS FOR CLASS DISCUSSION

1 What major elements would you include in an R&D organization's strategic plan?

2 Why is an understanding of external trends important?

3 Who all should be involved in developing a strategic plan?

4 How does one ensure that a strategic plan remains a flexible, dynamic, and useful document?

14.10 FURTHER READINGS

Hensey, M. (1991). Essential success factors for strategic planning. *Journal of Management in Engineering* (ASCE), **7**(2), 167–177.

Hax, A. C. and N. S. Majluf (1996). *The Strategy Concept and Process: A Pragmatic Approach.* Upper Saddle River, NJ: Prentice-Hall.

APPENDIX
RESEARCH, DEVELOPMENT, AND SCIENCE POLICY

We have added this appendix because we felt that researchers need to see how their activities relate to social goals and must be able to understand trends that shape science policy. Since they do have some choices of how they will spend their time, they must be able to influence funding for their research.

Science policy should not be shaped only in the nation's capital. Unfortunately, one of the major problems in the United States has been insufficient participation by scientists in shaping policy. In discussing research goals with colleagues, supervisors, subordinates, and science policy framers, researchers need to know something about the formulation of science policy. In a democracy, if government alone is left to decide such matters, it will result in second-rate policy. Participation by scientists is necessary for the sound development of such policy. To do this effectively, scientists need to understand relationships between

- Science and technology
- R&D expenditures and economic development
- R&D expenditures internationally and nationally
- R&D expenditures and science policy

Some of the information in this appendix provides a global view of international and national investment in research. While this may seem to be of little utility to a principal investigator or a research manager, there is no question that the information included here provides a broad understanding of the research enterprise and its implication for science policy and has a number of possible uses at certain times or at some levels. Some suggestions on how a principal investigator or a research manager can use this information follow.

A persuasive case has been made that investment in R&D plays a crucial role in the economic well-being of a nation, the profitability of a business enterprise, and the effectiveness of a technology-based governmental agency. There is evidence that the return on R&D investment in industry is higher than investment in other activities [Nadiri, 1980]. R&D managers can use this information to develop a strategy for making a case for their own program.

Using national information as a model, the R&D manager of a research group or a laboratory may want to conduct a study to check the effectiveness of its research activities. For example, return on investment (ROI) of selected completed research projects can be analyzed. To minimize biases, some research organizations have a third party conduct such ROI studies. In conducting such a study, tangible benefits accrued from the use of research output are documented and analyzed through direct discussions with the users while costs are determined through discussions with the research group. In one case, such an activity demonstrated that the ROI from completed research projects was, on the average, 30 to 1. As a note of caution, since only completed research projects can be analyzed and since there is a natural bias built into selecting more successful projects than others, the overall ROI may be somewhat lower than such a study has shown. In any case, such information can provide a powerful argument for convincing research sponsors to provide the necessary resources. The ROI in research, including the importance of basic research, was further discussed in the chapter on university research enterprise. (Chapter 13).

Research requires considerable resources; it is indeed an expensive activity. The data we present here may be used to support the need for realistic budgets that may seem inflated to those who are unfamiliar with R&D. For research excellence, it is necessary to attract talented scientists and have well-equipped laboratories. None of this is possible without sufficient funding. Seeking funds for research is also an excellent way to test the market and user response to previous research outputs. Making a successful case for research funding and convincing sponsors and customers of the considerable benefits that R&D output provides is, in the final analysis, an effective feedback mechanism that is healthy for all concerned.

A.1 RELATIONSHIP BETWEEN SCIENCE AND TECHNOLOGY

There is consensus in the literature and among knowledgeable scientists that technology stimulates science. As Bondi [1967] put it: "It is certainly a matter of experience that every time our experimental technique has taken a leap forward, we have found things totally unexpected and wholly unimaginable before. I see no reason whatever to expect that future improvements in experimental techniques will not have the same effect." This comment reminds us of the advances in astronomy after the development of the telescope, in bacteriology after the invention of the microscope, and so on. Innumerable other examples are possible.

On the other hand, in the opinion of some, it is not so clear that science stimulates technology. Price [1965] examined citation patterns in both scientific and technological journals and did not find a link; Langrish [1971] argued that the role of universities in industrial innovation is limited. While university research can be justified on other grounds, such as the training of students [Allen, 1977], it is not possible to justify it as a stimulant to innovation alone. On the other hand, Gibbons and Johnston [1974] found that one-sixth of the information needed to solve technological problems came from scientific journals.

The best conclusion, it seems, is to think of the scientific and technological systems as mutually supportive and interacting. Science, technology, and the market are interrelated [Freeman, 1982], and in developing science policy we ought to consider support for both science and technology.

A.2 TECHNICAL INNOVATION AND ECONOMIC DEVELOPMENT

Labor, capital investment, the availability of natural resources and raw materials, technical innovation, and management skills all contribute to high productivity and economic development in a nation. Here the discussion will focus on the role that R&D plays in technological innovation and economic development.

The discussion on the output of science and technology pointed out that R&D productivity to some degree can be measured by output in terms of scientific literature, patents, and the gross domestic product of each employed person. A general relationship or correlation between investment in R&D and these three items seems to exist. This general correlation can never be precise because factors other than R&D (such as capital, quality of the labor force, social, economic, and political factors, among other variables) also play a major role. There is also the phenomenon of economic cycles. Studies conducted by the Systems Dynamic Group at MIT suggest that economies move through long waves of approximately 50 years' duration. In the early part of the cycle, productivity per person increases. This upswing is due to an increase in capital investment per person, but after the accumulation of the physical facilities and the capital investment, adding more capital does not necessarily add to productivity [Rothwell and Zegwell, 1981, p. 39].

Technological innovation as discussed previously combines understanding and invention in the form of socially useful and affordable products and processes. To produce this basic understanding and invention, investment in basic research is required. One could argue that the investment in one country could be in basic research, but that the benefits might accrue to another country that combines the results of basic research with the capacity to produce useful and commercially profitable products and processes. In any case, looking at the world as a complete system, investment in R&D is necessary for innovation.

In examining the role of technology in enhancing productivity for the private, nonagricultural sector of the U.S. economy, we see that from 1909 to 1949 the cumulative percentage change in output per work-hour amounted to 80.9% [Rothwell and Zegwell, 1981, p. 24]. Of this, 87.5% of the total increase in output per work-hour was due to technical change; the rest was due to increased capital per work-hour. Another study conducted by Denison and reported by Rothwell and Zegwell [1981, p. 25] indicates that between 1929 and 1957 the contribution of technology to *total* growth of the economy was about 20%.

Commenting on the role of technological innovation in economic development, Charpie [1970, p. 3] states that in industrialized economies, all of the studies show that 30–50% of long-term economic growth stems from innovation that either improves productivity or leads to new products, processes, or completely new industries. It is further suggested [Charpie, 1970, p. 51] that technology affects international trade in several ways. International payments for technology such as patent royalties, payments for technical know-how, and so on, flow predominantly to the nations that invest heavily in research and development. According to OECD figures, for example, the United States receives 10 times as much in technological payments from abroad as it makes in such payments to other nations [Charpie, 1970, p. 5].

Possession of high-technology capacity has worked to the advantage of innovative industrialized nations in many other ways. When low-technology natural products are produced in low-labor-cost countries, industrialized countries can overcome the labor cost disadvantage by producing high-technology products for export, thus more than replacing the import balance. An example is the export of synthetic fiber from the United States to countries where low-technology, labor-intensive cotton and wool were produced for export to the United States [Charpie, 1970, p. 5]. Nations that have demonstrated the highest innovative performance and investment and that have the necessary infrastructure for R&D are most likely to encounter new opportunities that further improve their technological position in the world [Charpie, 1970, p. 5].

While in the short run economic progress may be made by applying the existing scientific knowledge base, in the long run this simply is not possible. Freeman [1982, p. 5] states the following:

> It is only to assert the fundamental point that for any given technique of production, transport or distribution, there are long-run limitations on the growth of productivity, which are technologically determined. No amount of improvement in education and quality of labor force, no greater efforts by the mass media, no economies of scale or structural changes, no improvements in management or in governmental administration could in themselves ultimately transcend the technical limitations of candlepower as a means of illumination, of wind as a source of energy, or iron as an engineering material, or of horses as a means of transport. Without technological innovation, economic progress would cease in the long run and in this sense we are justified in regarding it as primary.

It is important to note that factors such as the education and training of the labor force, an efficient industrial infrastructure (such as transportation and communication networks), capital investment, and management skills all contribute to economic productivity. But without new inventions and the scientific base to produce them, economic growth and productivity simply will not continue to increase.

When we examine the relative advantage of the United States versus Japan and other countries in terms of the rate at which new inventions are created and patents are registered, the United States shows a discernible reduction of its relative advantage (see Figure A.11). Thus the broad trends suggest that the United States is losing ground in relation to other industrial countries.

A.3 ANALYSIS OF INVESTMENT IN BASIC RESEARCH

Reviewing the basic research investment in the United States, we find that in 1960, 9% of total U.S. R&D support was applied for this purpose. By 1993, it had increased to approximately 16% of the total U.S. R&D effort [*Science and Engineering Indicators,* 1993, p. 92]. This indicates a significant increase in support of basic research.

In the United States, the federal government continues to be the primary source of basic research support, providing nearly two-thirds of the total funding [*Science and Engineering Indicators,* 1993, p. 94]. This is quite understandable. Results of basic research are a property of the commons, and these results are shared widely and without regard to commercialization. Since no property rights can normally be attached to the basic research output, significant investment by industry in basic research is unlikely. What investment industry does make in basic research serves the special needs of some high-technology companies. The reasons for such investment may relate to (1) industry recognition of the link between science and technology and, in turn, the link between basic research and innovation, (2) the need to have a diverse portfolio of activities to increase research productivity, (3) industry recognition that providing opportunities for conducting basic research is essential to keeping high-caliber scientists, and (4) attempts to reduce the time between basic research output and innovation by funding its own, rather focused, basic research to achieve a competitive advantage.

Federal government and industry investment in basic research is necessary to

- Support the science and engineering (S/E) education process and to train the needed S/E manpower
- Provide a vigorous link between invention and innovation
- Maintain international competitiveness for industry
- Provide technologies for critical national needs such as public health and national defense

National investment in basic research benefits industry and the public, and therefore it is an important element of national science policy. Sustained federal government investment in basic research was strongly supported and thoughtfully articulated by the Chief Executive Officers of major U.S. corporations as described in the chapter on university research enterprise (Chapter 13).

A.4 R&D EXPENDITURE

Investment in R&D in the United States and other countries increased rather rapidly during the decade from the mid-1970s to the mid-1980s. The trend in recent years has been mixed, with Japan, Germany, and France still showing increases in overall R&D expenditures from 1988 to 1991, while the United States and United Kingdom both recorded decreases in their R&D expenditures. These trends are shown in

Figure A.1. The expenditures as shown in Figure A.1 have been converted into constant 1987 U.S. dollars, taking into account inflation and differences in the power of the national currencies of the countries involved.

Studies show that firms receive a rate of return of 30% for the typical R&D project, while the rate of return to society is even higher [Nadiri, 1980]. Of course, if more firms engaged in R&D and at a higher level, it is very likely that the ROI would become lower. However, the fact that return on investment is now high suggests that we have not yet reached optimal levels of R&D activities. This does not imply that consulting engineering and professional firms or small industries could always profitably engage in internally funded basic or applied research. The size of the firm, the technology of the enterprise, the availability of resources for investment in research, the competitive market situation, the need for R&D activity, and an effective R&D organization within the firm should be evaluated before undertaking an investment in R&D.

Perhaps one of the best measures of R&D activity in a country is the number of scientists and engineers (S/E) employed in conducting R&D. Figure A.2 shows the relative R&D efforts of six countries as indicated by the proportion of the labor force employed as scientists and engineers in R&D. Figure A.3 shows national expenditures for performance of R&D as a percentage of GDP for selected countries. R&D expenditures as a percentage of GDP has become one of the most widely used indicators of a country's commitment to scientific knowledge growth and technology development. In the post–Cold War world of today, the ratio of nondefense R&D expenditures is probably a better yardstick of a country's true commitment to the advancement of science and technology than most other measures. International comparisons of R&D expenditures change dramatically when defense-related ex-

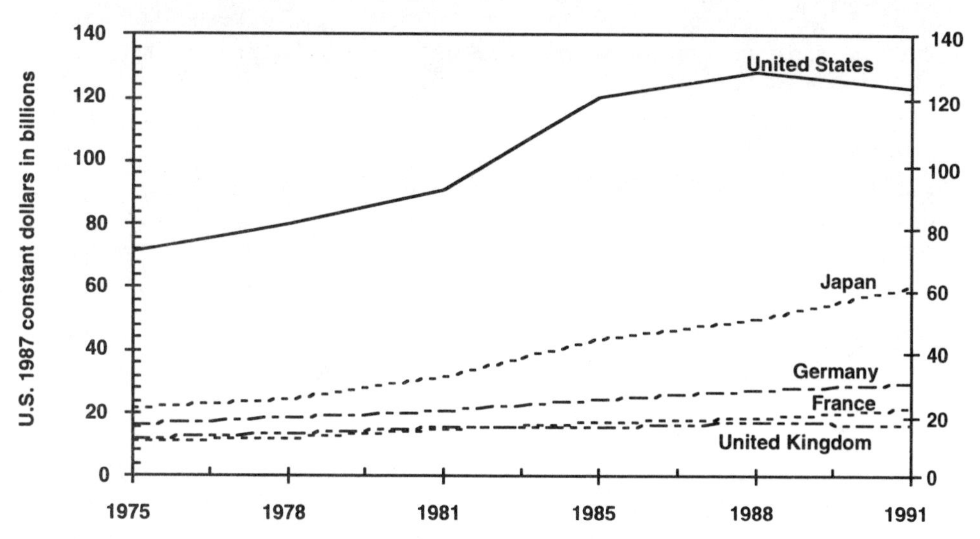

Figure A.1. Expenditures on research and development in selected countries. Foreign currency data are converted into U.S. dollars using current purchasing power parities. The data are then converted to 1987 dollars using the U.S. GDP implicit price deflators. (Source: *Science and Engineering Indicators*, 1993, p. 375.)

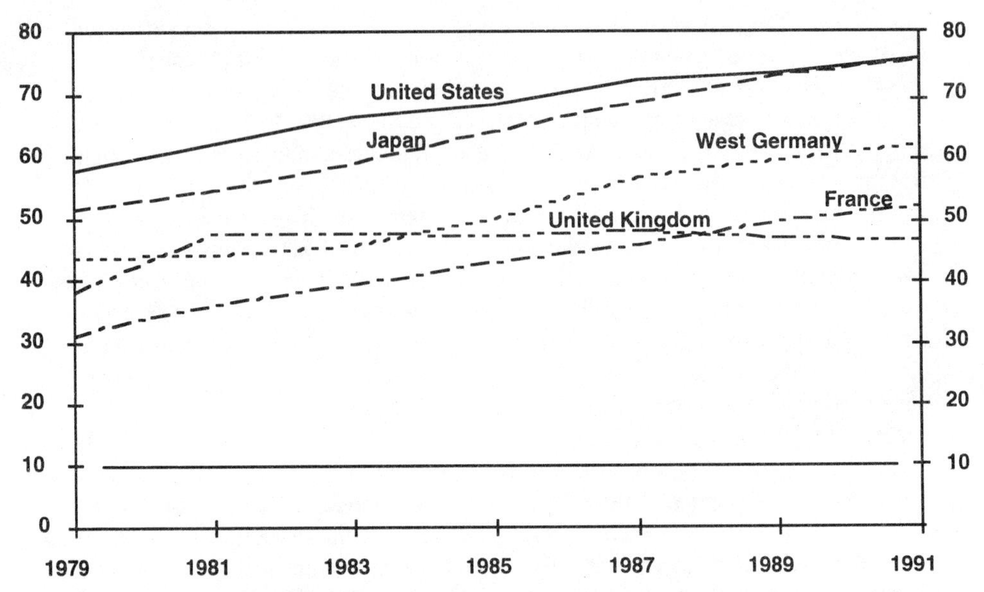

Figure A.2. Scientists and engineers engaged in R&D per 10,000 labor force population. (Source: *Science and Engineering Indicators,* 1996, p. 101.)

penditures are excluded. As we will see later, the nondefense R&D ratios of both Germany and Japan were considerably greater than that of the United States.

In the United States, R&D investment reached a level of about $171 billion in 1995. Information regarding trends in R&D investment in current and constant 1987 dollars is shown in Figure A.4. The relative distribution of R&D expenditures in 1995 by source, performer, and type of R&D is shown in Figure A.5. That year

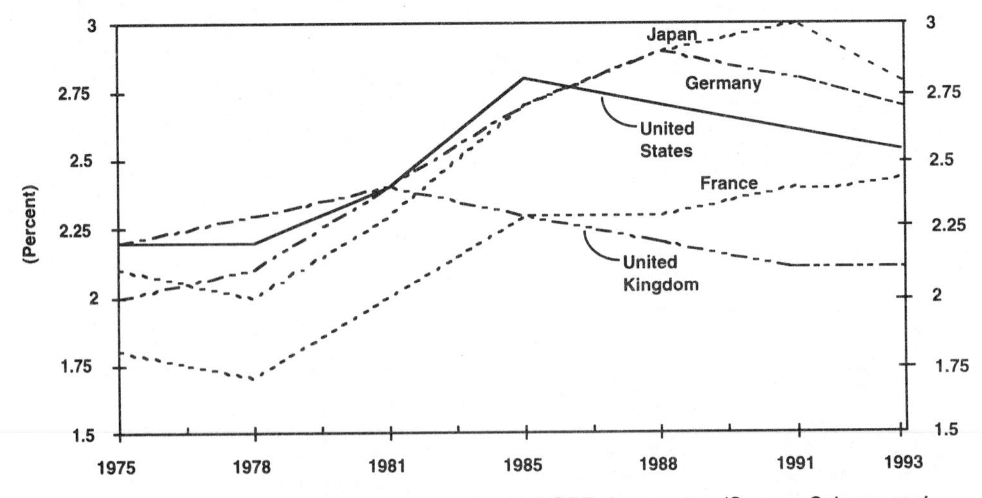

Figure A.3. R&D expenditures as a percentage of GDP, by country. (Source: *Science and Engineering Indicators,* 1996, p. 154.)

more than half of the R&D activities in the United States were supported by industry. As one would expect, industry-supported R&D focused on developing commercial products for marketing in the United States and overseas, whereas federally supported R&D focused on such areas as defense, health, space, energy, agriculture, and other noncommercial but nationally important areas.

The U.S. Office of Management and Budget (OMB) divides the federal budget into functional categories that reflect areas of U.S. federal government responsibility. Of the 16 major categories that contain R&D programs, national defense receives the largest share of U.S. federal investment in R&D, and health accounts for the next largest. Looking specifically at federal funding for basic research, the functional distribution is somewhat different, with health receiving the largest proportion, followed by general science and space. Basic research expenditures for defense are about 9% of the 1996 basic research total—down from 12% back in 1980. Relative distribution in 1994 of U.S. federal funds for R&D by selected budget function is shown in Figure A.6. U.S. government budget authority for national defense and nondefense purposes over the years in constant 1987 dollars is shown in Figure A.7. U.S. federal government support for defense R&D declined in constant dollars at an average annual rate of 3.1% between 1986 and 1992, following many years of continuous growth. Conversely, nondefense R&D support by the federal government showed an average annual increase of 5.5% over this same period. National defense (which includes DOD and DOE funds) remains the single largest budget category for the federal government, accounting for almost 59% of the total (down from its 69% peak share in 1987). In the future, defense R&D spending is expected to be reduced further to about 50% of the total federal budget. In addition, more emphasis on dual-use technologies and technology transfer activities from defense related activities to industry are expected to be major

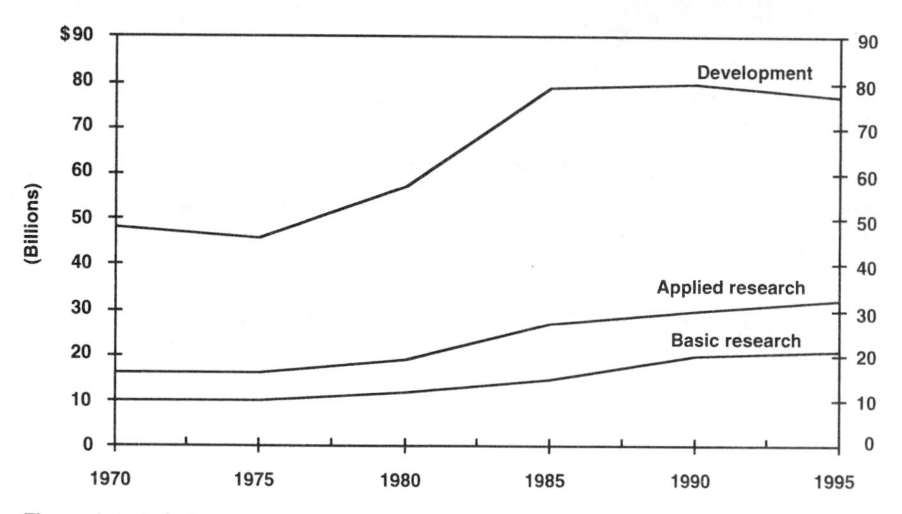

Figure A.4. U.S. R&D expenditures by character of work. GDP implicit price deflators are used to convert current dollars to constant 1987 dollars. (Source: *Science and Engineering Indicators,* 1996, pp. 108–110.)

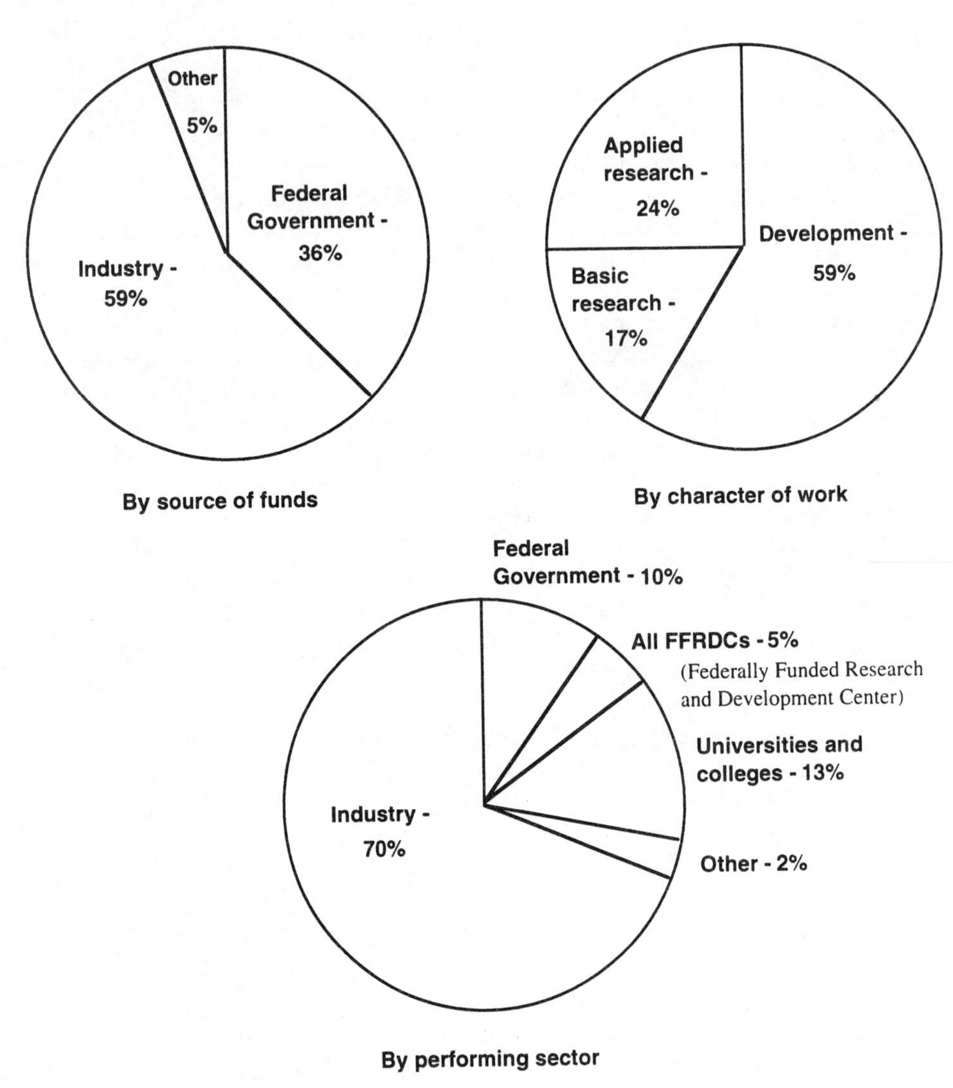

Figure A.5. Relative distribution of U.S. R&D expenditures by source, performer, and character of R&D: 1995. (Source: *Science and Engineering Indicators,* 1996, pp. 104–106)

DOD and DOE initiatives. In 1994, the following five functions accounted for 91% of the federal R&D budget [*Science and Engineering Indicators,* 1993, p. 102]:

- National defense—59%
- Health—15%. Incidentally, this is roughly equal to the nonfederal R&D support for health-related research.
- Space—9%
- Energy—4%
- General science—4%

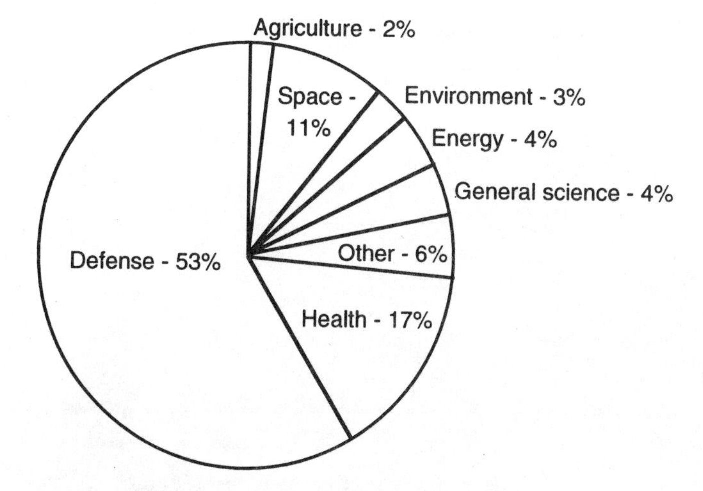

Total R&D: $70.5 billion current

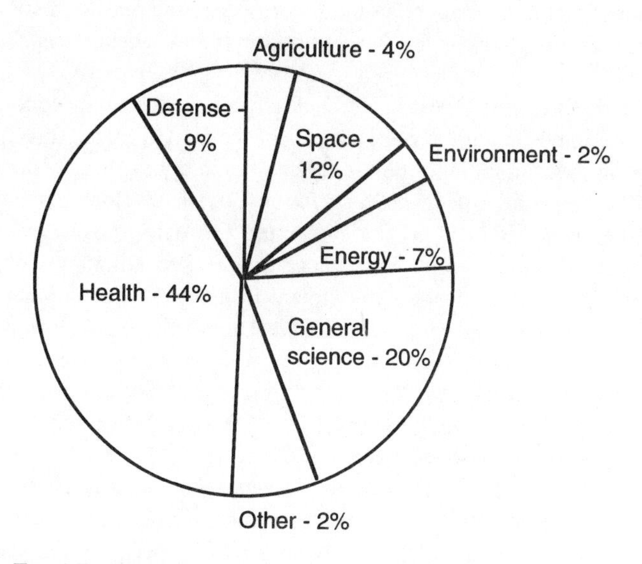

Total Basic Research: $14.3 billion current

Figure A.6. Federal R&D funds, by budget function: 1996. (Source: *Science and Engineering Indicators,* 1996, pp. 150–151)

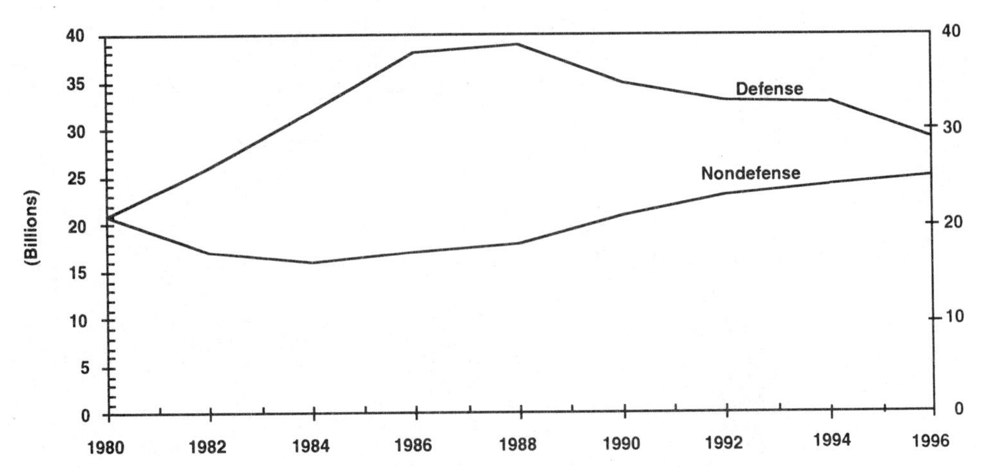

Figure A.7. Federal R&D funding for defense and nondefense—1987 dollars. (Source: *Science and Engineering Indicators,* 1996, p. 150.)

A comparison showing the federal obligations for research and development in constant 1987 dollars for 1980–1994 for selected major federal agencies is shown in Figure A.8.

How different governments invest in R&D clearly shows the socioeconomic objectives considered important at the policy level. In the United States, much of the federal government investment is for defense, health, and civil space, while investment in energy is moderate and investment for industrial development is almost nonexistent as shown in Figure A.9. For Japan and Germany, government R&D support patterns are much different. For France and United Kingdom, while the investment in defense is sizable, their investment in industrial development compares favorably with Japan and Germany. In the United States, historically, the direct role of government in industrial development or investing in R&D has not been well accepted by industry or science policy-makers. The question is often raised, Should the government pick "winners" and "losers" by supporting certain R&D projects for industrial development?

It is generally believed that, in the long run, government is not likely to allocate resources as efficiently as the market. As discussed earlier in the chapter on university research enterprise (Chapter 13), the government's proper role is investing in basic research, which provides a foundation for innovation without selecting winners and losers; and investing in nationally important missions such as defense, health, and civil space is a prudent course of action. Providing incentives to industry for taking a long-term view of investment in R&D could fulfill the need for increased industrial investment in R&D in the United States.

Since defense-related R&D is not primarily oriented toward a nation's trade competitiveness, its public health, or other nondefense objectives, a comparison of nondefense R&D expenditures to GDP may be of interest. Figure A.10 provides this information. It is interesting to note that the ranking of countries based on the

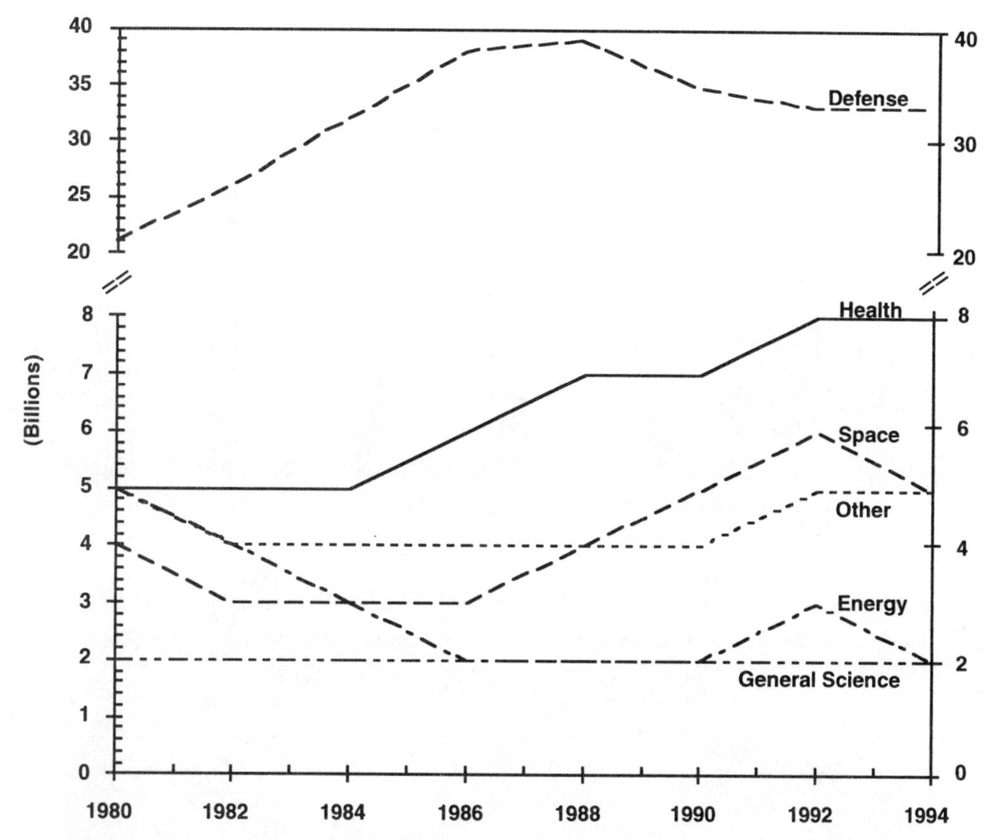

Figure A.8. Federal R&D funding, by major budget function—1987 dollars. (Source: *Science and Engineering Indicators,* 1993, p. 363.)

proportion of GDP devoted to nondefense R&D expenditure is similar to their ranking by the percentage of national R&D expenditure financed by industry. For instance, in 1991, between 43% and 50% of the national R&D effort was financed by private industry in the United States, the United Kingdom, and France, while in West Germany and Japan, private industry investments were 60% and 73% percent, respectively [*Science and Engineering Indicators,* 1993, p. 377].

The major market economies of the world spend similar proportions of their GDPs (between 2.2 and 2.7%) on all research and development. How this has varied over the years is shown in Figure A.3. Differences among the five nations are more dramatic when R&D expenditures for nondefense purposes are compared. For instance, Japan is spending in excess of 50% more as a percentage of GDP on nondefense R&D as compared to the United States, France, or United Kingdom. It is significant to note, however, that funding of defense-related R&D for 1992–1994 represented between 33% and 55% of total government R&D funding in the United States, the United Kingdom, and France. At the same time, this expenditure

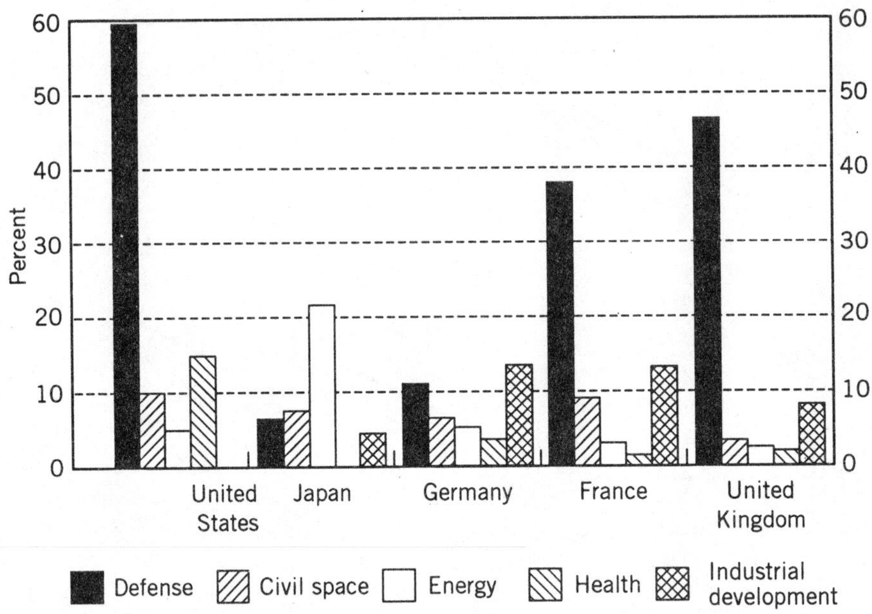

Figure A.9. Government R&D support, by country and socioeconomic objective: 1992. (Source: *Science and Engineering Indicators*, 1993, p. 106.)

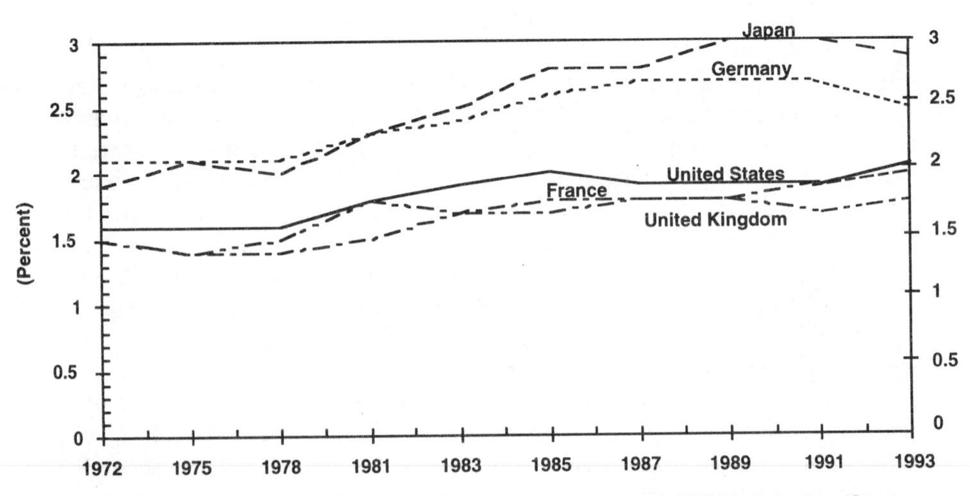

Figure A.10. Nondefense R&D expenditures as a percentage of GDP, by country. (Source: *Science and Engineering Indicators*, 1996, p. 155.)

was only 8.5% in West Germany and an even smaller 6% in Japan [*Science and Engineering Indicators*, 1995, p. 153]. Clearly, as the share of national R&D effort devoted by a country to defense-related activities increases, resources available to business-related research activities decreases. Historically, the reasons that West Germany and Japan spend less on defense-related R&D are the legal and constitutional constraints placed on these countries by the Allies at the end of World War II.

Data show that countries that spend proportionally less of their national R&D resources on defense enjoy the greatest GNP and productivity growth rate [Rothwell and Zegwell, 1981, p. 26]. Clearly, there are many technical spin-offs from defense-related R&D, but this is not an efficient way to develop new technology, and the *opportunity cost* to economic growth of defense-related R&D is clearly high [Rothwell and Zegwell, 1981, p. 26].

In developing national science policy, it is important to understand the critical need for R&D investment in defense-related activities. National policies in the United States and other industrialized countries with similar international responsibilities (e.g., United Kingdom and France) favor such expenditures. This is one area where the federal government, and not industry, must assume its proper responsibility and provide for an adequate national defense. Since R&D expenditures can increase productivity, R&D expenditures on defense-related activities can therefore reduce the proportion of GDP allocated to defense and thus lessen the nation's total defense burden.

A National Academy of Sciences [1995] report argues that the federal research budget is now defined so broadly that it is difficult to say how relevant it is to the health of U.S. science. The report suggests that a more meaningful definition of R&D that includes only activities that generate new knowledge or technologies and leaves out items which are related to mission oriented agency activities (e.g., new military hardware that consumes enormous R&D budgets) would provide a much sounder basis for understanding real national investment in science. In turn, this would provide a more meaningful basis for science policy-making. If this approach were used, the current federal R&D budget would be about half ($37 billion) of the total sum the government now claims to spend on R&D. Another benefit from this approach from a science policy point of view would be that mission-oriented agency programs that are now lumped in the total R&D budget would have to be defended on their own terms rather than for their contribution to science. Explicit decisions then could be made for these funds in terms of their tangible contribution to agency missions.

A.5 R&D PRODUCTIVITY

In this section, the emphasis is on R&D productivity at the national level. R&D productivity at the organizational level is discussed in Chapter 3, "Creating a Productive and Effective R&D Organization." R&D organization productivity is defined as an *organization effectiveness vector* that includes quantifiable and nonquantifiable organization outputs, and this vector reflects the quality and the relationship of outputs to organizational goals and objectives. Individual researcher productivity follows the general pattern of the organization effectiveness vector and is discussed in the chapter on performance appraisal (Chapter 10).

Figures A.1 through A.4 provide information about the magnitude of selected countries R&D investments. In addition to the monetary investment, in the long run, the ability of a nation to conduct research in science and technology is constrained by the ability of the nation's education system to produce the necessary science- and engineering-trained manpower.

It is difficult to compare the training of engineers and scientists from different countries because the curricula vary and the quality of training and the degrees conferred convey different meanings. Consequently, the number of engineers and scientists that a country trains may not predict its strength in science and technology. Nevertheless, the number of scientists trained is an indication of disciplinary emphasis by the country, and, over time, this is bound to affect the ability of the country to conduct research in science and technology. Therefore, it is important to note that in 1990 several countries awarded higher percentages of their first university degrees to candidates in the natural sciences and engineering than the United States. These include China (53%), Japan (26%), Mexico (34%), and Germany (47%), as well as several Central European nations. The figures for countries such as China and Mexico, where higher education is still a relatively elitist commodity, are somewhat misleading. China, for instance, has the lowest percentage of college-age cohorts obtaining a college degree (1%) compared with the United States, which has the highest percentage of 22-year-olds receiving college degrees (31%). Countries such as Japan and Germany, which both have high participation rates in university education combined with high percentages of degrees in natural science and engineering, produce greater numbers of graduates in these fields.

In addition to the number of engineers and scientists produced by a nation, R&D productivity is also a function of the quality of the training, effectiveness of R&D management, availability of modern research equipment, and computer technology. Some surrogates of R&D productivity could be the extent of the scientific literature, the number of patents, and the overall economic activity represented by the gross domestic product (GDP).

Scientific Literature

The relative strength of a country's R&D activities can be measured, to some degree, by its share of publications in the world's leading scientific journals. The United States share of world scientific and technical articles for selected fields for the years 1973, 1981, 1984, and 1991 is shown in Table A.1.

Patents

Data on patent activity permit some overall comparisons of the output of inventors in different countries. Significant inventions are patented in the host country and in other countries as well. Thus, the number of U.S. patents granted to foreign inventors provides an indicator of the level of invention in those countries, while the

TABLE A.1 U.S. Share of World Scientific and Technical Articles by Field[a]

Field	1973 (%)	1981 (%)	1984 (%)	1991 (%)
All fields	33	35	35	35.1
Clinical medicine	43	40	41	38.5
Biomedicine	39	39	39	38.9
Biology	46	38	37	37.6
Chemistry	23	20	21	23.1
Physics	33	28	27	29.8
Earth and space sciences	47	42	41	41.7
Engineering and technology	42	41	40	36.2
Mathematics	48	38	37	42.1

[a] Articles written by researchers from more than one country are prorated according to the number of author institutions in each country. Data for 1973–1980 are based on more than 2100 journals carried on the 1973 Science Citation Index Corporate Tapes of the Institute for Scientific Information. Data for 1981–1991 are based on more than 3500 U.S. and foreign journals on the 1981 Science Citation Index Corporate Tapes.

Source: CHI Research, Inc., Science & Engineering Indicators Literature Database, special tabulations, 1993.

proportion of foreign to U.S. inventors receiving U.S. patents provides a measure of the comparative inventiveness of foreign versus U.S. inventors. Figure A.11 shows an overall increase in all categories of patent activity. The number of U.S. patents granted to U.S. inventors increased to an all-time high of 51,000 in 1991. Foreign patenting also showed an increase over this period, accounting for 47% of the total number of U.S. patents issued in 1991. Five foreign countries account for 80% of U.S. foreign patents: Japan, Germany, Great Britain, France, and Canada.

Japan's share of *all* U.S. patents has increased steadily and dramatically, from 11% in 1978 to 22% in 1991. Japan accounts for 46% of all U.S. patents granted to foreign inventors. Given that the share of U.S. patents granted to the top three European patenting countries showed a general decline over this 14-year period and that Canada's average U.S. patenting rate has shown only a modest increase at best, Japan's ascendancy is all the more remarkable. It certainly cannot be ignored by U.S. science policy-makers.

Economic Activity

In a real sense, successful inventive and innovative activity over time should result in increases in an economy's ability to produce goods and services at low cost. Increased productivity and the resulting improvements in the standard of living for the nation's citizens are clearly the important benefits of an R&D activity. Consequently, one measure of the impact of science and technology on society is the value of the production accounted for by each employed person. Gross domestic product (GDP) measures the value added by industry and individuals in each coun-

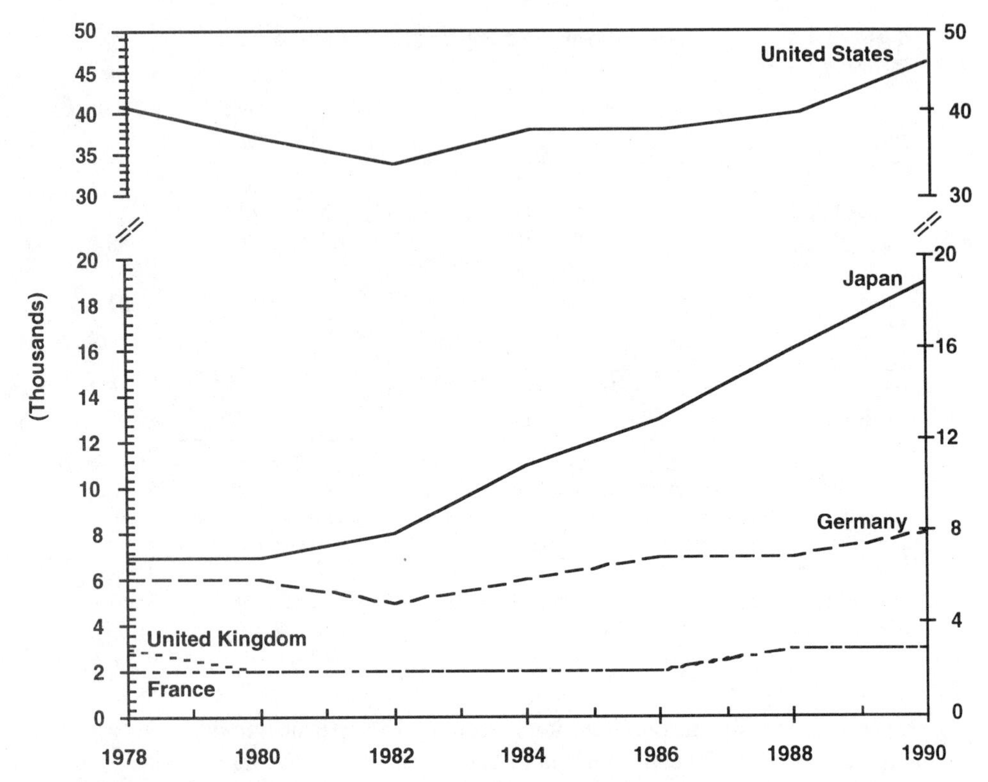

Figure A.11. U.S. patents granted by nationality of inventor. (Source: *Science and Engineering Indicators,* 1993, p. 172.)

try. In addition to the input provided by science and technology, *GDP per employed person* reflects factors such as availability of natural resources, expertise of the workforce, investment in industry, and many other social and economic factors. Among the many sources of productivity growth, science and technology have been identified as major causal factors. GDP, therefore, to some extent represents the strengths of science and technology of a nation. Real GDP per employed person in selected countries is shown in Figure A.12. With a major adjustment in West German and Japanese currency exchange rates and due to the continued growth of their economies, West German and Japanese GDP per employed person is expected to equal or surpass the United States before the end of this century.

A.6 GLOBAL PERSPECTIVES ON INNOVATION

We are finding that many of our competitors, especially in Japan and Germany, are able to go through each turn of the innovation cycle more quickly than their competitors in the United States. As one would expect, it takes only a few turns for the

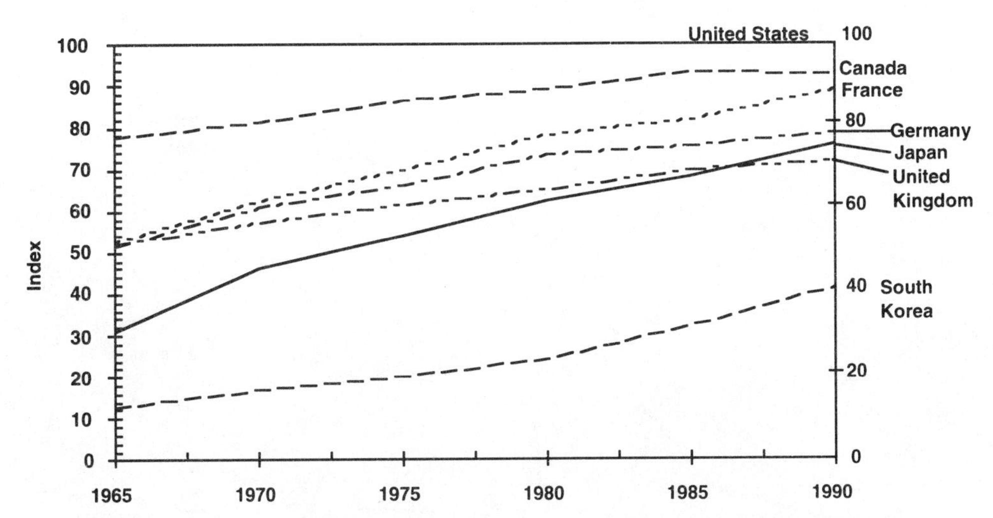

Figure A.12. Real gross domestic product per employed person in selected countries. Index: U.S. = 100. Country GDPs were calculated using 1985 purchasing power parities. (Source: *Science and Engineering Indicators,* 1993, p. 158.)

company with the shortest cycle time to build up a commanding lead in terms of productivity and the latest technology in the product. Strategies that allow a company to shorten the innovation cycle inevitably make them more competitive so they can be leaders in the global market.

Many Europeans, on the other hand, feel that Americans are better than they are at adapting and exploiting technology and that there is a good working relationship between the university community and industry in the United States. According to Goldsmith [1970, p. xvi], American skills lie in adapting and exploiting rather than in creating. To support this statement, he has given a number of examples of recent European developments that have had a major impact on American engineering techniques: the triple-deck railway wagon for transporting automobiles (German), the horizontal climbing crane (French), the flotation process for plate glass (British), the basic oxygen furnace (Austrian), the Hovercraft (British patent), and the fundamental digital computer (British).

Notwithstanding American ability to adapt and exploit technology developed elsewhere, let us examine American ability to create and invent. American share of world scientific and technical articles (Table A.1), patent applications (Figure A.11), and the number of prestigious awards (see Figure A.13 for Nobel Prizes for Scientific Discoveries) earned by American scientists would indicate that American scientists are indeed leaders in inventing and creating new knowledge. Perhaps many scientists and engineers in America feel, and rightfully so, that the emphasis on American investment in R&D needs to be strengthened further so as to sustain and enhance this leadership position. To the extent that this is not happening, many may feel that the United States is falling behind. Some examples cited are the relatively small number of engineers trained in the United States (as compared to

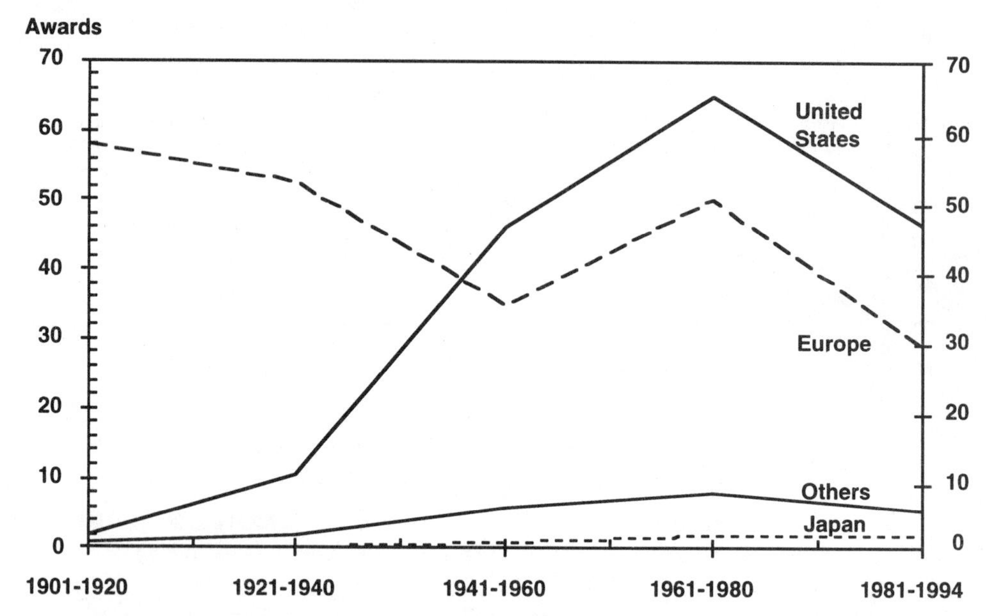

Figure A.13. Nobel Prize winners for scientific discoveries over the years. Includes awards in physics, chemistry, and physiology & medicine. (Source: *The World Almanac and Book of Facts,* World Almanac Books, 1996 edition, pp. 323–325.)

Japan and Germany), outdated laboratory facilities at universities, and little or no increase in the number of U.S. citizens who are full-time S/E graduate students in doctorate-granting institutions [*Science Indicators,* 1985, p. 102]. One factor in the recent decline of American innovation may be the improvement in the conditions of scientific work in other countries, which has reduced the "brain drain" to the United States.

Nevertheless, some Europeans feel that the climate for research, development, and innovation is more favorable in the United States than in most European countries. The following are some examples [Charpie, 1970, p. 9]:

- The U.S. tax code provides an assured flow of high-risk investment capital to those would-be innovators who are able to establish contact with adequate venture capital sources. The capital gains provisions of the U.S. tax law provides an incentive for taxpayers in high income brackets to assume substantial investment risks in anticipation of sharply reduced tax rates on high profits. With the 1987 tax reforms, this, of course, has changed.
- America has a tradition of publicly appreciating and recognizing the success of antiestablishment people. In addition, the mobile, less-structured American society is technologically sophisticated and is receptive to new ideas and inventions. American society is conditioned to give praise and prestige according

to accomplishments and it does not ignore a person of humble beginnings who chooses to fight the system and thus achieves success.

- One of the most important factors is that in America a would-be entrepreneur is surrounded by much evidence of successful entrepreneurship.

Billings and Yaprak [1995] argues that inventive (R&D) efficiency may be an important factor in the competition for global market shares and goods and services. R&D efficiency is compared for 14 industrial groups in the United States and Japan using multiple indices of inventive efficiency. Examples of these indices are:

- The marginal product of R&D capital
- The rate of change in total factor productivity
- The effect of R&D on future sales
- The effect of R&D on future value added

Findings show interesting differences in inventive efficiency across industrial groups and between the United States and Japan. U.S. food, textile, chemical, rubber, metals, fabricated metals, and other miscellaneous manufacturing industries appear to be relatively more efficient in inventive efficiency than their Japanese counterparts. In contrast, Japanese paper, petroleum, machinery, and scientific equipment industries display greater inventive efficiency than their U.S. counterparts [Billings and Yaprak, 1995].

According to a report in November 1995 by the Department of Commerce's Office of Technology Policy (OTP), foreign-owned companies are dramatically increasing R&D spending in the United States, with expenditures reaching $14.6 billion in 1993. At the same time, OTP says, U.S. companies have also increased R&D done abroad, nearly doubling spending between 1987 and 1993, reaching an investment of $9.8 billion in 1993 [Rotman, 1995].

A.7 R&D EXPENDITURE AND SCIENCE POLICY

The information and analyses presented here should provide an overview of the considerable investment industrialized nations are making in R&D. The importance of this investment in both basic and applied research cannot be overstated. Discussions related to the importance of basic research and the impact of technical innovation on economic development may provide some information useful for developing a national science policy.

Some interesting issues related to the global economy and science policy were raised in a recent publication by the National Academy of Engineering [1993]. In this report, relative roles of industry, the university community, and governmental agencies were discussed and goals and policy recommendations to further productivity and growth of the U.S. economy were articulated. Clearly, the performance of the nation's technology enterprise—that is, its total capacity for creating, developing, and deploying new technology—is a key factor in increasing the economic productivity and sustaining the growth rate of the U.S. economy [National Academy of Engineering, 1993].

Some goals and policy recommendations related to science and technology are [National Academy of Engineering, 1993]:

- Foster the timely adoption and effective use of commercially valuable technology throughout the U.S. economy.
- Increase civilian R&D investment in the U.S. economy and close emerging gaps in the nation's civilian technology portfolio.
- Access and exploit foreign technology and high-tech markets more effectively to advance the interests of U.S. citizens.
- Create a strong institutional framework for federal technology policy in support of national economic development, and integrate the planning and implementation of federal technology policy with that of national domestic and foreign economic policy.

In a general sense, these goals provide a blueprint for creating, developing, and deploying new technology to provide for increased productivity and sustained economic growth of the economy.

R&D Expenditures and Society

An analysis of R&D investment by the major industrialized countries gives an indication of the emphasis each nation places on its needs and the priorities each nation has established among competing social needs. Since investment in R&D affects the innovation process and thus the goods and services available to consumers, *the nature of R&D investment may to a large degree determine the real choices available to society*. Consequently, R&D investment at the national level cannot be viewed primarily in terms of profitability.

Freeman [1982, p. 201] argued that:

> The advance of science and technology must find its support and its justification, not merely in the expectation of competitive advantage, whether national or private, military or civil, but far more in its contribution to social welfare, conceived in a wider sense. The funding of R&D is extremely important for these basic goals.

Because of the problems of the commons and other associated issues, there is a tendency for R&D investments in industry and in government to focus primarily on short-term projects and not on problems of the environment, resource conservation, and the like. As Freeman [1982, p. 402] has stated:

> Present R&D project selection techniques are biased overwhelmingly towards technical and short-term competitive economic criteria. This is true of both capitalist and

> socialist economies. An extremely important problem for research and social implementation is the evolution of quite new techniques of selection and assessment which could be applied both in the private and public sectors. These should take into account aesthetic criteria, work satisfaction criteria, environmental criteria and other social costs and benefits which today are almost excluded from consideration.

Only when a national science policy explicitly recognizes that market rate of return cannot properly account for the investment needed in either basic or applied research can we begin to develop a broader and a more comprehensive approach to investment in research. This argument does not necessarily lead to more investment in research. Instead, it may lead to modifying some of our priorities and may also lead to analyzing both monetary and human capital investments in R&D explicitly.

R&D Expenditures on Space and Defense

Let us take the case of major research and development efforts that are undertaken to focus on national priorities such as space or defense. Without making any political judgments, although these can be important, it would be salutary to look at investments beyond the monetary aspects—for example, the investment of scientific talent incurred for these activities. Commenting on the desirability of the Strategic Defense Initiative, the Nobel Prize recipient John Bardeen stated that investment in this research program alone could be larger than the entire amount spent by the government for all nonmilitary research, including the National Science Foundation and the National Institute of Health. Perhaps, he stated, the country can afford this $30 billion investment in *monetary terms,* but not in the *diversion of the top talent* of scientific manpower to such activities [Bardeen, *Daily Illini,* April 21, 1986]. Further commenting on a similar experience in the early 1960s, Bardeen [*Daily Illini,* April 21, 1986] has stated that the Apollo Program in the early 1960s was a great technical success, but the real cost to the United States was far greater than the dollars spent might indicate, because it was during the 1960s that the aerospace industry expanded rapidly to meet military and space needs and drew scarce top technical talent from civilian and other human needs. This, in turn, provided nations such as Japan and Germany the opportunity to establish a lead in civilian-based industries and markets. Large undertakings such as the Apollo Program historically have caused perturbations in the supply and utilization of top technical talent, so that the aerospace or military industry grew faster than could be sustained in the long run. Within a decade of its expansion in the early 1970s, highly trained engineers, ill-suited for civilian needs, were being laid off by the aerospace industry.

No one questions the importance of national defense. Indeed, it is one of the most important priorities for a nation and especially so for a superpower like the United States. As a matter of national policy, however, it is important to consider not only the expenditures in major research programs of the magnitude of space and defense, but also the investment of top scientific talent in such activities. The impact that their programs have on the utilization of scientific talent is an important consideration. It would seem that a vigorous debate on these issues is essential so that various national priorities are appropriately balanced. Such a debate should bring into consideration a science policy that explicitly addresses the issue of top *talent diversion* to different national efforts in addition to the *monetary aspects*. Educational and training programs could then be geared to produce the necessary technical talent, perturbations could be minimized, and essential defense needs, along with other nonmilitary and human needs, could be effectively satisfied. This would produce a more sustainable defense policy, which would be less costly to the nation because engineers and scientists that are highly trained and experienced in addressing military needs would not be rendered useless when there are major funding shifts and the resources and manpower normally wasted in startups and winddown activities for large project undertakings for space and defense would be eliminated or minimized.

Choices among Competing Priorities

As the cost of specific projects (e.g., the supercollider) increases, some critics claim that "big science" threatens to eliminate "small science." Thus it seems that establishing science policy will eventually require making choices among competing national priorities. Is there a role for the science community in making these choices and formulating a national science agenda? In an address to the members of the National Academy of Science, Frank Press [1988] stated that "we must also be willing, for the first time, to propose priorities across scientific fields. . . . We can do so in a manner that is knowledgeable, responsible and useful. We should accept the challenge." If the scientific community does not get involved in proposing priorities and in formulating a national science agenda, others will set this agenda without critical analysis and proper knowledge. Such a national agenda is not likely to have the support of the science community, the confidence of the public, or the long-term commitment of the nation for financial support.

In making choices among competing national needs and proposing priorities across the scientific fields, a set of credible evaluation criteria is needed. The weight given to various elements of the criteria would naturally differ among scientists themselves and other decision-makers. It should, however, not be too difficult to reach an agreement on the major elements of such criteria.

Dutton and Crowe [1988] have proposed a set of evaluation criteria (see Table A.2) with a series of questions under each of the main elements. They propose that each major research program of national significance be evaluated using such criteria so that comparable judgments can be made across programs.

A.8 SUMMARY

In summary, science and technology are closely interrelated. Investment in R&D is essential for economic development and for sustaining economic growth. The level of support should be determined by the social benefits of competing usages of the funds. The few clues that we have at present suggest that in the United States the

TABLE A.2 Evaluation Criteria for Research Programs of National Significance

Scientific Merit

1. SCIENTIFIC OBJECTIVES AND SIGNIFICANCE

What are the key scientific issues addressed by the initiative?

Why are these issues significant in the context of science?

To what extent is the initiative expected to resolve them?

2. BREADTH OF INTEREST

Why is the initiative important or critical to the discipline proposing it?

What impact will the science involved have on other disciplines?

Is there a potential for closing a major gap in knowledge either within a discipline or in areas separating disciplines?

3. POTENTIAL FOR NEW DISCOVERIES AND UNDERSTANDING

Will the initiative provide powerful new techniques for probing nature? What advances beyond previous measurements can be expected with respect to accuracy, sensitivity, comprehensiveness, and spectral or dynamic range?

Is there a potential for insight into previously unknown phenomena, processes, or interactions?

Will the initiative answer fundamental questions or stimulate theoretical understanding of fundamental structures or processes related to the origins and evolution of the universe, the solar system, the planet Earth, or life on Earth?

In what ways will the initiative advance the understanding of widely occurring natural processes and stimulate modeling and theoretical description of these processes?

Is there a potential for discovering new laws of science, new interpretations of laws, or new theories concerning fundamental processes?

4. UNIQUENESS

What are the special reasons for proposing this initiative ? Could the desired knowledge be obtained in other ways? Is a special time schedule necessary for performing the initiative?

Social Benefits

1. CONTRIBUTION TO SCIENTIFIC AWARENESS OR IMPROVEMENT OF THE HUMAN CONDITION

Are the goals of the initiative related to broader public objectives such as human welfare, economic growth, or national security? Will the results assist us in planning for the future?

What is the potential for stimulating technological developments that have application beyond this particular initiative?

Will the initiative contribute to public understanding of the physical world and appreciation of the goals and accomplishments of science?

2. CONTRIBUTION TO INTERNATIONAL UNDERSTANDING

Will the initiative contribute to international collaboration and understanding?

Does the initiative have any aspects requiring special sensitivity to the concerns of other nations?

TABLE A.2 *(continued)*

3. CONTRIBUTION TO NATIONAL PRIDE AND PRESTIGE

   How will the initiative contribute to national pride and to the image of the United States as a scientific and technological leader?

   Will the initiative create public pride because of the magnitude of the challenge, the excitement of the endeavor, or the nature of the results?

Programmatic Concerns

1. FEASIBILITY AND READINESS

   Is the initiative technologically feasible?

   Are new technological developments required for the success of the initiative?

   Are there adequate plans and facilities to receive, process, analyze, store, distribute, and use data at the expected rate of acquisition?

   Is there an adequate administrative structure to develop and operate the initiative and to stimulate optimum use of the results?

2. SCIENTIFIC LOGISTICS AND INFRASTRUCTURE

   What are the long-term requirements for special facilities or field operations?

   What current and long-term infrastructure is required to support the initiative and the processing and analysis of data?

3. COMMUNITY COMMITMENT AND READINESS

   Is there a community of outstanding scientists committed to the success of the initiative?

   In what ways will the scientific community participate in the operation of the initiative and the analysis of the results?

4. INSTITUTIONAL IMPLICATIONS

   In what ways will the initiative stimulate research and education?

   What opportunities and challenges will the initiative present for universities, federal laboratories, and industrial contractors?

   What will be the impact of the initiative on federally sponsored science? Will new components be required? Can some current activities be curtailed if the initiative is successful?

5. INTERNATIONAL INVOLVEMENT

   Does the initiative provide attractive opportunities for involving leading scientists or scientific teams from other countries?

   Are there commitments for programmatic support from other nations or international organizations?

6. COST OF THE PROPOSED INITIATIVE

   What are the total direct costs, by year?

   What are the total costs, by year, to the federal budget?

   What portion of the total costs will be borne by other nations?

Source: Dutton, J. A. and L. Crowe (1988). Setting priorities among scientific initiatives. *American Scientist*, journal of Sigma xi, **76,** 600–601. Reprinted by permission.

benefits from increased support for R&D will be greater than the benefits that society will derive from competing usages of the funds.

Views on R&D expenditure and science policy, the topic discussed at the end of this appendix, naturally vary. Public policy decisions related to science policy and research expenditure affect resources available for conducting research and the emphasis on such research. Consequently, participation of engineers and scientists in the policy-making stage, where tradeoffs between competing demands are considered, needs to be reemphasized and institutionalized.

R&D scientists and engineers constitute a substantial portion of the American electorate. Over 1.4 million people, in a country where half the population usually does not vote, can be a critical force in some elections. The data presented here suggest that the United States lead in science is slipping. Furthermore, too much is spent on research that has little potential for technological innovation spin-offs, while other countries, such as Japan, are graduating more engineers than the United States, produce more patents per capita, and do more research in areas that have such spin-offs. These trends must be reversed. In a democracy, such changes come from concerned segments of the electorate. It is you, the reader, and others like you, that can initiate the needed policy changes.

REFERENCES

Albert, R. (1983). The cultural sensitizer or cultural assimilator. In D. Landis and R. Brislin (Eds.), *Handbook of Intercultural Training,* Vol. 2 pp. 186–217. New York: Pergamon.

Alderfer, G. P. (1972). *Existence, Relatedness and Growth.* New York: Free Press.

Allen, T. J. (1970). Communication networks in research and development laboratories. *R&D Management,* **1,** 14–21.

Allen, T. J. (1977). *Managing the Flow of Technology: Technology Transfer and the Dissemination of Technological Information Within the Research and Development Organization.* Cambridge, MA: MIT Press.

Allen, T. J., M. L. Tushman, and D. M. S. Lee (1979). Technology transfer as a function of position in the spectrum from research through development to technical services. *Academy of Management Journal,* **22**(4), 694–708.

Allio, R. J. and D. Sheehan (1984). Allocating R&D resources effectively. *Research Management,* **27**(3), 14.

Andrews, F. M. (1979). Motivation, diversity and the performance of research units. In F. M. Andrews (Ed.), *Scientific Productivity.* New York: Cambridge University Press.

Andrews, F. M. and G. F. Farris (1967). Supervisory practices and innovation in scientific terms. *Personnel Psychology,* **20,** 497–515.

Andrews, K. R. (1980). *The Concept of Corporate Strategy.* Homewood, IL: Richard Irwin.

Anthony, R. N. and R. E. Herzlinger (1984). *Management Control in Non-Profit Organizations.* Homewood, IL: Richard D. Irwin.

Argyris, C. (1985). *Strategy Change and Defensive Routines.* Marshfield, MA: Pitman.

Armstrong, D. J. and P. Cole (1995). Managing distances and differences in geographically distributed work groups. In S. E. Jackson and M. N. Ruderman (Eds.), *Diversity in Work Teams: Research Paradigms for a Changing Workplace,* pp. 187–215. Washington, D.C.: American Psychological Association.

Arrow, K. J. (1974). *Essays in the Theory of Risk-Bearing*. Amsterdam: North-Holland; New York: American Elsevier.

Atchison, T. and T. W. French (1967). Pay systems for scientists and engineers. *Industrial Relations*, **7**, 44–56.

Bailyn, L. (1984). *Autonomy in the Industrial R&D Lab*. Unpublished paper, Sloan School of Management, Massachusetts Institute of Technology, TR-DNR-#30.

Bailyn, L. and J. T. Lynch (1983). Engineering its difficulties. *Journal of Occupational Behavior*, **7**, 263–283.

Balkin, D. B. and L. R. Gomez-Mejia (1984). Determinants of R&D compensation strategies in the high tech industry. *Personnel Psychology*, **32**, 635–650.

Barron, F. (1969). *Creative Person and Creative Process*. New York: Holt, Rinehart & Winston.

Bass, B. (1985). *Leadership and Performance Beyond Expectation*. New York: Macmillan.

Baumgartel, H. (1957). Leadership style as a variable in research administration. *Administrative Science Quarterly*, **2**, 344–360.

Becker, L. J. (1978). Joint effect of feedback and goal setting on performance: A field study of residential energy conservation. *Journal of Applied Psychology*, **63**, 428–433.

Bell, J., R. Herman, and C. Sutton (1986). The acid test of innovation. *New Scientist*, **6**, 34 (March).

Bennis, W. (1984). The four competencies of leadership. *Training and Development Journal*, **38**(8), 14–19 (August).

Billings, B. A. and Yaprak, A. (1995). Inventive efficiency: how the U.S. compares with Japan. *R&D Management* **25**(4), 365–376.

Bishop, J. M. (1995). Enemies of promise. *Wilson Quarterly* **19**(3), 61–65 (Summer).

Black, J. S. and M. Mendenhall (1990). Cross-cultural training effectiveness: A review and theoretical framework for future research. *Academy of Management Review*, **15**, 113–136.

Blake, R. R. and J. S. Mouton (1986). From theory to practice in intergroup problem solving. In S. Worchel and W. G. Austin (Eds.), *Psychology of Intergroup Relations*, pp. 67–82. Chicago: Nelson-Hall.

Blake, S. P. (1978). *Managing for Responsive Research and Development*. San Francisco: W. H. Freeman.

Bok, D. C. (1982). *Beyond the Ivory Tower*. Cambridge: Harvard University Press.

Bondi, H. (1967). *Assumption and Myth in Physical Theory*. Tarner Lectures. New York: Cambridge University Press.

Boorstin, D. J. (1983). *The Discoverers*. New York: Random House.

Boring, E. G. (1950). *A History of Experimental Psychology*. New York: Appleton-Century-Crofts.

Bosomworth, C. E. (1995). How 26 companies manage their central research. *Research-Technology Management*, **38**(3), 32–40 (May–June).

Brickman, P., J. A. M. Linsenmeier, and A. G. McCarneins (1976). Performance enhancement by relevant success and irrelevant failure. *Journal of Personality and Social Psychology*, **33**, 149–160.

Brooks, H. (1968). *The Government of Science*. Cambridge: MIT Press.

Brooks, H. (1973). Knowledge and action: The dilemma of science policy in the '70s. *Proceedings of the American Academy of Arts and Sciences,* **102**(2), Spring 1973: The search for knowledge.

Brooks, H. (1994a). Current criticisms of research universities. In J. R. Cole, E. G. Barber, and S. R. Gaubard (Eds.), *The Research University in a Time of Discontent.* Baltimore: Johns Hopkins University Press.

Brooks, H. (1994b). *Science and Government Report,* **23**(3), Washington, D.C. (February).

Brown, J. K. and L. W. Kay (1987). *Tough Challenges for R&D Management.* Ottawa, Ontario: The Conference Board of Canada.

Bush, V. (1945). *Science, the Endless Frontier.* Report to the President on a program for postwar scientific research. Washington, D.C.: U.S. Government Printing Office.

Carson, J. W. and T. Rickards (1979). *Industrial New Product Development: A Manual for the 1980s.* New York: Gower Press.

Cetron, M. J. (1973). Technology transfer: Where we stand today. *Joint Engineering Management Congress 21st,* pp. 11–28.

Chakrabarti, A. K. (1974). The role of champion in product innovation. *California Management Review,* **XVII**(2), 58–62.

Chakrabarti, A. K. and A. H. Rubenstein (1976). Interorganizational transfer of technology—a study of adoption of NASA innovations. *IEEE Transactions on Engineering Management,* **23**(1), 20–34 (February).

Chakrabarti, A. K. and R. D. O'Keefe (1977). A study of key communicators in research and development laboratories. *Group and Organization Studies,* **2**, 336–346.

Chan, M. (1981). Intergroup conflict and conflict management in R&D divisions of four aerospace companies. *Dissertation Abstracts International,* **42,** 1767.

Chandler, A. A., Jr. (1962). *Strategy and Structure: Chapters in the History of American Industrial Enterprise.* Cambridge, MA: The MIT Press.

Charpie, R. A. (1970). Technological innovation and the international economy. In M. Goldsmith (Ed.), *Technological Innovation and the Economy,* pp. 1–10. London: Wiley-Interscience.

Cheng, J. L. C. (1984). Paradigm development and communication in scientific settings: A contingency analysis. *Academy of Management Journal,* **27,** 870–877.

Cialdini, R. B. (1985). *Influence.* Glenview, IL: Scott, Foresman.

Cohen, H., S. Keller and D. Streeter (1979). The transfer of technology from research to development. *Research Management,* **22**(3), 11–17 (May).

Cole, J. R., E. G. Barber, and S. R. Graubard (1994). *The Research University in a Time of Discontent.* Baltimore: Johns Hopkins University Press.

Cuadron, S. (1994). Motivating creative employees calls for new strategies. *Personnel Journal,* **73**(5), 103–106 (May).

Cunningham, J. B. (1979). The management system: Its functions and processes. *Management Science,* **25**(7), 657–670.

Daft, R. L. and K. E. Weick (1984). Toward a model of organizations as interpretation systems. *Academy of Management Review,* **9,** 284–295.

Dalton, G. W. (1971). Motivation and control in organizations. In G. W. Dalton and P. R. Lawrence (Eds.), *Motivation and Control in Organizations.* Homewood, IL: Dorsey Press.

Davis, E. E. and H. C. Triandis (1971). An experimental study of black–white negotiations. *Journal of Applied Social Psychology,* **1,** 240–262.

Davis, P. and M. Wilkof (1988). Scientific and technical information transfer for high technology: Keeping the figure in its ground. *R&D Management,* **18**(1), 45–58.

Dougherty, E. (1990). Career satisfaction: Would you do it again? *R&D,* **32**(7), 40–50 (July).

Dunnette, M., J. Campbell, and K. T. Jaastad (1963). The effect of group participation on brainstorming effectiveness for two industrial samples. *Journal of Applied Psychology,* **47,** 30–37.

Dutton, J. A. and L. Crowe (1988). Setting priorities among scientific initiatives. *American Scientists,* **76,** 599–603 (November–December).

Ellis, L. W. (1984). Viewing R&D projects financially. *Research Management,* **27**(2), 29 (March–April).

Evan, W. M. (1965a). Superior–subordinate conflict in research organizations. *Administrative Science Quarterly,* **10,** 51–64.

Evan W. M. (1965b). Conflict and performance in R&D organization: Some preliminary findings. *Independent Management Review,* **7,** 37–46.

Farris, G. F. (1982). The technical supervisor: Beyond the Peter Principle. In M. L. Tushman and W. L. Moore (Eds.), *Readings in The Management of Innovation,* pp. 337–348. Boston, MA: Pitman Publishing.

Fiedler, F. E. (1967). *A Theory of Leadership Effectiveness.* New York: McGraw-Hill.

Fiedler, F. E. (1986a). The contributions of cognitive resources and leader behavior to organizational performance. *Journal of Applied Psychology,* **16,** 532–548.

Fiedler, F. E. (1986b). Lecture at American Academy of Management. Chicago, August.

Fiedler, F. E., M. Chemers, and L. Mahar (1977). *Improving Leadership Effectiveness: The Leader-Match Concept.* New York: Wiley (2nd ed., 1984).

Fiedler, F. E., C. M. Bell, M. M. Chemers, and D. Patrick (1984). Increasing mine productivity and safety through management training and organizational development: A comparative study. *Basic and Applied Social Psychology,* **5,** 1–18.

Fiedler, F. E., W. A. Wheeler, M. M. Chemers, and D. Patrick (1987). Managing for mine safety. *Training and Development Journal,* pp. 40–43 (September).

Fineman, S. (1980). Stress among technical support staff in research and development. In C. L. Cooper and J. Marshall (Eds.), *White Collar and Professional Stress.* New York: Wiley, pp. 211–231.

Fisher, W. A. (1980). Scientific and technical information and the performance of R&D groups. In B. V. Dean and J. L. Goldhar (Eds.), *Management of Research and Innovation. TIMS Studies in the Management of Sciences,* Vol. 15, New York: North-Holland, pp. 135–150.

Foa, U. and E. Foa (1974). *Societal Structures of the Mind.* Springfield, IL: Thomas.

Francis D. and D. Young (1979). *Improving Work Groups.* San Diego, CA: University Associates (2nd ed., 1992).

Freeman, C. (1982). *The Economics of Industrial Innovation,* 2nd ed. London: Franes Pinter.

Freiberg, P. (1995). Creativity is influenced by our social networks. *Monitor,* American Psychological Association, p. 21 (Aug.).

French, J. R. P. and R. D. Caplan (1973). See T. Keenan [1980] for summary.

Friedman, L. (1992). Cognitive and interpersonal abilities related to the primary activities of R&D managers. *Journal of Engineering and Technology Management,* **9**(3), 211–242 (December).

Gibbons, M. and R. D. Johnston (1974). The roles of science in technological innovation. *Research Policy,* **3,** 220–242.

Gibson, J. E. (1981). *Managing Research and Development.* New York: Wiley.

Goldsmith, M. (1970). Introduction. *Technological Innovation and the Economy.* London: Wiley-Interscience.

Gordon, W. J. (1961). *Synectics.* New York: Harper & Row.

Griffiths, P. A. (1993). Science and the Public Interest, *The Bridge,* **23**(3), 3–14, National Academy of Engineering (Fall).

Guetzkow, H. S. and P. Bowman (1946). *Men and Hunger: A Psychological Manual for Relief Workers.* Elgin, IL: Brethen.

Hackman, J. R. and G. Oldham (1980). *Work Redesign.* Reading, MA: Addison-Wesley.

Hall, D. T. and R. Mansfield (1975). Relationships of age and security with career variable of engineers and scientists. *Journal of Applied Psychology,* **60,** 201–210.

Hanson, D. J. (1994). Academic earmarks scorned by lawmakers, defended by universities. *Chemical & Engineering News,* **72**(40), 22–24 (October).

Hax, A. C. and N. S. Majluf (1988). The concept of strategy and the strategy formation process. *Interfaces,* **18**(3), 99–109 (May–June).

Hax, A. C. and N. S. Majluf (1991). *The Strategy Concept and Process: A Pragmatic Approach.* Englewood Cliffs, NJ: Prentice-Hall.

Hax, A. C. and N. S. Majluf (1996). *The Strategy Concept and Process: A Pragmatic Approach,* 2nd ed. Upper Saddle River, NJ: Prentice-Hall.

Helmers, S. and R. Buhr (1994). Corporate story telling: The buxomly secretary, a pyrrhic victory of the male mind. *Scandinavian Journal of Management,* **10,** 175–191.

Hensey, M. (1991). Essential success factors for strategic planning. *Journal of Management in Engineering* (ASCE), **7**(2), 167–177.

Herold, D. M. and C. K. Parsons (1985). Assessing the feedback environment in work organizations: Development of the job feedback survey. *Journal of Applied Psychology,* **70,** 290–305.

Hersey, P. and K. H. Blanchard (1982). *Management of Organizational Behavior,* 4th ed. Englewood Cliffs, NJ: Prentice-Hall.

Holt, K., H. Geschka, and G. Peterlongo (1984). *Need Assessment.* New York: Wiley.

Howard, W. G. and B. R. Guile (Eds.) (1992). *Profiting From Innovation.* National Academy of Engineering. New York: Free Press.

Iansiti, M. (1995). Technology integration: Managing technical evolution in a complex environment. *Research Policy,* **24**(4), 521–542 (July).

Ilgen, D. R., C. D. Fisher, and M. S. Taylor (1979). Consequences of individual feedback on behavior in organizations. *Journal of Applied Psychology,* **64**(4), 349–371.

Isen, A. M., et al. (1985). The influence of positive effect on the unusualness of word associations. *Journal of Personality and Social Psychology,* **48,** 1413–1426.

Jabri, M. M. (1992). Job satisfaction and job performance among R&D scientists: The moderating effects of perceived appropriateness of task allocation decisions. *Australian Journal of Psychology,* **44,** 95–99.

Jackson, S. E. and R. S. Schuler (1985). A meta-analysis and conceptual critique of research on role ambiguity and role conflict in work settings. *Organizational Behavior and Human Decision Process,* **36,** 16–78.

Jain, R. K., L. V. Urban, and G. S. Stacey (1980). *Environmental Impact Analysis—A New Dimension in Decision Making.* New York: Van Nostrand Reinhold.

Janis, I. L. (1972). *Victims of Groupthink: A Psychological Study of Foreign-Policy Decisions and Fiascoes.* Boston: Houghton Mifflin.

Jaques, E. (1961). *Equitable Payment.* New York: Wiley (2nd ed., 1970).

Jones, O. (1994). Establishing the determinants of internal reputation: The case of the R&D scientists. *R&D Management,* **24**(4), 325–339 (October).

Kanfer, F. H. (1988). Contributions of a self-regulation model to the conduct of therapy. Invited address to the Midwestern Psychological Association, April 28, Chicago.

Katz, D. and R. Kahn (1980). *The Social Psychology of Organizations.* New York: Wiley.

Katz, R. and T. J. Allen (1982). Investigating the not invented here (NIH) syndrome: A look at the performance, tenure, and communication patterns of 50 R&D project groups. *R&D Management,* **12**(1), 7–19.

Katz, R. and T. J. Allen (1985). Project performance and the locus of influence in the R&D matrix. *Academy of Management Journal,* **28,** 67–87.

Katz, R. and M. Tushman (1979). Communication patterns, project performance, and task characteristics: An empirical evaluation and integration in an R&D setting. *Organizational Behavior & Human Performance,* **23**(2), 139–162 (April).

Katz, R. and M. Tushman (1981). An investigation into the managerial roles and career paths of gatekeepers and project supervisors in a major R&D facility. *R&D Management,* **11**(3), 103–110.

Keenan, T. (1980). Stress and the professional engineer. In D. C. Cooper and J. Marshall (Eds.), *White Collar and Professional Stress.* New York: Wiley, pp. 189–210.

Keeney, R. L., and H. Raiffa (1976). *Decisions with Multiple Objectives: Preferences and Value Tradeoffs.* New York: Wiley, pp. 6 and 68.

Keller, R. T. (1994). Technology-information processing fit and performance of R&D project groups: A test of contingency theory. *Academy of Management Journal,* **37,** 167–179.

Keller, R. T. (1995). "Transformational" leaders make a difference. *Research Technology Management,* **38**(3), 41–44 (May–June).

Keller, R. T. and W. E. Holland (1975). Boundary-spanning roles in a research and development organization: An empirical investigation. *Academy of Management Journal,* **18**(2), 388–393 (June).

Kennedy, D. (1994). Making choices in the research university. In J. R. Cole, E. G. Barber, and S. R. Graubard (Eds.), *The Research University in a Time of Discontent.* pp. 85–114. Baltimore: Johns Hopkins University Press.

Kennedy, D. (1985). Government policies and the cost of doing research. *Science,* **227,** 480–484 (February).

Kvande, E. and B. Rasmussen (1994). Men in male-dominated organizations and their encounter with women intruders. *Scandinavian Journal of Management,* **10**(2), 163–173.

Landis, D. and R. Bhagat (1996). *Handbook of Intercultural Training,* 2nd ed. Thousand Oaks, CA: Sage.

Landis, D. and R. Brislin (1983). *Handbook of Intercultural Training,* 3 vols. Elmford, NY: Pergamon.

Lane, N. (1996). *Thin Ice Over Deep Water: Science and Technology in a Seven Year Downsizing.* Remarks at the American Astronomical Society Meeting, January 15.

Langer, E. J. (1983). *The Psychology of Control.* Beverly Hills, CA: Sage.

Langrish, J. (1971). Technology transfer: Some British data. *R&D Management,* **1,** 133–136.

LaPorte, T. R. (1967). Conditions of strain and accommodation in industrial research organizations. *Administrative Science Quarterly,* **12,** 21–38.

Lawler, E. E. (1973). *Motivation in Work Organizations.* Monterey, CA: Brooks/Cole, p. 9.

Lawler, E. E., III (1986). *High Involvement Management.* San Francisco: Bass.

Lawrence, P. B. and W. J. Lorsch (1967). *Organization and Environment.* Boston: Harvard University Press (rev. ed., 1986).

Leonard-Barton, D. and W. A. Kraus (1985). Implementing new technology. *Harvard Business Review,* 102–110 (November–December).

Lewis, C. W. and M. J. Tenzer (1992). Political strategies for hi-tech development: The case of a university-related research park. *International Journal of Public Administration,* **15,** 1757–1801 (October).

Likert, J. F. (1967). *The Human Organization.* New York: McGraw-Hill.

Lincoln, J. F. (1951). *Incentive Management.* Cleveland: Lincoln Electric Co.

Locke, E. A. (1968). Toward a theory of task motivation and incentives. *Organizational Behavior and Human Performance,* **3,** 157–189.

Locke, E. A., K. N. Shaw, L. M. Saari, and G. P. Latham (1981). Goal setting and task performance: 1969–1980. *Psychological Bulletin,* **90,** 125–152.

Locke, E. A., E. Frederick, E. Buckner, and P. Bobko (1984). Effect of previously assigned goals on self-set goals and performance. *Journal of Applied Psychology,* **69,** 694–699.

Loher, B. T., R. A. Noe, N. L. Moeller, and M. P. Fitzgerald (1985). A meta-analysis of the relation of job characteristics to job satisfaction. *Journal of Applied Psychology,* **70**(2), 280–289 (May).

Long, J. (1992). Science funding: House drops 10 'pork-barrel' projects. *Chemical & Engineering News,* **70**(39), 6. (September 28).

MacKinnon, D. W. (1962). The nature and nurture of creative talent. *American Psychologist,* **17,** 7.

Mansfield, E. (1995). Academic research underlying industrial innovation: Sources, characteristics, and financing. *Review of Economics and Statistics,* **77**(1), 55–65 (February).

Marcson, S. (1960). *The Scientist in American Industry.* Princeton, NJ: Princeton University Press, pp. 78–151.

Marquis, D. G. and D. L. Straight (1965). *Organizational Factors in Project Performance.* Cambridge, MA: MIT Working Paper No. 133-65.

Maslow, A. (1992). *Motivation and Personality,* 3rd ed. New York: Harper.

Massey, W. E. (1994). Can the research university adapt to a changing future? In J. R. Cole,

E. G. Barber, and S. R. Graubard (Eds.), *The Research University in a Time of Discontent*. Baltimore: Johns Hopkins University Press, pp. 191–202.

McCain, G. (1969). *The Game of Science*. Belmont, CA: Wadsworth, p. 59.

McGregor, D. (1972). An uneasy look at performance appraisal. *Harvard Business Review*, pp. 133–138 (September–October).

Merten, U. and S. M. Ryu (1983). What does the R&D function actually accomplish. *Harvard Business Review* (July–August).

Merton, R. K. (1973). *The Sociology of Science: Theoretical and Empirical Investigations*. Chicago: The University of Chicago Press.

Mintzberg, H. (1973). *The Nature of Managerial Work*. New York: Harper & Row, pp. 56–58.

Mintzberg, H. (1975). The manager's job: Folklore and fact. *Harvard Business Review*, No. 75409, pp. 49–61 (July–August).

Misumi, J. (1985). *The Behavioral Science of Leadership*. Ann Arbor: The University of Michigan Press.

Morton, J. A. (1971). *Organizing for Innovation*. New York: McGraw-Hill.

Murphy, K. R., W. K. Balzer, M. C. Lockhart, and E. J. Eisenman (1985). Effects of previous performance on evaluations of present performance. *Journal of Applied Psychology*, **70**, 72–84.

Nadiri, M. I. (1980). Contributions and determinants of research and development expenditures in U.S. manufacturing industries. In G. M. von Furstenburg (ed.), *Capital Efficiency and Growth*. Cambridge, MA: Ballinger.

Nadler, D. A. (1982). Concepts for the management of organizational change. In G. L. Lippitt (Ed.), *Implementing Organizational Change*. Jossey-Bass, San Francisco.

Nadler, D. A. (1994). Collaborative strategic thinking. *Planning Review*, **22**(5), 30–31, 44.

Naisbitt, J. (1982). *Megatrends*. New York: Warner Books.

National Academy of Engineering (1993). *Prospering in a Global Economy—Mastering a New Role*, Washington, D.C.: National Academy Press.

National Academy of Sciences (1995). *Allocating Federal Funds for Science and Technology*, Washington, D.C.: National Academy Press.

Nelson, L. J. and D. T. Miller (1995). The distinctiveness effect in social categorization: You are what makes you unusual. *Psychological Science*, **6**, 246–249.

Newton, T. J. and A. Keenan (1985). Coping with work related stress. *Human Relations*, **38**, 107–126.

Newton-Smith, W. H. (1981). *The Rationality of Science*. London: Routledge and Kegan.

Organization for Economic Cooperation and Development (OECD). (1993). *The Measurement of Scientific and Technological Activities*. Paris: OECD.

Osborn, A. F. (1957). *Applied Imagination*, rev. ed. New York: Scribner.

Osborn, A. F. (1963). *Applied Imagination*. New York: Scribner.

Pelz, D. C. (1956). Some social factors related to performance in a research organization. *Administrative Science Quarterly*, **1**, 310–325.

Pelz, D. C. and F. M. Andrews (1966a). Autonomy, coordination, and simulation in relation to scientific achievement. *Behavioral Science*, **2**, 89–97.

Pelz, D. C. and F. M. Andrews (1966b). *Scientists in Organizations.* New York: Wiley.

Peters, T. J. and R. H. Waterman (1982). *In Search of Excellence—Lessons from America's Best Run Companies.* New York: Harper & Row.

Press, F. (1988). *The Dilemma of the Golden Age.* Address to the Members of the National Academy of Science, 26 April 1988.

Price, D. J. D. (1965). Is technology independent of science? *Technology and Culture,* **6,** 553–568.

Quinn, J. B. (1985). Managing innovation: Controlled chaos. *Harvard Business Review,* **63**(3), 73–84 (May–June).

Rahim, A. (1983). A measure of styles of handling interpersonal conflict. *Academy of Management Journal,* **26,** 368–376.

Regis, E. (1987). *Who Got Einstein's Office?* New York: Addison-Wesley.

Rheem, H. (1995). Improving productivity: The importance of R&D. *Harvard Business Review,* **73**(3), 12–13 (May–June).

Ritti, R. (1982). Work goals of scientists and engineers, in M. L. Tushman and W. L. Moore (Eds.), *Readings in the Management of Innovations,* pp. 363–375. Boston: Pitman.

Roberts, E. B. (1978). What do we really know about managing R&D? Interview with Michael Wolff. *Research Management,* **21**(6), 6–11 (November).

Roberts, E. B. (1995). Benchmarking the strategic management of technology—I. *Research-Technology Management,* **38**(1), 44–56 (January–February).

Roberts, E. B. and A. L. Frohman (1978). Strategies for improving research utilization. *Technology Review,* **80**(5), 32–39 (March–April).

Roberts, E. B. and A. R. Fusfeld (1981). Staffing the innovative technology-based organization. *Sloan Management Review,* **22**(3), 19–34 (Spring).

Rogers, E. M. (1983). *Diffusion of Innovations,* 3rd ed. New York: Free Press.

Rogers, E. M. (1995). *Diffusion of Innovations,* 4th ed. New York: Free Press.

Rosenbaum, M. E., D. L. Moore, J. L. Cotton, M. S. Cook, R. A. Hieser, M. N. Shovar, and M. J. Gray (1980). Group productivity and process: Pure and mixed reward structures and task interdependence. *Journal of Personality and Social Psychology,* **39,** 626–642.

Rosovsky, H. (1987). "Deaning." *Harvard Magazine* (January–February).

Ross, M. H. (1990). Opportunities for maximizing the effectiveness of the administrator/researcher relationship. *Journal of the Society of Research Administrators,* **22**(1), 17–22 (Summer).

Rothwell, R. and W. Zegwell (1981). *Industrial Innovation and Public Policy: Preparing for the 1980 and 1990s.* Westport, CT: Greenwood.

Rotman, D. (1995). Companies globalize research. *Chemical Week,* **157**(21), 41 (Nov. 29).

Ruzic, N. (1978). How to tap NASA developed technology. *Research Management,* **21**(6), 38–40 (November).

Saari, L. M. and G. P. Latham (1982). Employee reactions to continuous and variable ratio reinforcement schedules involving a monetary incentive. *Journal of Applied Psychology,* **67,** 506–508.

Sakamoto, N. (1982). *Polite Fictions: Why Japanese and Americans Seem Rude to Each Other.* Tokyo, Japan: Kinseido.

Salter, M. S. (1971). Management appraisal and reward systems. *Journal of Business Policy,* **1**(4), 41–51.

Saxenian, A. (1994). Lessons from silicon valley. *Technology Review,* **97,** 42–51 (July).

Schmitt, R. W. (1985). Successful corporate R&D. *Harvard Business Review,* **63**(3), 124–128.

Schneider, B., S. K. Gunnarson, and J. K. Niles (1994). Creating the climate and culture of success. *Organizational Dynamics,* **23,** 17–29.

Schneider, W. (1993). *Getting Smart Quicker: Training More Skills in Less Time.* Washington, D.C. Federation of Behavioral, Psychological and Cognitive Sciences.

Schriesheim, J., M. A. Von Glinow, and S. Kerr (1977). Professionals in bureaucracies: A structural alternative. In P. C. Nystrom and W. H. Starbuck (Eds.), *Prescriptive Models of Organizations.* New York: North-Holland, pp. 55–69.

Science and Engineering Indicators (1985). Washington, D.C.: National Science Board.

Science and Engineering Indicators (1993). Washington, D.C.: National Science Board.

Science and Engineering Indicators (1996). Washington, D.C.: National Science Board.

Science Indicators, The 1985 Report (1985). Washington, D.C.: National Sciences Board.

Scott, S. G. and R. A. Bruce (1994). Determinants of innovative behavior: A path model of individual innovation in the workplace. *Academy of Management Journal,* **37,** 580–607 (June).

Scott, W. B. (1994). NASA reshapes tech transfer. *Aviation Week & Space Technology,* **140**(20), 55 (May).

Shanklin, W. L., and J. K. Ryans, Jr. (1984). Organizing for high-tech marketing. *Harvard Business Review,* **62**(6), 164–171 (November–December).

Smith, G. C. (1970). Consultation and decision processes in an R&D laboratory. *Administrative Science Quarterly,* **15,** 203–215.

Snyder, M. (1979). Self-monitoring process. In L. Berkowitz (Ed.), *Advances in Experimental Social Psychology,* Vol. 12. New York: Academic Press, pp. 86–131.

Souder, W. E. (1975). Stage-dominant (S-D), process-dominant (P-D) and task-dominant (T-D) models of the new product development (NDD) process: Some straw-men models and their contingencies. *Technology Management Studies Group Paper,* November 1.

Souder, W. E. and A. K. Chakrabarti (1980). Managing the coordination of marketing and R&D in the innovation process. In B. V. Dean and J. L. Goldhar (Eds.), *Management of Research and Innovation. TIMS Studies in the Management Sciences,* Vol. 15, New York: North-Holland, pp. 135–150.

Spector, P. E. (1982). Behavior in organizations as a function of employee's locus of control. *Psychological Bulletin,* **91,** 482–497.

Sternberg, R. J. and J. E. Davidson (1995). *The Nature of Insight.* Cambridge, MA: MIT Press.

Sternberg, R. J. and T. I. Lubart (1995). *Defying the Crowd: Cultivating Creativity in a Culture of Conformity.* New York: The Free Press.

Sutton, C. (1986). Serendipity or sound science? *New Scientist,* **109,** 30–32 (February).

Szakonyi (1994). Measuring R&D effectiveness—I. *Research-Technology Management.* **37**(2), 27–32.

Szilagyi, A. D. and W. E. Holland (1980). Changes in social density. Relationships with function interaction and perceptions of job characteristics, role stress and work satisfaction. *Journal of Applied Psychology,* **65,** 28–33.

Thompson, J. D. (1967). *Organizations in Action.* New York: McGraw-Hill.

Thompson, P. H. and G. W. Dalton (1976). Are R&D organizations obsolete? *Harvard Business Review,* **54,** 105–116 (November–December).

Thomson, W. J. (1983). Effects of control on choice of reward and punishment. *Bulletin of Psychononomic Society,* **21,** 462–464.

Triandis, H. C. (1971). *Attitude and Attitude Change.* New York: Wiley.

Triandis, H. C. (1977). *Interpersonal Behavior.* Monterey, CA: Brooks/Cole.

Triandis, H. C. (1980). Values, attitudes, and interpersonal behavior. *Nebraska Symposium on Motivation, 1979.* Lincoln, NE: University of Nebraska Press.

Triandis, H. C. (1994). *Culture and Social Behavior.* New York: McGraw-Hill.

Triandis, H. C. (1995). *Individualism and Collectivism.* Boulder, CO: Westview Press.

Triandis, H. C., R. Hall, and R. B. Ewen (1965). Member heterogeneity and dyadic creativity. *Human Relations,* **18,** 35–55.

Triandis, H. C., L. Kurowski, and M. Gelfand (1994). Workplace diversity. In Triandis, H. C., M. Dunnette, and L. Hough (Eds.). *Handbook of Industrial and Organizational Psychology,* 2nd ed. Palo Alto, CA: Consulting Psychologists Press, pp. 769–827.

Tushman, M. L. (1988). Managing communication networks in R&D laboratories. In M. L. Tushman and W. L. Moore (Eds.), *Readings in the Management of Innovation,* 2nd ed. Cambridge, MA: Ballinger, pp. 261–274.

Twiss, B. C. (1992). *Managing Technological Innovation,* 4th ed. Marshfield, MA: Pitman.

Von Hippel, E. A. (1978). Users as innovators. *Technology Review,* Jan., pp. 31–37.

Vroom, V. and P. W. Yetton (1973). *Leadership and Decision Making.* Pittsburgh: University of Pittsburgh Press.

Wainer, H. A. and I. M. Rubin (1969). Motivation of research and development entrepreneurs: Determinants of company success. *Journal of Applied Psychology,* **53,** 178–184.

White, S. E., T. R. Mitchell, and C. H. Bell, Jr. (1977). Goal setting, evaluation apprehension, and social cues as determinants of job performance and job satisfaction in a simulated organization. *Journal of Applied Psychology,* **62,** 665–673.

Whyte, W. F. (1948). *Human Relations in the Restaurant Industry.* New York: McGraw-Hill.

Wicksteed, S. Q. (1985). *The Cambridge Phenomenon.* London: Brand.

Williams, C. W., R. S. Brown, P. R. Lees-Haley, and J. R. Price (1995). An attributional (casual dimensional) analysis of perceptions of sexual harassment. *Journal of Applied Social Psychology,* **25,** 1169–1183.

Wilson, D. K. (1994). New look at performance appraisal for scientists and engineers. *Research-Technology Management,* **37**(4), 51–55 (July–August).

Wilson, E. D. (1995). The molecular wars. *Harvard Magazine,* May–June, pp. 42–49.

Winchell, A. E. (1984). Conceptual systems and Holland's theory of vocational choice. *Journal of Personality and Social Psychology,* **46,** 376–383.

Worchel, S. and W. G. Austin (1985). *Psychology of Intergroup Relations.* Chicago: Nelson-Hall.

AUTHOR INDEX

SUBJECT INDEX